CHRIS STERLING
6 JAN 75

TELECOMMUNICATIONS

TELECOMMUNICATIONS

THE TELEPHONE SYSTEMS OF THE CONTINENT OF EUROPE

A. R. Bennett

•

THE DEVELOPMENT OF THE TELEPHONE IN EUROPE

Herbert Laws Webb

ARNO PRESS
A New York Times Company
New York • 1974

Reprint Edition 1974 by Arno Press Inc.

The Telephone Systems of the Continent of Europe was reprinted from a copy in The Columbia University Library.
The Development of the Telephone in Europe was reprinted from a copy in the University of Illinois Library.

TELECOMMUNICATIONS
ISBN for complete set: 0-405-06030-0
See last pages of this volume for titles.

Manufactured in the United States of America

Library of Congress Cataloging in Publication Data

Bennett, Alfred Rosling, 1850-1928.
The telephone systems of the continent of Europe.

(Telecommunications)
Reprint of 2 works, the 1st published in 1895 by Longmans, Green, London; the 2d published in 1910 by Electrical Press, London.
1. Telephone--Europe--History. I. Webb, Herbert Laws. The development of the telephone in Europe. 1974. II. Title. III. Series: Telecommunications (New York, 1974-)
TK6055.B46 1974 384.6'094 74-4666
ISBN 0-405-06033-5

TELEPHONE SYSTEMS

THE GENERAL TELEPHONE COMPANY'S STEEL TOWER AT THE BRUNKEBERG EXCHANGE, STOCKHOLM

THE

TELEPHONE SYSTEMS

OF THE

CONTINENT OF EUROPE

BY

A. R. BENNETT

Member of the Institution of Electrical Engineers; Divisional Engineering Superintendent in London
to the United Telephone Company, Limited, 1880
Engineer to the Commercial Telephone Exchange, Glasgow, 1881-3
Chief Engineer for Scotland and Ireland to the National Telephone Company, Limited, 1883
General Manager and Chief Engineer in Scotland and the North-west of England
to the National Telephone Company, Limited, 1883-90
General Manager and Chief Engineer to the Mutual Telephone Company, Limited, 1890-2
General Manager and Chief Engineer to the New Telephone Company, Limited, 1892-5

WITH 169 ILLUSTRATIONS

LONDON
LONGMANS, GREEN, AND CO.
AND NEW YORK
1895

CONTENTS

III. BELGIUM

IV. BOSNIA-HERZOGOVINA

V. BULGARIA

VI. DENMARK

VII. FINLAND

VIII. FRANCE

IX. GERMAN EMPIRE

(exclusive of bavaria and würtemberg)

X. GREECE

XI. HOLLAND

XII. HUNGARY

XIII. ITALY

XIV. LUXEMBURG

XV. MONACO

XVI. MONTENEGRO

XVII. NORWAY

XVIII. PORTUGAL

XIX. ROUMANIA

XX. RUSSIA

XXI. SERVIA

XXII. SPAIN

XXIII. SWEDEN

XXIV. SWITZERLAND

XXV. TURKEY

XXVI. WÜRTEMBERG

THE TELEPHONE SYSTEMS

OF THE

CONTINENT OF EUROPE

INTRODUCTION

DURING the discussions on the existing state and future conduct of telephony in the United Kingdom which have taken place pretty continuously during the last few years, many references have cropped up to foreign and, more especially, to continental practice. Statements have frequently been made as to the existence of what to the British public have appeared fabulously low rates in Holland, Switzerland, Norway, Sweden, and elsewhere—statements to which support was given, from time to time, by various consular reports. The facts set forth, the believers in, and advocates of, low rates in this country have endeavoured occasionally to turn to their advantage in argument, but, owing to lack of exact information and the denials of their opponents, with little result. The apologists of the existing monopoly have either traversed *in toto* the truth of the statements or have declared that the conditions under which such rates exist are radically different from those which obtain in the United Kingdom. They have asserted, for example, that the low rates are not inclusive of all charges; that the subscribers have to pay the cost of their lines or instruments, or both, and, after connection, for any repairs that may be necessary; that foreign telephone companies are not burdened with such payments to the Government as are imposed on the National Telephone Company here; that foreign administrations (this has been specially said of Germany) have an absolute right to fix supports and wires wherever they please, underground or overhead, without payment ; that labour is less costly on the Continent than with us ; that foreign workmen and operators are not only badly paid, but mercilessly sweated ; that the cheap systems are ill-constructed and worse managed ; that the low rates, if they exist, are only applied in small towns ; that they do not pay ; together with various other assertions intended, and tending, to create doubt, and confuse the advocates of telephonic reform.

The points at issue were so numerous and involved, and the question so interesting and replete with importance to the British commercial community, particularly in view of a possible Post Office acquisition, partial or complete, of the telephone systems, that the author determined to ascertain the truth for himself, and that by the best of all methods, personal inspection and investigation. Controversy had chiefly centred on the Scandinavian and German countries, Holland, Belgium, and Switzerland. All these, together with France, have been visited by the author, and the most minute inquiries as to the tariffs, rules, laws, technical practice, and other matters of interest conducted on the spot, the points enumerated above as being specially in dispute and indicated for examination receiving more than ordinary attention. The results of this inquest are now presented to the public in a form that, it is hoped, will facilitate reference to particular points and enable an accurate idea of the true state of matters to be readily arrived at.

It will be found that no two nations have solved, or attempted to solve, the problem in exactly the same manner. In some cases the divergencies are wide, but in most great intelligence, combined with solicitude for the public weal, has been brought to bear, often with the most satisfactory results.

It will be seen that except in two Russian towns, St. Petersburg and Moscow, which are in the hands of a monopolist company and where the rates are 25*l.* per annum, no continental subscription comes up to the 20*l.* rate with which we are familiar in London. On the other hand, subscriptions in some places descend to 2*l.* 10*s.* and 2*l.* 9*s.* 7*d.* per annum, everything included, and are made to pay. The contention of the high-rate apologists that the low rates are *not* inclusive will be found, for the purposes of their argument, to be untrue and delusive. The fact is that practice in this respect varies, even in the same countries, as in Norway and Denmark, some of the rates being inclusive and others not. Full particulars are given of these variations under the headings of the several countries, but it may be as well to state definitely here that inclusive rates, covering the supply and maintenance of all wires, instruments and accessories, of 2*l.* 10*s.*, 2*l.* 15*s.* 7*d.*, and 3*l.* 6*s.* 8*d.* exist in Norway; of 2*l.* 15*s.* 7*d.* and 3*l.* 6*s.* 8*d.* in Denmark; of 2*l.* 9*s.* 7*d.* and 2*l.* 17*s.* 10*d.* in Holland; of 3*l.* 4*s.* in Finland; and of 2*l.* 16*s.* and 3*l.* 12*s.* in Italy; while rates of 4*l.* and 5*l.* are of frequent occurrence elsewhere. In refutation of the assertion that low rates mean bad workmanship, the author would direct special attention to the installation at Zutphen, a town of 17,000 inhabitants (where the Zutphen Telephone Company applies an inclusive rate of 2*l.* 17*s.* 10*d.*), which is fitted throughout with metallic circuits of stouter bronze wire than the National Telephone Company habitually uses in this country; with the very best of transmitters, receivers, and instruments, together with an expensive switch-board by one of the leading manufacturers, and all the usual complement of lightning-guards, cross-connecting apparatus, and testing instruments. The outside construction consists of standards, poles, insulators, and general fittings of the best description, the work throughout being thoughtfully designed and well carried out. An all-night service is provided, and the company pays 4 2 per cent. on the capital invested. In proof of this a translation of the last balance-sheet, dated February 1895, is given. A translation of the last accounts of the Co-operative Company at Aarhus, which has an inclusive rate of 4*l.* 3*s.* 3*d.* for local connections and of 5*l.* 16*s.* 7*d.* for those who wish to speak to the other towns within a radius of about 20 kilometers, will also be found in the Danish section. Finally, in order to dispose of the assertion that low rates are only applicable to small towns, the accounts of the Christiania Telephone Company for 1893 are printed at the end of the Norwegian section. This company, operating in a capital city, has nearly 5,000 subscribers, and has regularly paid dividends of from 5 to 5½ per cent. since 1885, besides keeping its system up to date, providing ample reserve funds, and liberally contributing to the benevolent funds of its male and female employees, all on an *inclusive* subscription of 4*l.* 8*s.* 11*d.*

What can be done in the direction of municipal telephones is instanced by the example of Trondhjem, the third town of Norway, where the telephone exchange is in the hands of the town council. The exchange is well built of good material, provided with the most expensive instruments in the market—those of Ericsson & Co., and earns a profit of 4 per cent. for the ratepayers, on an inclusive rate of 2*l.* 10*s.* for business connections and 1*l.* 5*s.* for private houses, rates which apply to lines not exceeding 1½ kilometers in length! Is there a valid reason why a British municipality should not do as well in a town of corresponding size?

Switzerland is another country in which low rates prevail, soon to give way (see page 378) to lower; but the system adopted of charging per call or connection renders comparison with the foregoing rates, which cover any desired number of local calls, difficult. For the subscriber who makes but little use of his telephone the new Swiss tariff will be the cheapest of all, while the busy firm's contribution may exceed the highest rates mentioned in this book. Thus, a man calling only once per working day will pay (after having been a member of the exchange for two years) only 2*l.* 4*s.* 6*d.* per annum, while a subscriber calling 20 times a day will pay 14*l.* 2*s.*, and one originating 30 talks per day as much as 20*l.* 7*s.* 7*d.* This plan is unquestionably the most rational one, but experience shows that it tends to reduce the number of calls, so that the average of the daily connections asked for at Zürich is only two per subscriber. The Swiss plan, therefore, restricts the volume of business and the usefulness of the telephone, while the lines and exchange apparatus must be as expensive and perfect as in the busiest centre.

In the tabulated statements in the Danish and Norwegian sections will be found many particulars of, and results obtained in, the smaller towns.

Some information as to the way-leave facilities enjoyed by the telephone administrations or companies in most of the countries is given. The author's inquiries tend to show that the autocratic privileges talked of are mostly imaginary. The French Government possesses greater power over private property than any other, and, unluckily for those who seek to establish a connection between rates and way-leaves, the French rates are amongst the dearest on the Continent. In Germany the Government has no more power to put a standard and wires on a man's house without his permission than it has to burn it down.

It will be seen that many foreign companies are burdened with far more onerous payments to their governments or municipal authorities than is the National Telephone Company. In Madrid, 20 per cent.; in Bilbao, 34 per cent.; in Barcelona, 33·75 per cent.; and in Valencia 31·5 per cent. of the gross receipts are payable to the Government. In Italy a uniform tax of 10 per cent. on the gross receipts and 2*l.* per annum for every public telephone station (call office) is levied; in Russia the tax is also 10 per cent. of the gross takings, while the Portuguese get off with 3 per cent.

During one public discussion on the subject of telephone rates it was stated as justifying a 10*l.* rate in Manchester that subscribers in Amsterdam have to pay practically the same—9*l.* 14*s.* 2½*d.* But the apologist, probably because he knew no better, omitted to say that the Amsterdam company has to pay 2*l.* 1*s.* 9*d.* per subscriber per annum to the town council; and that, while the 10*l.* rate in Manchester is limited to a distance of one mile, the Dutch subscription applies to the whole of Amsterdam proper.

Workmen's wages, according to the particulars supplied to the author by the officials of the various administrations and companies, are not invariably noticeably lower, nor the hours worked much longer, than in this country. As a rule, the female operators are better paid, in some cases markedly so, than those of the National Telephone Company.

On the other hand, where low wages prevail, their effect on cost of production is sometimes neutralised more or less by the Customs import duties. For instance, Norway possesses no iron, and the author found English iron on the roofs of Christiania in the form of telephone supports. Norway, too, either imports her telephone apparatus or makes it from imported materials. Switzerland and Holland, which produce no iron and import instruments or raw material, are in the same case.

The author did not undertake a foreign tour for the purpose of convincing himself of the feasibility of low rates, but in order to obtain authoritative evidence to help him to convince others of the fact. Personally, he required no convincing, as his experience in Scotland and Manchester rendered any further evidence unnecessary. The author knows that many of the National Telephone Company's exchanges absorb less than 50 per cent. of the subscriptions collected in them for upkeep and contingencies, so that a municipality or company putting into the business only the capital actually required for establishment could earn a fair profit on not more than half the present rates.

That it must be so is evident from a consideration of the National Company's capital and regular 5 per cent. dividend. It has been stated frequently in print, and at public meetings, in the presence of the company's directors and officials ; [1] and—so far as the author is aware never seriously contradicted—that the amount of 'water' to paid-up capital is as two or three to one ; that is to say, that out of a capital of four millions for which dividends must be found, only one million, or at most one and a third millions, have been put into the business. To pay 5 per cent. on four millions this one million must earn 20 per cent.

That it actually does so is unquestionable : in fact, telephony in the United Kingdom is really conducted to-day as cheaply as on the Continent, the only difference being that each sovereign invested has to find interest for two or more unproductive companions. Actual experience affords this assertion ample confirmation.

From 1880 to 1885 the National Telephone Company was opposed in Dundee and its vicinity by the Dundee and District Telephonic Company, Limited, which company had commenced business with a rate of 10*l.* designed to oppose the rate of 20*l.* which the National Company had established in the same town. Finding that it could not hold its own, the National determined to ruin the opposition by a war of rates, and suddenly came down from 20*l.* to 5*l.* per annum at one swoop. The Dundee and District replied with a reduction to 5*l.* 10*s.*, below which they considered

[1] *Truth*, August 21, 1890, and March 10, 1892. Councillor Southern's speech to the Manchester Town Council, *Manchester Guardian*, March 8, 1894.

it inexpedient to go, as the telephones they used, owing to patent complications, were costing from 20*l.* to 25*l.* each, and were difficult to procure even at those prices. At the first annual meeting (February 1883) after the reduction, the Dundee and District had only a balance of 400*l.* to the good, which was carried to a reserve fund ; but in February 1884, after nearly two years' experience at 5*l.* 10*s.*, it declared a dividend of 10 per cent. per annum, besides adding 200*l.* to the reserve fund. In February 1885 the dividend was 5 per cent., and 200*l.* to reserve. But this victorious career proceeded but a little further, as the National made up its mind that the speed at which the Dundee Company was ruining itself was not rapid enough, knowing besides, from its own experience, now of considerable duration, with a 5*l.* rate, that the 5*l.* 10*s.* of the opposition was sufficient, and more, to permit it to live and prosper ; and so made an offer to buy the Dundee Company, which was ultimately accepted. At the final meeting the chairman congratulated the shareholders on having received an average dividend of 9 per cent. per annum for the four and a quarter years of the company's existence, and on the return of their capital with 40 per cent. by way of bonus. That was how ruin had spelled with them !

On its side the National had not done badly. It professed to have lost the difference between the original rate of 20*l.* and the fighting rate of 5*l.* ; but that was no real loss, since its subscribers at 20*l.* would have been very few, while, as matters stood, its exchange had grown out of all knowledge. After the purchase the combined systems numbered some 1,200 subscribers, and constituted together the largest exchange in the United Kingdom, excepting, perhaps, London. Subsequently, when the rate was put up to 10*l.*, it dwindled away to about half. This great increase in Dundee showed, as it did afterwards in Manchester, under the Mutual Telephone Company, that a 5*l.* rate taps a class of subscribers which cannot afford, or will not give, 8*l.* or 10*l.* for the accommodation. In Dundee a considerable number of small shopkeepers, grocers and others, came on at 5*l.* and went off when the rate was increased to 10*l.* ; in Manchester numbers of packing-case makers, sign-writers, plumbers, &c., who had never thought of joining the National exchange at 10*l.*, subscribed to the Mutual at 5*l.* as soon as the opportunity was afforded them. The National had found in Dundee, much to its surprise, that a 5*l.* rate was not only sufficient to cover expenses, but to leave a profit into the bargain, even after debiting Dundee with its due proportion of directors' fees, Post Office royalty of 10 per cent. on the gross receipts, and London office general expenses, provided that the patent royalty of 2*l.* per annum per subscriber were set aside. The author believes that the United Telephone Company, the owners of the patents, eventually agreed to abrogate the Dundee royalty, so that the National really made no loss during the competitive period. But there are no patent royalties now, and the Dundee Town Council or a local company would not have any London office burden to bear, so that the author is quite sure that an exchange with metallic circuits, underground wires in the centre of the town to each block of buildings, and all modern improvements, could readily be made remunerative at 5*l.* per annum, Post Office royalty included. The experience gained since the days of the Dundee and District Company renders it possible to provide an improved system at still lower rates than it did.

Other competitors had arisen and were still to arise in different parts of the kingdom. Messrs. D. and G. Graham in Glasgow at 12*l.*; Charles Moseley in Manchester at 8*l.*; Tasker & Co. in Sheffield at 7*l.*; the Globe Telephone Company in London at 10*l.*; and Mr. Sharples in Preston at 6*l.*; all of which were, after shorter or longer combats, ultimately bought out—some at extravagant premiums—because they, having no need to die, steadfastly declined to do so. In not one instance did the National run a competitor to a standstill, although in the cases of Sharples and Tasker the contest went on for years. Their rates were sufficient to gain a livelihood, and the National knew it.

But the most recent home proof of the sufficiency of low rates is that afforded by the Mutual Telephone Company, Limited, of Manchester, which started with a 5*l.* rate for its shareholders and 6*l.* for non-shareholders. The Mutual Company's case is different from all the others, inasmuch as its exchange was constructed entirely on the metallic circuit principle and comprised all the latest improvements. The Mutual exchange was opened on February 28, 1891, but owing to the determination of the directors to charge nothing until a large circle of subscribers had been put in communication, no rentals were made payable until July 1. From this date until October 31, the end of the financial year, the accounts showed a revenue of 3,906*l.* 5*s.*, only 1,145*l.* 1*s.* of which was applicable to the four months dealt with; nevertheless a credit balance of 378*l.* 11*s.* 7*d.* was available, which was carried forward. In the third half-year of the exchange's existence the receipts, after deducting Post Office royalty, averaged 4*l.* 12*s.* 2*d.* per line per annum, the annual revenue being 6,401*l.* 2*s.* 4*d.* and the number of lines 1,389. The actual working expenses for the half-year were at the rate of 3,244*l.* 17*s.* 6*d.* per annum. Adding to this 500*l.* for directors' fees, 250*l.* for general expenses, and 1,650*l.* (5 per cent. on 33,069*l.*, the capital actually expended) for deterioration, the gross expenses were 5,644*l.* 17*s.* 6*d.*, or 4*l.* 1*s.* 3*d.* per line, leaving a profit of 10*s.* 11*d.* per line per annum, or a total profit of 758*l.* 3*s.* 3*d.* This is only 2·29 per cent. per annum on the capital expended; but the 33,069*l.* included the cost of two trunk lines to Bolton, of some 500 spare metallic circuits, of a central switch-table fitted complete for 2,000 lines and with ultimate accommodation for 4,000, and of standards, poles and general fittings of capacity far in excess of immediate requirements, so as to leave room for future expansion. Less the cost of the trunk lines the actual expenditure on the system had been 31,939*l.*, not quite 23*l.* per line. Deducting the cost of the extra accommodation provided everywhere, the cost per line was only some 16*l.* But for a town of the size of Manchester with Salford (population 703,507) the rate proposed by the author (see page 25) is 5*l.* 15*s.*, which would materially increase the net revenue and obviously give a municipality or an unburdened company a handsome margin of profit.

Most unfortunately for the Manchester public and its own shareholders the Mutual Company was induced to sell its business to the New Telephone Company, Limited. As general manager and chief engineer of the Mutual Telephone Company, the author is glad of an opportunity to state definitely that the sale was in no wise justified in any way by the position of the company. Its business was rapidly increasing; the proportion of net revenue was growing every month; many initial difficulties, including an attempt to prevent the company's wires being run, had been triumphantly overcome; the most flattering opinions of its service had been given in writing by its subscribers, many of whom were also members of the National Manchester exchange, and so peculiarly qualified as judges. Moreover, the Lancashire County Council had granted permission to the company to erect poles and wires on every road in the county of Lancaster, so leaving the way clear for the connection of every town by trunk lines sooner or later. In fact, the company's success had been phenomenal, and its prospects at the date of sale were of the brightest.

But the directors became persuaded that the company's object of winning cheap telephony for the nation would be furthered by transferring the business to a powerful fighting organisation such as the New Telephone Company was supposed to be, and it would certainly be unfair to blame them for not foreseeing the extraordinary turn which that company's affairs subsequently took. In a few months it had fallen completely under the control of the National. The rates in Manchester were shortly afterwards raised and alterations effected which rendered a realisation of the Mutual Company's programme impossible. But the superiority of its service and the sufficiency of its rate had been nevertheless conclusively demonstrated.

The Mutual campaign of course confirmed the author's previous experience at Dundee; and the two cases together will probably be accepted as conclusive evidence of the exorbitant character of the existing rates. That being so, it will surely not be contended that the commercial community has not a right to demand that its business intercourse shall not be burdened with avoidable expense, or in any way, or through any cause, be rendered more costly than that of its trade competitors abroad.

In Belgium, one of England's keenest competitors, a merchant at any town receiving an inquiry by mail or telegraph from, say, South America can put himself in almost instantaneous communication with the chief manufacturers at Liége, Verviers, or elsewhere, and with the shippers of Antwerp, each communication costing 1 franc ($9\frac{1}{2}$*d.*), and in the course of half an hour is in a position to forward a quotation for the desired shipment. Similarly, a German merchant can telephone all over the country for 1*s.* per connection. A British trader receiving the same inquiry would be at a great disadvantage: the delay and uncertainty in getting through would probably deter him from using the telephone at all; if not, he would have to pay the Belgian or German charges many times over. This is not what the public wants or ought to be called upon to submit to.

It has been said that the British public does not care for telephony, and that it would not in any case take advantage of cheap rates to the same extent as continental peoples do. The author considers that this constitutes a most unfair and unwarrantable prejudgment of what the British public would do if it were placed on an equality as regards facilities with other peoples.

What has the telephone service, even in the best conducted exchanges, hitherto meant, and what does it mean to-day, to the British subscriber? Simply that he may call up, and be called up by, other subscribers in his own town and, to a limited extent, other towns also. He may also be called up by non-subscribers speaking from public stations (call offices) established, not at the post and telegraph offices, where people naturally expect to find them, but scattered anywhere where room for an instrument can be found. Dealing with these facilities in the same manner as the services rendered to the public by foreign administrations and companies are dealt with in this present book, it may be said that the British subscriber enjoys for his money four services, to wit:—

1. Local exchange intercourse. 2. Internal trunk line intercourse. 3. Public telephone station intercourse. 4. Forwarding and receiving his telegrams by telephone (in some of the large towns only).

Now, let it be thoroughly grasped what foreign subscribers obtain for subscriptions which sometimes amount to a third or less of the British.

AUSTRIA.—1. Local exchange. 2. Internal trunks. 3. International trunks. 4. Telephoning of telegrams. 5. Local telephonogram[1] service (ten words for *2d.*). 6. Telephoning of

[1] In Austria and Switzerland a message telephoned by a subscriber to the central office to be written down and delivered by messenger to non-subscriber is officially known as a phonogram, a word which, without official authority, has also been adopted in the same sense in several other countries. In the United Kingdom and the United States, at least, phonogram means the record of the

messages to be written down at the central office and mailed as letters or post-cards. 7. Messages calling a non-subscriber to a distant public station to converse. 8. Public telephone stations.

BAVARIA.—1. Local exchange. 2. Internal trunks. 3. International trunks. 4. Telephoning of telegrams (free). 5. Local telephonogram service (ten words for 2*d*.). 6. Telephoning of mail matter as above. 7. Public telephone stations.

BELGIUM.—1. Local exchange. 2. Internal trunks. 3. International trunks. 4. Telephoning of telegrams (free). 5. Public telephone stations. 6. Messages calling strangers to distant public stations.

DENMARK.—1. Local exchange. 2. Internal trunks (often free). 3. International trunks. 4. Telephoning of telegrams. 5. Local telephonogram service (ten words for 1·99*d*.). 6. Public telephone stations. 7. Messages calling strangers to distant public stations.

FRANCE.—1. Local exchange. 2. Internal trunks. 3. International trunks. 4. Telephoning of telegrams. 5. Local telephonogram service. 6. Public telephone stations. 7. Municipal telephone stations. 8. Special wayside exchange service.

GERMANY.—1. Local exchange. 2. Internal trunks. 3. International trunks. 4. Public telephone stations. 5. Telephoning of telegrams. 6. Local telephonogram service (ten words for 2*d*.). 7. Telephoning of matter to be mailed.

HOLLAND.—1. Local exchange. 2. Internal trunks. 3. Public telephone stations. 4. Telephoning of telegrams. 5. Time service.

HUNGARY.—1. Local exchange. 2. Internal trunks. 3. International trunks. 4. Telephoning of telegrams. 5. Public telephone stations. 6. Rural or village intercourse.

ITALY.—1. Local exchange. 2. Internal trunks. 3. Public telephone stations (in some towns only). 4. Telephoning of telegrams.

LUXEMBURG.—1. Local exchange. 2. Internal trunks (included in the local subscriptions). 3. Telephoning of telegrams. 4. Local telephonogram service. 5. Public telephone stations. 6. Messenger service. 7. Parochial telephone stations. 8. Telephoning of mail matter.

phonograph; so, to avoid possible confusion, the author substitutes the word telephonogram wherever necessary throughout the book.

NORWAY.—1. Local exchange. 2. Internal trunks. 3. International trunks. 4. Telephoning of telegrams. 5. Local telephonogram service. 6. Public telephone stations. 7. Messenger service.

PORTUGAL.—1. Local exchange.

ROUMANIA.—1. Local exchange. 2. Internal trunks. 3. Public telephone stations. 4. Telephoning of telegrams. 5. Local telephonogram service.

SPAIN.—1. Local exchange. 2. Internal trunks. 3. Public telephone stations. 4. Telephoning of telegrams. 5. Local telephonogram service (twenty words for 1·92*d.*).

SWEDEN (State administration and General Telephone Company).—1. Local exchange. 2. 70-kilometer free radius. 3. Internal trunks. 4. International trunks. 5. Telephoning of telegrams. 6. Local telephonogram service (forty words for 3·3*d.*). 7. Messenger service. 8. Public telephone stations.

SWITZERLAND.—1. Local exchange. 2. Internal trunks. 3. International trunks. 4. Telephoning of telegrams. 5. Telephonogram service. 6. Parochial telephone stations. 7. Public telephone stations. 8. Special wayside exchange service.

WÜRTEMBERG.—1. Local exchange. 2. Internal trunks. 3. International trunks. 4. Telephoning of telegrams. 5. Public telephone stations. 6. Local telephonogram service (ten words for 1*d.*). 7. Telephoning of matter to be mailed.

It will be seen from this summary that only one country—Portugal—has an inferior list of facilities. One other—Italy—has the same number ; but all the rest enjoy superior advantages. In the countries noted for the widest spread of telephony it will be found that subscribers have at their command seven or eight different applications of the telephone ; thus—Sweden, 8 ; Switzerland, 8 ; Austria, 8 ; Germany, 7 ; Bavaria, 7 ; Würtemberg, 7 ; Norway, 7 ; Denmark, 7. Is it fair, therefore, to reproach the British public for being slow to subscribe ? Should it not be recognised that the telephone is one thing in Britain and another in Sweden or Austria ?

Had the telephoning of telegrams been free and unrestricted during the past fourteen years ; had it been within the power of subscribers to despatch telegrams to non-subscribers in the same town, twenty words for 1·92*d.* as in Spain, or even ten words for

2d. as in Germany and Denmark ; had they been free to telephone messages to be written down and posted as letters or post-cards, the tale might have been quite different.

It would not be correct to blame the National Telephone Company for not giving these facilities, for, indeed, it has not been in its power to accord them. No ; the blame must be borne by the Post Office, which, under the mistaken idea that the best way to serve the public interest is to curtail such facilities as are not provided by itself, has denied the public these advantages. As a consequence, its revenues have suffered by the competition of the telephone trunk lines. During the author's continental tour of investigation nothing was made clearer to him by the foreign officials than that the encouragement given to the subscribers to forward telegrams by telephone for transmission has compensated to a very large extent, if not altogether, for the telegraphic traffic lost by the rivalry of the trunk lines. In every country the tale is the same : the telegraph revenues have not suffered by the competition of the trunk lines because the extension of the telephone system has provided new feeders to the telegraph in every direction, and these newly-found feeders have provided traffic enough to outweigh the loss on certain long distance lines. Thus, to cite an example, the extensive telegram traffic which formerly prevailed between the Bourses of Brussels and Paris, and which necessitated the constant employment of several direct telegraph wires, has been entirely wiped out by the telephone circuits ; but, notwithstanding this, the telegraph receipts continue to grow. During these fourteen years, therefore, the Post Office has been engaged in cutting off its nose to spite the companies, and has voluntarily cast away a source of income which would have rendered unnecessary the wails made over revenue lost through the competition of telephone trunks. The author is of course aware that the telephone exchanges in some of the principal towns are, and have been for some years, in connection with the Postal Telegraph Office ; but what is wanted is not a partial, but a universal and unrestricted, application of the service. Obstacles are thrown in the way of establishing connection with telegraph offices. For instance, the Mutual Telephone Company applied for, but could not obtain, a connection in Manchester, while its rival, the National, was permitted to provide its subscribers with the service. In Edinburgh, after long agitation, the telephone exchange was joined to the telegraph office in 1888 or 1889, but childish regulations were made which greatly impaired the usefulness of the service, it being permissible, under them, for a man on one side of a street to have his telegrams telephoned to him, while his opposite neighbour could not. There is no parallel to such things on the Continent.

It is true that the Post Office now proposes, under its agreement with the National Telephone Company, to give facilities more commensurate with foreign practice, which is distinctly news to be rejoiced at; but why has the community been forced to wait fourteen years for them?

The charges scheduled in respect to the new services in the agreement compare most unfavourably with those in vogue elsewhere. Thirty words, *if they can be telephoned and written down by a possibly inexpert clerk in three minutes*, are to cost 3*d.* in a message intended for local delivery; but ten words for 2*d.* without any time limit would be better. Few people require to send thirty-word messages, and those who do may without injustice be left to pay extra for them. No provision is made for allowing the replies to such messages to be prepaid by the senders, nor for the messenger who delivers them to bring back the replies for immediate telephoning, which is a facility that is enjoyed in several countries abroad. The foreign telephonograms operate both ways; apparently the British message is to be *from* the subscriber only. Then it is restricted to subscribers only. In Denmark and Spain such messages, *written down*, may be handed in at any public telephone station, telephoned by the attendant to the central office, and thence delivered by messenger. In Germany a ten-word message of this description costs 2*d.*; in Copenhagen 1·99*d.*; while in Madrid one of twenty words can be sent for 1·92*d.* Is there any valid reason why the Londoners or Glaswegians should be denied a parallel privilege, or why the Post Office should discriminate against the general public in favour of subscribers to a monopoly like the National Telephone Company? Then the Post Office charge of 3*d.* is liable to be increased by a terminal charge on the part of the company. This should

Country	Minutes allowed for unit charge	Miles																		
		20	40	80	120	160	200	240	280	320	360	400	440	480	520	560	600	640	680	720
United Kingdom †.	3	6*d*	9*d*	1/3	1/9	2/4	2/10	3/4	4/-	4/6	5/-	5/6	6/-	6/6	7/-	7/6	8/-	8/6	9/-	9/6
Austria	3	6	10	1/2	1/8	and any distance beyond, 1/8														
Bavaria *	5	5	5	1/-	and any distance beyond, 1/															
Belgium *	5	generally free	9·6	any distance																
Bulgaria	5	9·6	any distance																	
Denmark—State	3	—	—	1/1¼	1/8															
„ Companies	—	Usually included in local subscription; if not, it is covered by a small increase thereon																		
Finland	5	1·12*d*.	1·21	2·3	2·7	4·3	5·4	6·5	7·6											
France * ††	5	4·8	4·8	9·6	9·6	1/2·4	1/7·2	1/7·2	2/-	2/4·8	2/4·8	2/9·6	3/2·4	3/2·4	3/7·2	4/-	4/-	4/4·8	4/4·8	4/9·6
Germany	3	5	1/-	and any distance beyond, 1/																
Holland **	3	9·9	9·9	and any distance beyond, 9·9*d*.																
Hungary	3	1/8	and any distance beyond 1/8																	
Italy	5	2/5	2/5	2/5	2/5	2/5	2/5	2/5	2/5	2/10·7	2/10·7	3/4·5	3/10·2	3/10·2	4/4	4/9·8	4/9·8	5/3·5	5/3·5	5/9·3
Luxemburg	—	Free	any distance																	
Norway	5	Varies capriciously from 3·3*d*. to 6·5*d*.																		
Roumania *	3	1/2·4	1/2·4	2/-	2/-	2/9·6	3/7·2	3/7·2	4/4·8	5/2·2	5/2·2									
Spain	3	6·7	6·7	1/-	1/-	1/4·8	1/9·6	1/9·6	2/2·4	2/7·2	2/7·2	3/-	3/4·8	3/4·8	3/9·4	4/2·2	4/2·2	4/7	4/7	5/1·8
Sweden	3	Free	Free	4	4	6·6	6·6	6·6	6·6	6·6	6·6	9·9	9·9	9·9	9·9	1/1·2	1/1·2	1/1·2	1/1·2	1/1·2
Switzerland	3	2·88	4·8	and all distances beyond, 4·8*d*.																
Würtemberg	5	5	any distance																	

NOTE.—The charges in this table are given in shillings, pence, and decimals of pence.

REMARKS.—In the countries marked * the unit charges are reduced when a number of talks is subscribed for in advance. ** In addition to an annual charge of 16*s*. 6½*d*. † Including the terminal charges which the company is authorised to levy. †† Rates reduced to nearly one-half between the hours of 9 P.M. and 7 A.M.

not be, or, at least, the terminal fee ought to be ascertained before the agreement becomes law, and the gross cost to the customer fixed.

The proposed trunk rates, even without the terminal charges which the agreement authorises the company to levy, are, without exception and by far, the dearest in Europe.

The table on p. 16 drawn up from official data contrasts the proposed British with the trunk rates of all countries in which trunk lines are actually working.

It will be seen that, saving for a few of the shorter distances, the British rates are far higher than any of the others, with the single exception of Roumania's. In that country all telephone rates, local as well as trunk, are phenomenally dear, and the natural result is shown in the fact that Bucharest, the capital, possesses only 100 subscribers after several years' exploitation. Italy and Spain are the next dearest, but in neither country has any considerable experience in trunk work yet been gained. The lines opened are short and of recent origin. The tariffs have been made in advance, and are not, consequently, of the same value as guides as those of Sweden or Germany, which have been in operation over long distances for several years. The French rates average about half the British and are yet amongst the dearest on the Continent.

At four hundred miles, say the length of a trunk from London to Glasgow or Edinburgh, the British charge is 5*s.* 6*d.*, against 1*s.* 8*d.* Austria, 1*s.* Bavaria, 2*s.* 9½*d.* France, 1*s.* Germany, 3*s.* 4½*d.* Italy, 3*s.* Spain, and 10*d.* Sweden. At six hundred miles Britain is 8*s.*, Austria 1*s.* 8*d.*, France 4*s.*, Germany 1*s.*, Italy 4*s.* 10*d.*, Spain 4*s.* 2*d.*, and Sweden 1*s.* 1½*d.* In fact, the British tariff, it is to be feared, will not give telephonic traffic a chance to develop at the longer distances : it is likely to prove prohibitive for all but stockbrokers, a class of the community very estimable no doubt in its way, but not sufficiently so to entitle it to the monopoly of lines erected at the public expense. In considering the trunk question it should be borne in mind that in several countries large reductions on the tariff rates may be obtained by subscribing for a number of talks in advance. This is the case even in Roumania, Britain's only rival in dearness.

There is no indication of any intention to reduce trunk

rates at night and so encourage communication during the off hours. In France and between France and Belgium rates are reduced about one half between 9 P.M. and 7 A.M., with satisfactory results.

Indeed, there is no assurance that the trunks will be open at night at all. At present they are; but when they terminate, as it is intended that they shall, at the post offices, which mostly close at 9 P.M., a retrograde step in this respect is to be feared. Then the greater part of the capital invested in the trunk lines will lie fallow during ten hours or so out of the twenty-four.

It may be well to point out in connection with the trunk line question that in Norway and Denmark, where independent companies exist in nearly every town, trunk line communication is established and worked without friction by the adoption of a very simple plan—that of allowing each company to erect and maintain the trunks within its own territory, and to keep all the money it can take at its own end.

The author must confess inability to understand the proposal of the Post Office to pay a commission to the company on telegrams telephoned. No such commission is paid anywhere on the Continent, for the very good reason that the mere existence of the facility of telephoning telegrams constitutes a valuable aid to the company in securing new subscribers. The usual practice (when the service is not perfectly free, as in Belgium and Bavaria) is to require a payment *from* the company or subscriber. The Post Office should afford connection to its telegraph offices in all towns where the facility is asked for, and abolish all vexatious restrictions and regulations; but it has no call to pay the company for doing what it is glad and anxious to do wherever permitted. At least, if a commission *is* paid to the company it should be stipulated that it, on its part, must impose no charge of any description on its subscribers in connection with the telegram service.

The proposal of the Post Office to withdraw its veto on the establishment of public call offices in the houses or shops of sub-postmasters is only what it ought to have done years ago. In fact, the veto should never have been imposed. On the Continent, call offices or public telephone stations at the post and telegraph offices are generally provided (in Germany they usually exist nowhere else), and are found a great convenience. The duty ought to be imposed on the Post Office of finding room for a call-box at all its chief branches, and to recoup itself, not by charging a rent which might prove prohibitive to the company, but by retaining, say, half the receipts. It would then be to the interest of both Post Office and company to develop the traffic. In Italy the Government imposes a tax of 2*l.* per annum on all public telephone stations, with the result that they are few and far between, several of the largest towns not possessing even one.

The proposal to allow railways, canals, &c., to be used by the company at a nominal charge is only reasonable. The monopoly given by Parliament to the Post Office in respect to the erection of wires on railways was conferred before the existence of telephony as a practical art was dreamed of, and was never intended to act as a bar to legitimate public requirements. Any powers in connection with railways or canals conferred on the National Company ought to be extended to any other companies, municipalities, or persons who may hereafter become licensees; and also to those who may require to erect private telephone lines.

A table is given on pp. 20, 21 of the charges made in the various continental countries for the different services rendered. The exceptions and variations are so numerous that it is a little difficult to make comparisons at every point; but by taking the most commonly used unit charges in each country it is nevertheless possible to compress a mass of information into a small compass.

One feature in the table will doubtless strike the observer. It is the column headed 'Entrance fee,' and it refers to a practice which has enabled wonders to be wrought in the direction of cheap telephony on a modest amount of capital, for practically it works out that the subscriber finds, in the shape of 'entrance fee,' 'admission charge,' or 'contribution,' as it is named in various countries, the capital, or the greater part of it, required for the installation of his line, instrument, and share of exchange apparatus. The custom prevails in Austria, France, Monaco, Roumania, and Sweden, on the part of the respective States, and in Denmark (partially), Finland (partially), Norway (partially), and

—	Country	Entrance Fee	Annual Subscription for one exchange line and instrument	Charge for a second connection
1.	Austria	4*l.* 3*s.* 4*d.*	4*l.* 3*s.* 4*d.* 500 meters	4*l.* 3*s.* 4*d.*
2.	Bavaria		7*l.* 10*s.* 5 kilometers	3*l.* 15*s.*
3.	Belgium	—	From 5*l.* to 10*l.* Usually 3 kilometers	Variable
4.	Bulgaria	—	8*l.* first year ; 6*l.* subsequently	—
5.	Denmark	—	Copenhagen, 8*l.* 6*s.* 8*d.* Provinces, from 1*l.* 18*s.* 8*d.* to 4*l.* 8*s.* 11*d.*	Copenhagen, 6*l.* 13*s.* 4*d.*
6.	Finland	Companies, nil	Companies, 3*l.* 4*s.* to 4*l.* 16*s.*	—
7.	France	Co-operative Societies, 8*l.* to 10*l.*	Co-operative Societies, 2*l.* to 2*l.* 16*s.*	—
		Paris, nil	Paris, 16*l.*	6*l.* 8*s.*
		Lyons, nil	Lyons, 12*l.*	4*l.* 16*s.*
		12*s.* per 100 meters of single wire other towns	Other towns over 25,000, 8*l.* Other towns under 25,000, 6*l.*	
8.	Germany	—	7*l.* 10*s.* 5 kilometers	5*l.*
9.	Holland	—	Amsterdam, 9*l.* 14*s.* 2½*d.* Dordrecht, 4*l.* 3*s.* 11*d.* Breda, 2*l.* 17*s.* 10*d.* Alkmaar, 2*l.* 9*s.* 7*d.*	—
10.	Hungary	—	Buda-Pesth, 12*l.* 10*s.* Other towns, 5*l.*	—
11.	Italy	—	2*l.* 16*s.* to 8*l.*	—
12.	Luxemburg	—	3*l.* 4*s.*, including right to use all trunks	—
13.	Monaco	12*s.* per 100 meters of single wire	6*l.*	—
14.	Norway	—	Christiania, 4*l.* 8*s.* 11*d.* Provinces, 1*l.* 8*s.* to 3*l.* 6*s.* 8*d.*	1*l.* 13*s.* 4*d.*
15.	Portugal	—	7*l.* 10*s.*	5*l.* 12*s.* 6*d.*
16.	Roumania	6*l.*	8*l.* to cover 1,000 talks per annum ; 16*s.* per 100 afterwards	—
17.	Russia	—	St. Petersburg and Moscow, 25*l.* Other towns, 10*l.* to 12*l.* 10*s.*, 2½ miles	—
18.	Spain	—	5*l.* 12*s.* to 12*l.*, according to population	—
19.	Sweden	Company, 2*l.* 15*s.* 7*d.* any distance State, 2*l.* 15*s.* 7*d.*	Company, 5*l.* 11*s.* 1*d* State, 4*l.* 8*s.* 11*d.*	Company, 4*l.* 8*s.* 11*d.* State, 3*l.* 6*s.* 8*d.*
20.	Switzerland	—	1st year, 4*l.* 16*s.* 2nd year, 4*l.* 3rd year, 3*l.* 4*s.* Covers 800 calls only per annum	—
21.	Würtemberg	—	5*l.* 3 kilometers	2*l.* 10*s.*

Erratum

Page 20.—Table of Rates. In first column the word 'France' should be opposite 'Paris' in second column. The information *re* Co-operative Societies refers to Finland.

—	Internal Trunk Rates		Fee for telephoning telegrams	Fee for telephonograms	Fee for telephoning mail matter	Public Telephone Station charges
	Minimum	Maximum				
1.	6*d.* 3 minutes	1*s.* 8*d.* 3 minutes	1*d.* + ·1*d.* per word	1*d.* + ·1*d.* per word	1*d.* + ·1*d.* per word	2*d.* 3 minutes
2.	5*d.* 5 minutes	1*s.* 5 minutes	Free	1*d.* + ·1*d.* per word	1*d.* + ·1*d.* per word	2·5*d.* 5 minutes
3.	9·6*d.* 5 minutes	9·6*d.* 5 minutes	Free	—	—	2·4*d.* 5 minutes
4.	9·6*d.* 5 minutes	9·6*d.* 5 minutes	—	—	—	4·8*d.* 5 minutes
5.	Cos., free State, 1*s.* 0½*d.* 3 minutes	Cos., free State, 1*s.* 8*d.* 3 minutes	Companies ·133*d.* per word ; State, 2·6*d.* per message	10 words for 1·99*d.*	—	2*d.* 5 minutes
6.	1*d.* 5 minutes	9*d.* 5 minutes	—	—		—
7.	4·8*d.* per meters, 5	100 kilo- minutes	Paris and Lyons, 2*l.* per annum ; other towns, free	4·8*d.* per 5 minutes occupied	—	Paris, 4·8*d.* Provinces, 2·4*d.* 5 minutes
8.	5*d.* 3 minutes	1*s.* 3 minutes	1*d.* + ·1*d.* per word	1*d.* + ·1*d.* per word	1*d.* + ·1*d.* per word	2·5*d.* 3 minutes
9.	16*s.* 6½*d.* and per 9·9*d.* 3 minutes	per annum ; talk 9·9*d.* 3 minutes	8*s.* 3*d.* per annum, and per message ·99*d.*	—	—	4·95*d.* 5 minutes
10.	1*s.* 8*d.* 3 minutes	1*s.* 8*d.* 3 minutes	2*d.* per message	—	—	2*d.* 5 minutes
11.	2*s.* 5*d.* not exceeding 500 kilometers, and 5·76*d.* per 100 kilometers beyond		1·92*d.* per message	—	—	·96*d.* to 2·4*d.* 5 minutes
12.	Included subscri	in local ption	·98*d.* per message	·98*d.* per message + 3·36*d.* cost of special messenger	·98*d.* per message + postage	3·36*d.* 5 minutes
13.	—	—	Free	4·8*d.* per 5 minutes occupied	—	2·4*d.* 5 minutes
14.	3·3*d.* 5 minutes	6·5*d.* 5 minutes	2·6*d.* not exceeding 20 words ; ·66*d.* per 10 words after	30 words for 4*d.*	—	1·3*d.* 5 minutes
15.	—	—	—			—
16.	14·4*d.* fir meters ; 100	st 100 kilo- 9·6*d.* per after	·96*d.* per message + ·48*d.* for each 5 words	4·8*d.* first 20 words, and 1·92*d.* for each 20 after	—	9·6*d.* 5 minutes
17.	—	—	—		—	
18.	5·3*d.* 3 minutes	2*s.* 7*d.* 3 minutes	—	1·92*d.* 20 words ; ·48*d.* each 5 after	—	1·92*d.* 3 minutes
19.	Free up to meters ; 2*d.*	70 kilo- beyond, 1*s.* 1½*d.*	Company, free ; State, ·66*d.*	40 words for 3·3*d.*	—	Company, 1·3*d.* within radius 70 kilometers ; State, 1·3*d.* within Stockholm ; 1·99*d.* within 70 kilometers
20.	2·88*d.* 3 minutes	7·2*d.* 3 minutes	·96*d.* per message	1·92*d.* + ·096*d.* per word	—	·96*d.* 3 minutes
21.	5*d.* 5 minutes	5*d.* 5 minutes	·1*d.* per word ; minimum, 1*d.*	·1*d.* per word, minimum 1*d.* + cost of messenger	·1*d.* per word ; minimum, 1*d.*	2*d.* 5 minutes

Sweden, on the part of the various telephone companies. The contributions exacted vary greatly. In France they are high, amounting to 12*s.* per 100 meters of single wire, or 1*l.* 4*s.* per 100 meters of double wire, which, in most cases, is in excess of the real cost, so that the State makes a profit out of the subscriber at the first onset. The Austrian is less, being 4*l.* 3*s.* 4*d.* for 500 meters of double wire, against the French 6*l.* In Sweden, however, the contribution is only 2*l.* 15*s.* 7*d.*, just half of one year's (company's) rental, irrespective of the length of the line so long as it does not extend beyond the limits of the town. Such an amount once paid is not felt by the subscriber, but is of enormous importance to a company or individual concessionary, as it provides funds wherewith to construct the exchange. This is the secret of the existence and success of many of the small Norwegian and Danish exchanges, and the author is aware of no valid reason why it should not be practised in the United Kingdom too. It would operate admirably in aid of the smaller municipalities desiring to start their own exchanges, for it would obviate the necessity of drawing on the rates for the purpose, a method to which objection has been expressed in certain quarters. Nobody could demur to municipalities establishing exchanges with the subscribers' own money, which might be returned gradually in the shape of reduced rentals after the business had begun to yield a profit.

It is this contribution system, together with the profits remaining after paying its maximum dividend of 8 per cent., which has helped the General Telephone Company of Stockholm, with a paid-up capital of only 32,966*l.*, to cover a radius of 43·49 miles of country round the capital, with a network of trunk lines comprising 121 switch-rooms and 10,346 subscribers' instruments, and to evolve a property valued, at the end of 1894, after eleven years' working, at 205,648*l.*, besides building up substantial reserves, employees' accident and benevolent funds, and paying for the conversion of the whole of its Stockholm system from single to double wire.

It is such results as these which should command the attention of the British public. Let those interested—and who is not?—in the serious question of trade depression and want of employment for the masses ask themselves why similar results, which would find occupation both for idle capital and for thousands of workmen, clerks, and female operators, cannot be achieved in our own country. Where one telephone employee now exists, five or six years' vigorous development would call fifty into being.

The proposal of the Post Office to buy the existing trunk lines at 'cost price as shown by the company's books, together with a further sum of 10 per cent.,' should be jealously examined. The Post Office officials have a standing complaint that in 1870 the telegraphs were acquired at twice or three times their proper value, and anxiety is professed to avoid a similar extravagance in the case of the telephones. But in the author's opinion the Post Office officials are on the eve of tumbling into as grave an error now as did their predecessors of 1870. Many of the existing trunk lines are ten years old at least, and consequently, even when built of good materials, are far on the road towards the natural life limit of creosoted telegraph poles. But, as a matter of fact, many of the lines were not built of creosoted timber at all, but of wood unimpregnated with any preservative compound. The author himself erected trunk lines in the years 1885–89 with poles that were of insufficient diameter and otherwise unsuited for such purposes, but which were the best the company could be induced to provide. To buy these to-day at cost price plus 10 per cent. would be a transaction as improvident as any concluded in 1870.

In connection with the acquisition of the trunk lines by the Government, another point requires to be considered: viz., can the trunks be worked under the new conditions as promptly and satisfactorily as at present? According to accepted interpretations of the Post Office intentions, it is proposed to terminate the trunk lines in the post offices of the various towns, communication being had with the telephone exchanges by means of junction wires. This means that each telephonic call from one town to another will have to be dealt with by four operators instead of two, and consequently double the time will be taken in getting a connection through; the cost in wages and in wear and tear of apparatus will be also doubled, while the earning capacity of the trunks will be materially reduced, which may bring about a tendency to compensate for reduced carrying power by the imposition of higher rates. In trunk switching it is necessary, in order to obtain maximum speed, that a branch from each subscriber's wire shall be present on the trunk switch-board, so that the trunk operator may be able to put a trunk in connection with a subscriber's line directly, without the intervention of another person. To give effect to this plan after the acquisition of the trunks by the Post Office, it will be necessary to extend *all* the subscribers' wires from the telephone exchanges to the local post offices. In Manchester, as in Liverpool, the two institutions are some quarter of a mile apart, while in each town there are some 2,500 subscribers, any one of which may be asked for at any moment over a trunk line. It will be requisite, therefore, if the present speed of trunk switching is to be maintained, to construct 2,500 wires, each a quarter of a mile long, in Manchester and the same number in Liverpool, or a total length of 1,250 miles of new wires for those two towns alone. In towns worked on the metallic circuit system the mileage required would be doubled. But it is understood that it is not proposed to adopt this plan; consequently the switching speed, together with the earning power of the trunks, must be inevitably reduced.

It is generally believed in telephonic circles that, Parliament consenting, the Post Office will acquire the entire business of the National Telephone Company at December 31, 1897, the next break in the licence. It behoves the public, and, above all, the commercial community, to watch that the transfer is only allowed to take place under conditions which will assure a good service and an uninterrupted development at reasonable rates, to be set forth and fixed beforehand. It is useless to attempt to disguise the fact that the Post Office has always opposed low rates, no matter to what applied. The twopenny post, the penny post, the newspaper post, the parcel post, post-cards, reply post-cards, sixpenny telegrams; in short, every improvement without exception had to pass the gauntlet of official obstruction before it could attain the stage of useful existence. It may safely be predicted, therefore, that the Post Office will seek, whenever the acquisition of the whole telephonic business of the country comes up for settlement, to induce Parliament to sanction rates far in excess of those current on the Continent. That should in no wise be permitted. The preceding pages have amply demonstrated that rates of 2*l.* 10*s.* in the smaller and of 5*l.* in the larger towns are made remunerative abroad. The author's view is that, following the example of the French and Spanish Governments, Parliament should impose a scale of rates varying with the populations of the towns. After much consideration and analysis the author has satisfied himself that municipalities could establish and efficiently work exchanges on the metallic circuit plan, constructed underground in the centres of the towns and overhead in the suburbs as in Vienna and Zürich (see Austrian and Swiss sections), on the following rates, which are inclusive of Post Office royalty. These rates being possible for municipalities, should be possible for the Post Office also, and accordingly imposed on that department. No article is worth more than it can be bought for, and the commercial community is entitled to purchase what it wants in the cheapest market.

PROPOSED SCALE OF INCLUSIVE RATES TO BE CHARGED IN THE UNITED KINGDOM BY THE POST OFFICE OR BY FUTURE LICENSEES FOR LINES NOT EXCEEDING ONE MILE IN LENGTH.

	£	*s.*	*d.*
Towns up to 10,000 inhabitants	4	0	0
,, of 10,000 to 25,000 inhabitants . .	4	5	0
,, of 25,000 to 50,000 ,, . . .	4	10	0
,, of 50,000 to 100,000 ,, . . .	4	15	0
,, of 100,000 to 150,000 ,, . . .	5	0	0
,, of 150,000 to 250,000 ,, . . .	5	5	0
,, of 250,000 to 500,000 ,, . . .	5	10	0
,, of 500,000 to 750,000 ,, . . .	5	15	0
London	8	0	0

Of course the Post Office would not willingly accept such rates, in, the author believes, the perfectly sincere and honest conviction that they would not pay. But still the fact remains that they have been made to pay and *are* made to pay. A telephone engineer fetched over from Trondhjem, where a population of over 30,000 souls is successfully catered for on a 2*l.* 10*s.* rate, would no doubt be of a different opinion and anxious for an opportunity of showing how it's done.

But a conflict of views is inevitable, and the author would propose the following plan as being both practicable and calculated to bring conviction in its train.

As before stated, the Post Office cannot possibly acquire the whole business of the National Telephone Company before December 31, 1897. In the interim period there is plenty of time for a municipality to show what can be done in the direction of low rates and improved service. Let two or three municipalities be licensed on the condition that metallic circuits are employed throughout, so that, in the event of an ultimate Post Office purchase, the municipal exchanges will fit in properly with, and make part and parcel of, the postal system. By the end of 1897 such experience will be gained, if the municipalities go wisely to work, as will put an end to all quibbles as to the sufficiency of a 5*l.* rate. Such a test should be welcomed by all parties, whether for or against low rates, really wishing for a settlement of the question.

But the author doubts whether the Post Office realises the importance of the subject of national telephony. Speaking in the House on March 1, 1895, the Postmaster-General ('Daily Chronicle,' March 2, 1895) said that 'the telephone could not, and never would be, an advantage which could be enjoyed by the large mass of the people. He would go further and say if in a town like London or Glasgow the telephone service was so inexpensive that it could be placed in the houses of the people, it would be absolutely impossible. What was wanting in the telephone service was prompt communication, and if they had a large number of people using instruments they could not get prompt communication and yet make the telephone service effective.'

What can be expected from a department whose chief entertains opinions such as these? What hope can be entertained when the fountain of knowledge is thus found frozen at its source? Let the reader turn to the Swedish, Norwegian, and Swiss (with its parochial telephone stations) sections of this book, and judge whether Mr. Arnold Morley really knows so much of what is passing in the world as to justify his assumption of the rôle of prophet. The 'could not and never would be' is strongly suggestive of the predictions about railways and telegraphs and steamboats which used to be made when those inventions were in their infancy—of the late Dr. Lardner's rash undertaking to eat the first steamer, cargo and all, that succeeded in crossing the Atlantic. The author believes that Mr. Morley will live to be wiser. If not, then Stockholm with its 11,534 exchange telephones and Berlin with its 25,000 and odd subscribers exist in vain.

We are all addicted to accept our own individual experiences as guides, and the fact probably is that Mr. Morley, not unnaturally perhaps, but still with a limitation of vision rather amazing in a Minister of Posts and Telegraphs, is basing his belief on home, nay London, experience. He believes that the presently existing system is the best possible, and he deduces (and with infinite correctness) that no possible modification of it can bring the telephone home to the masses. Pursuing the same line of arguement, but substituting provisions for telephones, Mr. Morley would be equally safe in declaring that the large mass of the population of London or Glasgow could not, and never would be, provided with daily bread. And he would be right, assuming that the distribution of food were carried out on a plan analogous to that on which telephones are now supplied. If all provisions brought into a large town were carried to one central site and thence distributed *direct* to the house of each individual consumer, without the intervention of markets, of shops, of costermongers, or any of the usual intermediaries, the task involved would border on the impossible. Division of labour is imperative in such a case. When the labourers are many and work intelligently on an organised plan, each in his own sphere, a very minute sphere perhaps too, the bread and the milk and the meat will find its way almost, to appearances, automatically to the remotest capillaries of the city's anatomy. So it is with telephones.

Take a town, however immense, and realise that at no very remote period telephones will be numerous in many parts of it and totally wanting in none, and the task of devising a plan for an exchange to meet all possible requirements becomes an easy one. Such a plan the author laid before the British Association in 1891,[1] and such a plan has recently been adopted in the reconstruction of the General Telephone Company's system at Stock-

[1] *On the Telephoning of Great Cities*, pamphlet by the present author. Whittaker & Co. 1*s*.

holm (see Swedish section). With it each telephone ordered drops into its place naturally and economically. A large portion of that traffic which Mr. Morley fears, would not pass beyond the local exchange (or shop) at all; and there exists no difficulty whatever in dealing with the whole, however extensive it may be. With such a plan in operation, whole suburbs of London would not be totally cut off from telephone exchange intercourse as at present.

In the speech already quoted the Postmaster-General ('Daily Chronicle,' March 2, 1895) told the House of Commons that the charge for telephones in London is only 10*l.*, exactly one half of the actual figure, and that rates rose as high as 40*l.* and 50*l.* in America, meaning, no doubt, the United States. These high American rates are confined to a few towns, and there are special, although not very satisfactory, reasons for their existence; but why should the Postmaster-General, when instructing the House of Commons, mention high rates, which are exceptional, and omit all reference to the low rates which are almost universal elsewhere than in Britain? Is the British Post Office really unaware of the existence of these last? The author thinks not, and for the following reason. In October 1894 the author in the course of his continental tour of investigation made formal application through the British Consulate at Berlin (a procedure he was advised was necessary) for permission to inspect the Berlin telephone system. He was informed that this could not be permitted without an introduction from the British Postmaster-General. At the same time the German Government wrote to the British Post Office inquiring whether it approved of the application or had any objection to its being complied with. It may read strange that the German Government imagines that a British electrician is necessarily in the leading strings of his Post Office, and stranger still—although somewhat flattering to the national vanity—that the Imperial German Post Office considers itself under the orders of St. Martin's-le-Grand; but so it is. What the tenor of the reply was the author knows not, but the result was a refusal to allow any inspection or to impart any information. It would not be complimentary to the intelligence or patriotism of the Post Office to imagine that it would, without an object, deliberately obstruct a

British subject in a quest for legitimate information abroad on a question in which he is known to be specially interested. The author shrinks from even verging on the uncomplimentary, so it is necessary to at least imagine a reason. Can it be that, knowing the author's consistent advocacy of low rates, the British Post Office feared that he would learn that in Germany the maximum local rate, even in Berlin with its 25,000 subscribers, is only 7*l.* 10*s.* per annum, everything included; and that a three-minute conversation can be had between any two points of the Imperial German Post Office territory—even when six hundred miles or more apart—for one shilling?—a facility for which the British Post Office proposes to charge 8*s.* If this was *not* the reason, it is of course open to the Post Office to make known its real motive.

Fortunately, this unpatriotic obstruction did not prevent the author from eventually obtaining all the information he sought, as will appear from a perusal of the German section.

The book is not entirely devoted to tariffs and regulations. Such matters are indissolubly bound up with technical questions, for cheap rates with bad construction and indifferent service are to be deprecated, and indeed disallowed altogether, for the author holds them to be intolerable, and only less acceptable than the combination of dear rates and a bad service. The service of a telephone exchange should be the first consideration. This opinion has always led the author to advocate the universal use of metallic circuits, without which privacy of conversation and speech undisturbed by strange noises, together with effective long-distance talking, is unattainable. Prompt and correct switching, with no uncertainty between signals intended to have different meanings, are also essential to a good system; and the operators' voices should never be heard on the wires. The familiar 'Have you finished?' and other intrusive cries with which London operators break in upon one's conversation every few seconds are totally unnecessary in a well-ordered exchange. In large towns the main routes of wires should be laid underground, at least in the central parts. These preliminaries and essentials having been attended to, and the best of material and workmanship employed in carrying them out, attention may be profitably given to the rates. The author's contention has always been, at least for the past ten years, that all these things are compatible with the scale of charges given above.

It is hoped that the analyses of the facilities, regulations, and methods of dealing with traffic given in the book will prove of interest, and even profit, to telephone managers. The result of the working of many intelligent minds separately striving after a solution of the same problem must be always worthy of contemplation; and none are so wise as to be independent of the experience of others. The details given in the various sections make it abundantly evident that telephone managers and engineers may learn much from each other, for the facilities given to the public vary considerably in different countries, while some methods are obviously superior to others in vogue elsewhere.

The author has endeavoured, as far as possible, to avoid describing well-known apparatus and methods. In respect to the technical portions a familiarity on the part of the reader with ordinary telephone exchange work and management is throughout assumed.

It will be readily understood that a book like the present would be impossible without the cordial co-operation of many friends, and the author has pleasure indeed in acknowledging his indebtedness to the gentlemen of all nationalities with whom it was his good fortune to come in contact during his continental tour. Everywhere (except at Berlin) officials, whether of State administrations or of companies, permitted, and even courted, the fullest inspection, and placed the most ample information, documentary and otherwise, at the author's disposal. Specially he would like to place on record his thanks to the following gentlemen:—M. J. BANNEUX, Director, and M. H. FRENAY, Engineer, of the Belgian Posts and Telegraphs, Brussels; Mr. E. B. PETERSEN, General Manager of the Copenhagen Telephone Company; Mr. F. ROSBERG, Telephone Engineer, Helsingfors; M. SELIGMANN, Chief Engineer, French Telephone Administration, Paris; Dr. H. F. R. HUBRECHT, Managing Director, and Mr. N. HEINZELMANN, Engineer, Netherlands Bell Telephone Company, Amsterdam; Mr. A. E. R. COLLETTE, Engineer, Dutch Administration of Posts and Telegraphs, The Hague; Mr. C. J. VAN BUEREN, Managing

Director, Zutphen Telephone Company, Zutphen ; Messrs. RIBBINK & VAN BORK, Telephone Engineers, Breda and Amsterdam ; Signor E. GEROSA, Manager, Società Telefonica Lombarda, Milan ; Mr. KNUD BRYN, Manager, Christiania Telephone Company, Christiania ; Mr. H. T. CEDERGREN, Managing Director, General Telephone Company, Stockholm ; Mr. AXEL HULTMANN, late Chief Engineer, Swedish State Telephone Administration, Stockholm ; Dr. T. ROTHEN, Director of the Bureau International des Administrations Télégraphiques, Berne ; Dr. WIETLISBACH, Director, Swiss Telegraphic Administration, Berne ; Mr. A. HOMBERGER, Local Telephone Manager, Zürich ; Mr. MAX HAHN, Vienna ; Mr. C. SIEGEL, St. Petersburg ; M. BERTHON, Société Industrielle des Téléphones, Paris ; Mr. SPRINGBORG, Manager, Aarhus Telephone Company ; Mr. L. M. ERICSSON, Stockholm ; M. F. NEUMAN, Director of Posts and Telegraphs, Luxemburg ; Mr. C. G. NIELSON, Chairman of the Drammen Uplands Telephone Company ; Mr. NORSHUUS, Manager of the Bergen Telephone Company.

The following works have been occasionally used in writing the Belgian, French, Italian, and Dutch sections respectively :— 'La Téléphonie,' E. Piérard ; 'Téléphonie Pratique,' L. Montillot ; 'Telefono,' Domenico Civita ; 'Het Plaatselijke Telephoonnet te Zutphen,' Aug. Collette. The 'Journal Télégraphique,' the official organ of the telegraph administrations, edited by Dr. T. Rothen, has been freely drawn upon, especially Dr. Wietlisbach's description of the Zürich exchange, which he has very kindly allowed the author to use.

22 ST. ALBAN'S ROAD,
HARLESDEN, LONDON, N.W.
March 9, 1895.

I. AUSTRIA

HISTORY AND PRESENT POSITION

THE history of the telephone in Austria dates from 1880, when the Government granted a concession for the city of Vienna to the Vienna Private Telegraph Company. This was soon followed by concessions to various persons and firms for several of the principal towns, the most valuable of which were acquired by an English association, the Telephone Company of Austria, Limited. Some of these concessions were, however, burdened by impracticable conditions owing to the desire of the Government to leave the settlement of details to the local authorities most interested. For instance, it is related that the Cracow municipality required of the concessionary for that town that all wires should be run horizontally, immediately beneath the projecting eaves of the houses, and always at the same height above the ground (incompatible conditions since the heights of the buildings varied); that the wires should never cross a street, and that a sum of money should be deposited out of which the municipality could satisfy any claims for damages that might arise. It is perhaps needless to say that the Telephone Company of Austria did not touch *that* licence; and, in fact, the municipality of Cracow had to wait for its telephones until 1887, when the State began the construction of exchanges on its own account, and then, strange to say, obtained the fulfilment of none of its conditions. In addition to that of the capital, the Vienna Private Telegraph Company undertook the exchange at Brünn; the Telephone Company of Austria constructed from time to time, until its acquisition by the State on January 1, 1893, the exchanges at

Prague, Trieste, Lemberg, Graz, Czernowitz, Pilsen, Reichenberg, and Bielitz-Biala; and a company called the Linz-Urfahr Undertakers (Unternehmung) established an exchange system in Linz-Urfahr, which was also absorbed by the Government on the first day of 1893. After that date, the only company left was the Vienna Private Telegraph, which maintained an independent existence until January 1, 1895, when the State finally became the possessor of the whole Austrian system.

The rates charged by the companies varied from 8*l.* 6*s.* 8*d.* in Vienna and 7*l.* 10*s.* in Prague and Trieste to 5*l.* in the smaller towns, out of which 10 florins or 16*s.* 8*d.* per subscriber had to be paid annually to Government.

In 1887 the State began to open exchanges and construct trunk lines in accordance with the provisions of a law promulgated on October 7 of that year. Its first ventures were at Baden, Vöslau, and Wiener-Neustadt, which were connected to Vienna by single 3 mm. bronze wires. Soon afterwards, State exchanges were opened in Aussig, Teplitz, and Carlsbad, while Brünn was joined to Vienna by two telegraph wires fitted with the Van Rysselberghe apparatus. Subsequently, the extension of the State system went on rapidly until, on December 31, 1892, the day before the absorption of the first two companies, it comprised sixty-one exchanges and twenty-nine metallic circuit trunk lines, including seven international. At the date of writing (February 1895) practically all the Austrian towns of any note are in possession of exchanges, and in the enjoyment of trunk line communication. The Van Rysselberghe system has not been persisted in, so that the trunks are invariably metallic circuits intended exclusively for telephony.

The law referred to was a most important one, as it specified the services to be rendered to the public by the Imperial Post and Telegraph Department, the tariffs to be levied, and the general rules to be observed, both by the State officials and the subscribers. In future it will constitute the groundwork of Austrian telephony. The late companies' regulations will be brought into line with it as soon as existing agreements will permit, and in a few years absolute uniformity will prevail. The development of the Austrian system is likely to be rapid and extensive, for the law is conceived in a most liberal spirit. The facilities placed at the disposal of subscribers and of the general public are not only numerous, but the charges are extremely moderate, and that in spite of the adoption of the principle—first introduced, the author believes, by Mr. H. T. Cedergren of Stockholm, and subsequently adopted by the French Government—of causing the subscribers to pay for the installation of their lines and instruments by a 'contribution' as it is called in Austria, or 'admission fee' as it is termed in Sweden. This plan obviates, of course, the necessity of finding a heavy capital; each unit brings its initial cost with it, and the annual subscription has to cover only maintenance and working expenses, and not interest on capital. The 'contribution' in Austria is 4*l.* 3*s.* 4*d.* for lines not exceeding 500 meters in length, and 16*s.* 8*d.* for each additional 100 meters, making the initial cost to the subscriber of a 1-kilometer line 8*l.* 6*s.* 8*d.*, payment of which, in accordance with the law, may be extended over five years if desired. But the annual subscription is only 50 florins, or 4*l.* 3*s.* 4*d.*, so that the contribution to first cost is a bagatelle to a subscriber who comes on, as most of course do, for the term of his business life. By spreading payment over five years, a line not exceeding 500 meters in length costs only $4l.\ 3s.\ 4d. + \frac{4l.\ 3s.\ 4d.}{5} = 5l.$ per annum for the first five years, and 4*l.* 3*s.* 4*d.* per annum thereafter. Similarly, a 1-kilometer line costs $4l.\ 3s.\ 4d. + \frac{8l.\ 6s.\ 8d.}{5} =$ 5*l.* 16*s.* 8*d.* for the first five years, and 4*l.* 3*s.* 4*d.* thereafter. One good effect of the contribution system is that the line, whatever its length, being paid for, the State can afford to make the annual subscription uniform for all distances. Actually, in Austria the unit subscription of 4*l.* 3*s.* 4*d.* covers all distances up to fifteen kilometers. These facts constitute a lesson which British municipal authorities would do well to study, for it teaches how a telephone exchange may be started without capital and supported on very slender subscriptions. The trunk tariff, while not so liberal as that of Germany, is still most commendably moderate, as under it 1*s.* 8*d.* franks a three-minute conversation from one end of Austria to the other.

But there is one thing in Austria which the author would like to see remedied with all practicable despatch. Except in Vienna, where many of the lines are already doubled, although not always used as metallic circuits, the system employed is single wire with earth return. If the Austrians are prudent, they will discard this while the change is yet comparatively easy. As already stated, the development under this wise telephone law will be rapid and practically boundless. There is no finality in telephone exchange work when conducted on liberal and far-seeing principles—the horizon ever recedes as progress is attained, and new and unexpected channels for usefulness ever present themselves. It would be a pity, therefore, beyond expression, if the system were allowed to drift into such a muddle as that which already exists in Germany. Let the Austrians open no more exchanges except on the metallic circuit plan, and address themselves to the task of altering—gradually if expense is a grave consideration, but still methodically altering—all the existing single-wire ones. Otherwise in a few years' time they will find themselves in possession of a system altogether behind the age, and which, to the humiliation of the national pride, will not bear comparison with those of France, Sweden, Belgium, or Switzerland, nor, the author hopes, with that of Great Britain either.

The gist of the law will be given under the various headings.

SERVICES RENDERED TO THE PUBLIC

1. **Intercourse between the subscribers and public telephone stations of the same town.**—Subscribers are held responsible for all damage to their instruments, or to the connecting wires within their premises, arising from malice or want of proper care. They have to pay the actual cost of shifts consequent on removals as determined by the Government engineers. The State reserves power to suppress any connection, temporarily or permanently, at any time without notice ; if this is done before the expiration of five years, money paid in advance as contribution to the cost of the line will be refunded for the unexpired period ; if after five years, no refund will be made. The State accepts no responsi-

bility for interruptions, and no subscriptions will be refunded on account of failure of service. No hard and fast radius within which local subscriptions apply has been fixed, and in practice the privileged area comprises a town, and the suburbs and surrounding districts which naturally group with it. This is a wise and liberal measure, which frees the people from the restrictions imposed on suburban intercourse in France, Germany, and Würtemberg. The use of instruments is restricted to the subscribers, their servants, and to friends staying with them.

2. **Internal trunk line communication.**—The trunk system is already very extensive. At the end of 1893 forty metallic circuits, with a length of 3,302 kilometers, were in operation, and considerable extension has taken place since. The longest lines are those between Vienna and Prague (354 kilometers) ; Vienna and Trieste (505 kilometers) ; and Prague and Asch (230 kilometers). The Vienna-Prague route comprises three metallic circuits.

3. **International trunk line communication.**—With the exception of the line to Hungary, which gives Vienna communication with Buda-Pesth, Szegedin, Temesvar, Arad, Raab, Pressburg, and other towns in the sister kingdom, the most important line by far is the Vienna-Berlin (660 kilometers), opened in January 1895. The others are with Switzerland, Bavaria, Würtemberg, and Saxony (two circuits), but their use is restricted to the towns adjacent to the frontiers. The Italian Government has proposed a connection between the two countries, but nothing has yet been settled on the subject.

4. **Telephoning of telegrams.**—Every facility is given for the exercise of this privilege, the State recognising the utility of creating a branch telegraph station in every subscriber's office or house, thereby encouraging the use of the telegraph and tending to compensate for any evil influence exercised by the telephonic trunk lines on the telegraphic revenue. The telephone exchanges are usually located at a telegraph office ; when this is not the case the two are joined by wire, and clerks are always in attendance to write down messages from subscribers, or telephone those arriving for them. Messages are accepted in any ordinary language, but when the clerks are not acquainted with the tongue used, subscribers must number the letters of their messages according to a

preconcerted plan, and dictate them, by the aid of German numerals, letter by letter. The code, which is printed in the subscribers' lists, provides for forty-three different letters, including the accented ones of the French, German, and Hungarian languages. Copies of telegrams telephoned to subscribers are not afterwards delivered by messenger, but, on demand, are posted free to the addressees by the next mail. This plan saves messengers' wages, uniforms, and boot-leather to no inconsiderable amount.

5. **Telephoning of messages (telephonograms) for local delivery.**—These are of several classes, viz. :—

(1) Telephoning of written messages addressed to subscribers handed in at any public telephone station.

(2) Written messages in the form of letters or post-cards forwarded to the central telephone exchange office, by letter post or pneumatic post, in order to be telephoned to subscribers. These must bear postage-stamps to the amount of the tariff charge.

(3) Messages telephoned by subscribers to the central office to be written down and forwarded to non-subscribers by (*a*) messenger ; (*b*) post-cards or letters, by letter post ; or (*c*) pneumatic post.

(4) Message calling a non-subscriber in the same or another town to a specified public station in order to hold a conversation with the sender.

6. **Public telephone stations.**—These are fairly numerous, there being thirty-one in Vienna and ten in the suburbs, generally situated at the post and telegraph offices. The provincial towns are proportionally well served. Users of public stations can avail themselves of any of the privileges open to subscribers, telegrams and telephonograms being accepted and trunk talks allowed. No distinction is made at the public stations between subscribers and non-subscribers.

TARIFFS

1. **Rates for local exchange communication.**—Payments come under two headings—(*a*) contribution to the cost of the line and instrument ; (*b*) annual subscription. The 'contribution,' which in most cases will cover the entire cost of the line, is 4*l.* 3*s.* 4*d.* for wires not exceeding 500 meters in length, after

which it is increased at the rate of 16*s.* 8*d.* per 100 meters, up to a maximum length of fifteen kilometers. Lines exceeding fifteen kilometers in length are to be specially arranged for. The contribution for a 1-kilometer line is consequently 8*l.* 6*s.* 8*d.* Contributions may be paid down, or divided into five equal annual payments, at the subscriber's option.

The annual subscription consists nominally of two parts, 2*l.* 10*s.* for the subscriber's station and 1*l.* 13*s.* 4*d.* for the exchange apparatus ; but actually the subscriber has only to concern himself with one payment of 4*l.* 3*s.* 4*d.* This annual subscription covers all distances up to fifteen kilometers. The unit subscription of 4*l.* 3*s.* 4*d.* is, however, doubled for instruments located in railway stations, hotels, or theatres, where they can be used by travellers, guests, or spectators. Clubs and kindred institutions must also pay double rates if they wish their members to be free of the instrument.

No reduction is made for second, third, or multiple instruments. When a person takes several lines the contribution is calculated on the sum of their lengths, and the unit annual subscription is collected for each instrument.

When a subscriber cannot be joined up without the use of cables or other special works, the State reserves the right to fix his contribution at a higher rate.

Government offices pay only half of the above-named rates, and, on the recommendation of the Minister of Commerce, the same reduction is accorded to municipal and other public offices.

Subscribers who only use their instruments for six months or less in each year are also admitted to the benefit of half rates.

Subscriptions are payable half-yearly, in advance, during the first fortnights respectively of January and July.

2. **Rates for internal trunk communication.**—The time unit is three minutes.

	s.	d.
0 to 50 kilometers	0	6
51 to 100 ,,	0	10
101 to 150 ,,	1	2
Over 150 ,,	1	8

When conversation is required between two towns which can

only be joined by the connection of several trunk lines, the rate levied is the sum of the charges ordinarily made for the use of each trunk separately, provided the total does not exceed 2*s.* 6*d.*, which is the maximum.

Urgent conversations, i.e. talks which take precedence of all others, are allowed at triple the usual charge. Annual subscriptions are not admitted in connection with trunks. Users of trunks must keep a deposit of 2*l.* 1*s.* 8*d.* with the State.

3. **Rates for international trunk communication.**—Time unit, three minutes.

	s.	*d.*
Vienna—Berlin	2	6
Vienna—Buda-Pesth and the other Hungarian towns	1	8
Bregenz—Bavaria	1	0
Bregenz—Würtemberg	1	0
Bregenz—Switzerland	1	0

Express or urgent talks are admitted on all lines except to Switzerland, at triple unit rates.

4. **Rates for the telephoning of telegrams.**—For each telegram received or delivered through the telephone exchange the charge is 1*d.* plus ·1*d.* per word, fractions of a kreuzer (·2*d.*) being inadmissible in the total. A ten-word message consequently costs to telephone, 1*d.* + 10 × ·1 = 2*d.*; and an eleven-word, 1*d.* + 11 × ·1 = 2·1 + ·1 = 2·2*d.* Charges on telegrams must be covered by deposit.

5. **Rates for messages telephoned for local delivery.**—The rates for this service, as defined on page 37, are the same as for the telephoning of long-distance telegrams, viz., 1*d.* per message plus ·1*d.* per word, fractions of ·2*d.* being inadmissible. This service is restricted to subscribers who keep deposits with the State. When the message is posted as a letter an extra charge of ·2*d.* is made for the paper and envelope.

6. **Rates levied at public telephone stations.**—Talks with local subscribers, per three minutes, 2*d.* Trunk talks, forwarding of telegrams and of telephonograms, are charged as in the preceding sections.

Alarms of fire or flood or notices of accidents may be telephoned from any public station without charge.

WAY-LEAVES

The State enjoys no absolute right of way. Local authorities and proprietors are constrained from offering vexatious opposition to the passage of wires by the Telegraph Acts, but on the other hand the State must do nothing without previous consultation. Fixtures on private buildings must be negotiated with the proprietors.

SWITCHING ARRANGEMENTS

Much of the old companies' work, of course, still remains. The Vienna Private Telegraph Company in 1888, and the Telephone Company of Austria, at Prague in 1889, fitted up multiple boards designed by Mr. Otto Schäffler, of Vienna, and manufactured in that town. At Trieste the latter company placed a 1,200-line non-multiple board manufactured by the Consolidated Telephone Construction and Maintenance Company, Limited, London, which firm also supplied boards of smaller capacity for the other towns worked by the Telephone Company of Austria. Both the Schäffler and Consolidated boards are highly spoken of, and are all still in use. At Vienna there is only one central station, and there are collected (March 1895) some 7,700 lines, mostly double wires, representing subscribers, trunks, and public stations. The switching arrangements are peculiar, and probably even unique. On the ground floor are installed two Schäffler multiples of the respective capacity of 2,400 and 3,000 lines, and on the first floor another of 3,000 lines. Each is complete in itself, but connections between the respective sets of subscribers have to be made by junction wires and jacks, just as non-multiple boards were worked in the old days. Roughly, two-thirds of the calls have to be transferred in this way, a fact which naturally militates against the attainment of the highest degree of rapidity in switching (intercourse between the boards being conducted by indicators, and not *viva voce*), although each operator looks after only fifty subscribers. It must be allowed, however, that the Vienna service is markedly better and quicker than that of Paris or Berlin ; on this point there seems to be unanimous agreement.

Fig. 1

On receiving a call for a line not at her command an operator switches the caller through to the proper board, where he must repeat his order to a young lady who has duplicate jacks before her for the whole of that board. Talking is done through two ring-off drops, which both fall when the end of a connection is signalled. The Schäffler boards have jacks in series. The test is managed by completing a circuit through one of the ring-off drops, and not by the ordinary click. If a line asked for is engaged, the application of the calling plug to the jack tumbles the drop. This would be by no means a bad plan, were it not that indicator flaps are so many Humpty Dumptys, unable to pick themselves up after a fall. Every operator has before her fifty signalling drops with answering jacks for the subscribers, together with transfer jacks and nine ring-off drops with their corresponding cords, plugs, and switches. The switches have black and white handles for operating the right and left cords respectively; the cord in connection with the white handle is short, and will reach only to the answering jacks; the other is three meters long, and is used for testing and connecting the lines called for. The jacks are in rows of twenty-five, thirty rows making a vertical division, and four divisions comprising a repeat of 3,000 jacks, of which there are fifteen in the latest board installed, a view of which is given in fig. 1. When full, the boards will contain 134,000 spring-jacks and seat a total of 168 operators. The wiring is effected by twenty-six cables containing wires of thirteen different colours, each twisted with a white one. The calls dealt with are said to sometimes amount to fourteen per subscriber per day. The cost of the 3,000-line board last installed is stated, with its cables and all fittings, to have been 19,537*l*., and exclusive of these, 15,000*l*. The workmanship is undoubtedly good and substantial, and so, happily, is in thorough accord with the price.

HOURS OF SERVICE

These coincide, as a rule, with the hours of telegraphic service, which in Vienna, Trieste, Prague, and other chief towns is continuous day and night. In the smaller towns the exchanges open at 7 or 8 A.M. and close at 8 or 9 P.M.

SUBSCRIBERS' INSTRUMENTS

There has been considerable variety in these since the early days of telephony in Austria. The Vienna Private Telegraph Company commenced with a modification of the Blake as a transmitter in Vienna and Brünn, and ordinary Bell receivers; while the Telephone Company of Austria adopted the Gower-Bell with a magneto ringer, in Prague, and Blake-Bells in their other towns, all their instruments being supplied by the Consolidated Telephone Construction and Maintenance Company of London. The Linz-Urfahr Undertakers went in for a modified Edison lamp-black button transmitter. It speaks well for the foresight of all the Austrian telephone engineers that they strictly avoided battery ringing, adopting magnetos from the outset, and have thus saved themselves from the embarrassment and expense now being experienced in connection with batteries in France and Germany. Latterly the Vienna Company has adopted the set shown in fig. 2. The transmitter is sometimes of the Schäffler and Körner types, but is now generally the well-known Deckert, which was introduced into the United Kingdom by the General Electric Company of London in 1891 as the 'Hunnings Cone,' and adopted by the author for the Mutual Telephone Company's exchange in Manchester with happy results. It has since been largely used by the National Telephone Company in London. The Vienna set of instruments cannot be commended as comprising the best possible arrangements. The magneto has no automatic cut-in for the generator coils, so the button g must be pressed when a

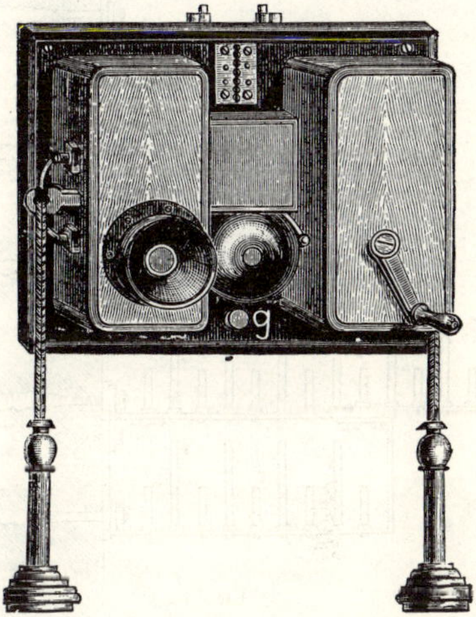

FIG. 2

subscriber would ring, thus needlessly occupying both hands and rendering the use of papers or pencil difficult, especially as no scribbling desk is provided. The magneto crank is inconveniently placed on a level with the subscriber's mouth, and in a position which renders it liable to be knocked against and damaged. There must be a separate battery-box, on the floor or elsewhere, with the expense of long connecting wires. The phones are hung up by looped cords in a manner calculated to fray both the cords and the users' tempers. Trembling bells are employed in conjunction with magnetos; in fact, every practicable sin against convenience and teachings of experience is committed. The combination is the more extraordinary, seeing that the Vienna Company's engineers have had the Consolidated Company's sets, comprising magneto bell, desk, battery-box, crank at right-hand side, automatic cut-in, forked lever for holding phone—all on one back-board—before their eyes for years, in Prague, Trieste, and other towns.

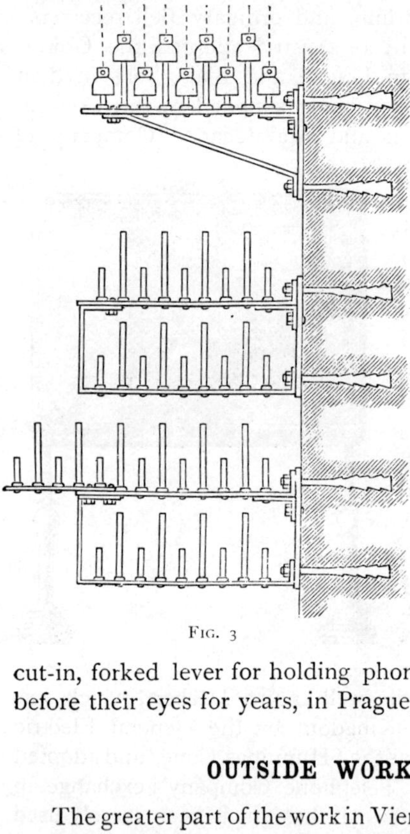

Fig. 3

OUTSIDE WORK (LOCAL)

The greater part of the work in Vienna is underground, the cables extending in some directions as far as four and a half kilometers from the exchange. The subscribers, however, in the immediate neighbourhood of the central station are served by overhead wires: these number some 300 only. The underground conductors are of 1 mm. copper, insulated with gutta-percha covered with cotton. They are spiralled together, and made up into cables, containing 5, 10, 15, and 20 pairs, by being wound with waterproofed, and then with tarred, tape. The cables are laid in larch troughs which are filled in with a mixture of asphalt and hydraulic lime, and

FIG. 4

then closed with strips of wood. The asphalt mixture never completely hardens, and forms no fissures through which moisture can reach the cables. The success of this method is reported to be complete, the cables suffering no appreciable deterioration after several years' service. There has certainly been plenty of opportunity for deciding the point, for at the end of 1893 the cables—many of which have been manufactured by Mr. O. Bondy, of Vienna—measured 154 kilometers, while the conductors reached a total of 35,493 kilometers. The wires are led to the subscribers overhead by the aid of distributing poles or standards on which the cables terminate, and the aërial lines (which are of 1·25 mm. silicium bronze, supported on double-shed insulators) commence. The immunity of the cables is the more remarkable inasmuch as there are no lightning guards at the junctions with the open wires, although protectors are provided at the exchange and on the

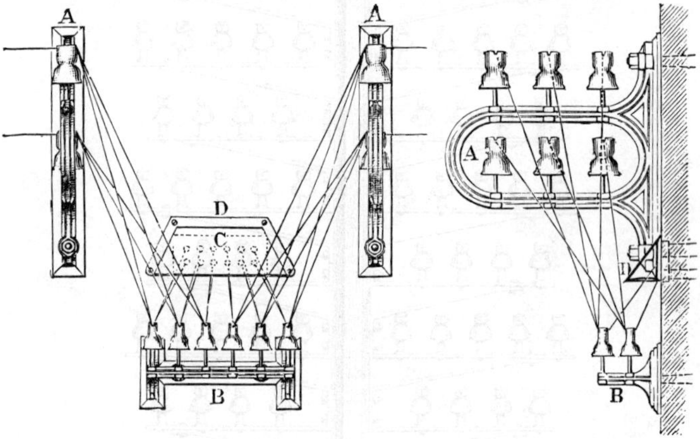

FIG. 5

subscribers' instruments. The overhead work is extensive in the suburbs and down by and across the river, attaining a total length of wire (in 1893) of 6,000 kilometers. Wall-bracket supports of the forms shown in fig. 3 are extensively used. The same style of bracket is also attached to poles, and makes a very presentable design. Along the river at Vienna a handsome route of octagonal poles so fitted (fig. 4) exists. A form of wall-bracket used by the State is shown at fig. 5, together with a method of leading wires into a house, which is largely practised in Austria and Germany. From the terminal insulator A the wire goes to a smaller bracket and insulator B, whence it is taken through a hole, C, in the wall, a cover

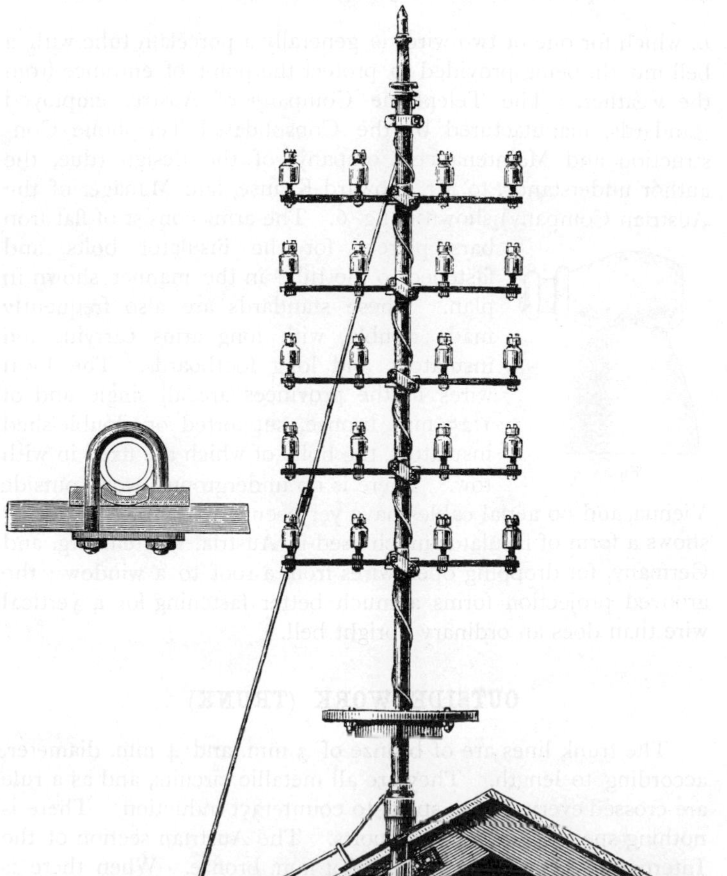

FIG. 6

D, which for one or two wires is generally a porcelain tube with a bell mouth, being provided to protect the point of entrance from the weather. The Telephone Company of Austria employed standards, manufactured by the Consolidated Telephone Construction and Maintenance Company, of the design (due, the author understands, to Mr. Howard Krause, late Manager of the Austrian Company) shown in fig. 6. The arms consist of flat iron bars pierced for the insulator bolts, and fastened to the tube in the manner shown in plan. These standards are also frequently made double, with long arms carrying ten insulators, and long footboards. The local wires in the provinces are all single and of 1·25 mm. bronze, supported on double-shed insulators, the bolts of which are fixed in with tow. There is no underground work outside Vienna, and no aërial cables have yet been used in Austria. Fig. 7 shows a form of insulator much used in Austria, Würtemberg, and Germany, for dropping open wires from a roof to a window; the grooved projection forms a much better fastening for a vertical wire than does an ordinary upright bell.

FIG. 7

OUTSIDE WORK (TRUNK)

The trunk lines are of bronze of 3 mm. and 4 mm. diameter, according to length. They are all metallic circuits, and as a rule are crossed every sixteen spans to counteract induction. There is nothing special about the supports. The Austrian section of the International line to Berlin is of 4 mm. bronze. When there is more than one metallic circuit between the same points they generally follow different routes; thus Vienna has three loops to Prague, measuring respectively 307, 308, and 354 kilometers.

PAYMENT OF WORKMEN

Foremen receive 1*l.* 8*s.* 4*d.* per week; skilled wiremen, 1*l.*; and labourers, from 15*s.* to 16*s.* 8*d.* The sleeping allowance is 10*d.* per night. A day's work is ten hours in summer and eight in winter.

PAYMENT OF OPERATORS

When first taken on, girls receive 1*l.* 13*s.* 4*d.* per month ; which is increased to 2*l.* 1*s.* 8*d.* when passed as quite competent. Subsequently they are advanced by stages to a maximum of 2*l.* 18*s.* 4*d.*, attained in three years. Lady superintendents receive 4*l.* 3*s.* 4*d.* per month. The girls' duty never, except at night and under very special circumstances, exceeds six hours per day. One watch takes duty from 8 A.M. till 2 P.M. ; the second, thence till 9 P.M. At that hour the night staff, consisting of six young ladies, arrives and continues the service until 8 A.M. They watch and sleep by turns. The Vienna staff, all told, comprises 334 girl operators.

STATISTICS

In March 1895 the subscribers in Vienna numbered 7,700. For the other towns no figures are obtainable later than December 31, 1893, when there were 80 exchanges belonging to the State (including 10 taken over from the companies on the preceding January 1), comprising 177 public stations and 7,483 subscribers. Vienna thus possesses a good half of the total number of subscribers. At the same date there were 40 metallic circuit trunks, of a total length of 3,302 kilometers, in operation. The principal exchanges were as follow :—

Town	Number of subscribers	
Prague	1,070	Taken over from the companies
Trieste	692	
Graz	598	
Brünn	568	
Lemberg	518	
Reichenberg	427	
Linz-Urfahr	221	
Bielitz-Biala	191	
Pilsen	182	
Czernowitz	113	

Town	Number of subscribers	
Cracow	214	Commenced by the State
Carlsbad	176	
Salzburg	133	
Aussig-on-Elbe	131	
Teplitz	110	
Troppau	109	
Warnsdorf	107	

The capital expenditure, receipts, and working expenses for 1892 are given as follow :—

STATE

	£
Capital to date	41,289
Receipts for 1892	33,875
Expenses ,,	10,700

VIENNA PRIVATE TELEGRAPH COMPANY

	£
Capital expenditure to date	498,000
Receipts for 1892	64,290
Expenses ,,	36,989

II. BAVARIA

HISTORY AND PRESENT POSITION

LIKE Würtemberg, Bavaria has preserved the autonomy of its Posts and Telegraphs, and consequently conducts its telephonic business without interference from Berlin. In the early days of telephony it steadily declined all applications for concessions, and everything has been done by the State itself since, in 1882, it opened the first Bavarian exchange at Ludwigshafen-on-Rhine. Until recently the opinion was held that single wires were adequate for local connections, but it is satisfactory to learn that a complete change of opinion in this respect has been brought about, and that all new work is now designed with a view to the ultimate adoption of metallic circuits. Munich, Nuremberg, and Würzburg are the three chief telephonic centres of Bavaria, each being surrounded by quite a galaxy of satellite switch-rooms. Lesser groups are Hof, Münchberg, and Berchtesgaden, while Augsburg stands by itself. In the detached left-Rhine palatinate, Ludwigshafen forms the centre of a group consisting of Speÿer, Kaiserslautern, Neustadt, and Lambrecht. With the exception of this last, with which communication can only be had *viâ* Stuttgart and Mannheim, or *viâ* Frankfort-on-Main and Mannheim, the different groups are joined by trunk lines belonging to the Bavarian Government. When it is stated that Munich (population 350,594) has close on 5,000 instruments connected to its exchange, and that Nuremberg (population 142,590) has over 2,500, while Würzburg (61,039). Augsburg (75,629), Fürth (43,206), and Bamberg (35,815) have 800, 750, 620, and 400 respectively, it will be understood that Bavaria is a very long way in advance of the United Kingdom in respect to its telephones. The fact is due, no doubt, in the first place to the facilities given, and in the second to the moderate tariff, which, although somewhat high (7*l.* 10*s.*) for a first connection, is remarkably low (3*l.* 15*s.*) for second and subsequent instruments. A consequence is that a larger proportion of the subscribers go in for more than one instrument than in any other country with which the author is acquainted. The length of line allowed for the subscription is very liberal—5 kilometers (3·1 miles). One objection to the rate is that it is uniform for all places, capital and village alike—treatment which is neither economically just nor calculated to encourage development. The obstacles imposed in the neighbouring kingdom of Würtemberg, in the Imperial postal territory and in France, to free communication between a town and its suburbs are absent in Bavaria, there being but two classes of charges for internal trunk communication, viz., between towns of the same telephonic group, and between one group and another.

SERVICES RENDERED TO THE PUBLIC

1. **Local exchange communication between the subscribers and public stations of the same town.**

2. **Trunk communication between towns of the same group.** The distances separating towns of the same group are often considerable, especially in the case of the Nuremberg group, which comprises Fürth, 5 miles ; Anspach, 25 miles ; Bamberg, 33 miles ; and Amberg, 35 miles off. The joining of the Amberg and Bamberg trunks therefore produces a circuit of sixty-eight miles, for which the charge is 5*d.* per five minutes.

3. **Trunk communication between towns of different groups.** All the groups are joined, there being only one isolated exchange, Kempten, which has not been reached by the trunks. Munich and Nuremberg are connected by two widely-differing routes, *viâ* Ratisbon and *viâ* Weissenberg, with the view of diminishing the chance of total interruption.

4. **International trunk communication.**—The principal international line is that between Munich and Berlin, over which the other chief Bavarian towns also obtain connection ; but the line to

Ulm and Stuttgart is also an important one. Besides these, there is communication with Frankfort-on-Main and Southern Germany, including Baden. The Bavarian lines also cross the Austrian frontier at Salzburg and Lindau, but in these cases talking is restricted to towns not far removed from the border.

5. **Telephoning of telegrams.**—This is restricted to the German and French languages. No charge is made for the service, the State taking the sensible view that the telephone constitutes a natural feeder of the telegraph, and as such should be encouraged as much as possible. The facilities given are very good, as reply-paid messages may be forwarded, and paid replies to telegrams received by messenger may be telephoned to the telegraph office. No deposits in advance are required, a signed promise to pay monthly the accounts rendered being considered sufficient.

6. **Telephoning of messages for local delivery.**—This service is confined to subscribers, and not extended, as in some countries, to the users of public stations. The sender of a telephonogram may undertake, when dictating his message, to pay for a reply, in which case the messenger who delivers it to the addressee will, if possible, bring back the answer to the central office, whence it is immediately telephoned to the sender. This service is an important one, for it makes a subscriber's telephone a channel which leads not only to every other subscriber, *but to every non-subscriber as well.* Dictation of difficult words is helped by a code of numbered letters published in the subscribers' lists, but this code is not so comprehensive as the Austrian, as it provides for only twenty-eight letters.

7. **Telephoning of mail matter.**—Subscribers may dictate messages to the central office to be mailed as letters or post-cards. In the former case they are written in pencil on telegram forms, enclosed in an envelope, addressed, stamped, and posted immediately. A post which would be missed in the ordinary way may thus frequently be saved. A slight drawback is that such letters cannot be registered; but then it is certain that they do not often contain bank-notes or other valuables.

8. **Public telephone stations.**—These are almost invariably located at post or telegraph offices, and are fairly numerous, there

being thirty-five in Munich, thirteen in Nuremberg, eight in Würzburg, and at least one in every town. An attendant is always provided, who collects the fees and obtains the connections asked for.

9. **Fire service.**—In such towns as do not enjoy a night service the lines of those subscribers who pay a small extra annual subscription are switched through to the fire station at closing time. A full description of this service will be given in the Würtemberg section, at page 422.

TARIFFS

1. **Local exchange rates.**

	Per annum £	s.	d.
An ordinary subscriber's station within 5 kilometers	7	10	0
Excess charge for distances beyond 5 kilometers, per 100 meters	0	3	0
A second instrument on the same line, but not in the same building	3	15	0
Second and subsequent instruments in connection with the same line, and in the same building	1	0	0
An instrument used by a tenant which can be switched on to a line rented by the proprietor of a building let off in flats or workshops	2	10	0
An extra bell	0	5	0

Government and municipal offices enjoy a reduction of one half. All distances measured as the crow flies. Agreements for lines not exceeding five kilometers, one year; exceeding that distance, two years. This tariff applies to all towns, irrespective of size.

2. **Trunk communication between towns of the same group.** Subscribers may pay per conversation, or by annual subscription.

	£	s.	d.
Per conversation of 5 minutes	0	0	5
The right to call any subscriber in any town of a group, per annum	2	10	0

3. **Trunk communication between towns of different groups.**

	s.	d.
Up to 100 kilometers, per 5 minutes	0	5
All distances beyond, ,, ,,	1	0

Express or urgent talks are admitted at triple fee.

4. **International trunk communication.**

	s.	d.
Munich and other chief towns to Berlin, per 3 minutes	2	0
Bavaria to Würtemberg, per 5 minutes	1	0
,, Austria, per 3 minutes	1	0
,, Switzerland, *viâ* Austria, per 3 minutes	1	2
,, towns in the south-west of the Imperial Post Office territory, per 3 minutes	1	0

There are a few rates of 3*d.* and 5*d.* in operation between towns situated close together, but on different sides of the frontier, as Ludwigshafen and Mannheim, Lindau and Bregenz, and Bad Reichenhall and Salzburg.

5. **Telephoning of telegrams.**—This service is free.

6. **Telephoning of messages for local delivery.**—For each telephonogram delivered by messenger the charge is 1*d.* plus ·1*d.* per word. Thus a ten-word message costs 2*d.*, and a twenty-word 3*d.*

7. **Telephoning of mail matter.**—The charge for this service is the same as for telephonograms, plus the value of the post-card or postage-stamp required.

8. **Public telephone station rates.**—Time unit, five minutes.

Local talks: A subscriber, member of his family, partner, or employee	1*d.*
All other persons	2·5*d.*

A non-subscriber may, however, put himself on an equality with a subscriber by buying a book containing fifty penny tickets, each of which will entitle him to a local talk if presented within one year from date of purchasing.

Trunk talks, as from subscribers' offices.

9. **Fire service charge.**—For connection with the fire station after an exchange is closed for the night, per annum, 10*s.*

WAY-LEAVES

The Government has no right to fix supports and wires on private property without the owner's permission. Subscribers can only give leave to attach wires intended for their own use to premises they lease or rent.

SWITCHING ARRANGEMENTS

Hitherto 'standard' boards of the Western Electric Company have been used at the principal switch-rooms, so it may be imagined that smart management has been requisite at Munich, with its 5,000, and Nuremberg, with its 2,500 subscribers, in order to provide an acceptable service. But multiple switch-boards have been ordered for, and will soon be fitted at, both these centres. They are of the same company's manufacture, with self-restoring drops of the type already installed at Zürich, and which is described in the Swiss section (p. 390). Fig. 8 is a plan of a recent Bavarian switch-board for small centres, showing how both single and double subscribers' lines and trunk wires are dealt with. The bar commutator is for cross-connecting and joining any wire temporarily to the testing apparatus. This bar commutator is sometimes replaced in the larger exchanges by a cross-wire commutator invented by Mr. J. Baumann, an engineer of the Royal Bavarian Telephone Department, which, for a large number of lines, is far cheaper to construct, while it occupies less space and is simpler to manipulate. Mr. Baumann's cross-connecting board consists of a strong rectangular iron frame encased in beechwood, and arranged to receive a number of silicium bronze wires of 8 mm. diameter, strung, some horizontally and some vertically, so as to cross each other at right angles at a distance of some two centimeters. The wires are insulated at the frames, and provided with tightening screws, similar to those of a violin (a tension of from thirteen to sixteen kilogrammes is kept normally on the wires), and connection terminals, by which the horizontal wires are joined to the subscribers' lines and the vertical wires to the switch-board. Under the tension applied, the wires remain so taut that it is not found necessary in practice to allow a greater clearance between parallel conductors than from three to five millimeters. The necessary connections between the horizontal and vertical wires are effected by small brass plates, each bearing two hooks about one centimeter apart, one hook adjusted to hold a horizontal, the other a vertical, wire. When two such wires,

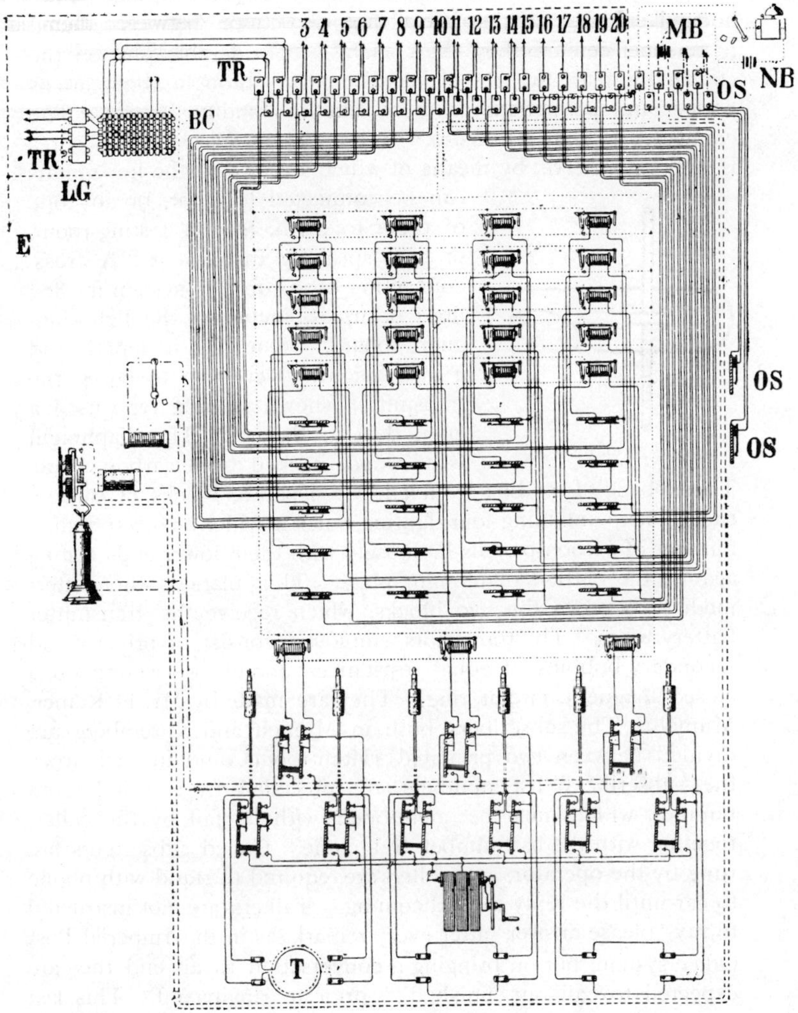

FIG. 8.—BC, bar commutator; T, translator; TR, trunks; MB, transmitter battery; NB, night bell; OS, transfer jacks; LG, lightning guards; 3 to 14, subscribers' double lines; 15 to 20, subscribers' single lines.

tightly stretched and separated by a space of two centimeters, are hooked together so that the clearance between them is reduced to one centimeter, they exert such a considerable pull on the connecting plate that the electrical contacts brought about are as perfect and permanent as those due to binding screws. The connector is shown in fig. 9. A number of spare vertical wires is kept in reserve, by means of which any two of the horizontals can be connected together, or any one of them to earth, to the testing-room, or to a speaking instrument. A cross-connecting board on this system for 800 lines occupies, including the lightning-guard board, 5·6 meters in length and 2·4 in height. Reverting to fig. 8, the transmitter shown is of a type used a good deal in Bavaria. The diaphragm is of wood, backed by a carbon plate.

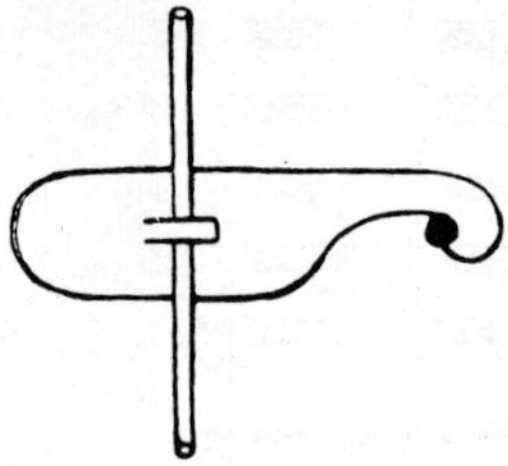

FIG. 9

To the back of the box are fixed two separate blocks of carbon, each block containing four slanting holes in which a corresponding number of carbon pencils lie loosely with their lower ends resting against the carbon diaphragm plate. This plate is then intermediate between the two blocks, which receive the transmitter battery wires. The translators employed consist of primary and secondary bobbins of equal resistance—200 ohms—wound on a closed magnetic circuit ring. They are made by Mr. F. Reiner, Munich. The subscribers, both in Munich and Nuremberg, are divided between two principal switch-rooms, and in each town the subscribers' list numbers are preceded by a switch-room number, which must be mentioned without fail by the caller, together with the list number and name. Called subscribers are rung by the operator, and callers are required to stand with phone to ear until the reply is forthcoming. Talkers are not instructed to say 'please answer' after every remark, as in the Imperial Post Office system, but on bringing a conversation to an end they are expected to call out 'finished!' prior to ringing off. This last signal has nothing to differentiate it from a ring through, so that the Bavarian subscribers, in common with all others on the Continent, cannot leave their instruments during a talk. During a thunder-

FIG. 18.—V, lightning guard ; M, transmitter ; *a*, magneto crank ; *a'*, generator cut-in ; *j*, magneto box ; W, bell ; T, receiver ; h, automatic switch ; h', hook for spare phone ; B, battery box.

storm operators retire from the switch-tables, and subscribers are instructed not to touch their instruments. The operators use watches to time trunk talks, and not sand-glasses. It is worthy of remark that Bavarian operators are invariably of the male sex, although it is in contemplation to introduce girls at Munich and Nuremberg simultaneously with the multiple switch-boards now on order.

HOURS OF SERVICE

Munich, Nuremberg, and Fürth are open day and night; the nine next most important towns from 7 A.M. till 11 P.M.; nine more from 7 A.M. till 9 P.M.; and the rest according to the duration of the duty at their respective telegraph offices.

SUBSCRIBERS' INSTRUMENTS

These usually comprise the modified De Jongh transmitter already described, manufactured by F. Reiner, of Munich; Bell receivers, magneto, trembling bell, battery box, and backboard; and, although not so neat and businesslike in appearance as the English, American, or Swedish sets, are well and substantially made, and give good results. The chief drawback is that the generator coils have to be cut into circuit by means of a push-button instead of by an automatic contact, an arrangement which compels the use of both hands in ringing, and causes the subscriber to incontinently transfer to his mouth, as to a third hand, any papers or pencils he may be carrying. Figs. 10 and 11 give a good representation of the subscribers' wall and table sets respectively.

OUTSIDE WORK (LOCAL)

All local wires are now run with silicium bronze of 1·5 or 2 mm. diameter. In the towns the supports are generally on the houses. Three designs of Bavarian standards are shown in figs. 12, 13, and 14. The first is a single wooden pole bolted to the

FIG. 11.—M, transmitter; W, bell; *j*, magneto box; T, receiver; h, automatic switch; B, battery box; *a*, magneto crank
a', generator cut-in.

roof timbers and provided with an angle-iron frame carrying a number of angle-iron arms. The second is also of wood, but double, and supporting angle-iron arms of greater capacity. The third, which is entirely of angle-iron, bears a strong resemblance to, without being identical with, the Belgian design of standard, which is not excelled anywhere for ability to withstand success-

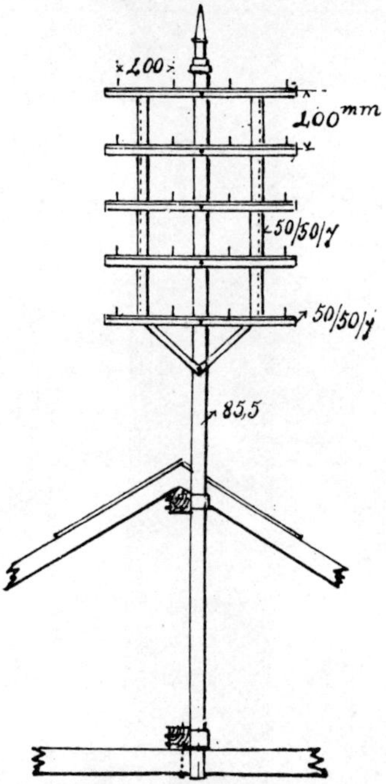

FIG. 12

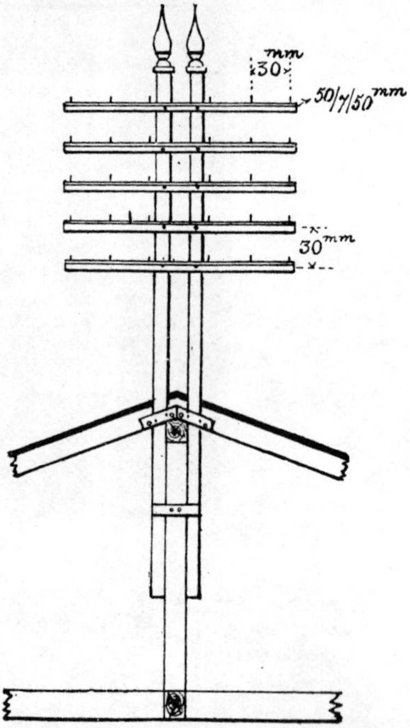

FIG. 13

fully the many vicissitudes to which roof-supports of large capacity are subject. Ground poles are usually of wood and in no wise noteworthy for size or design; but there are also a few iron columns of the Zürich type (see Swiss section, fig. 147). A few aerial cables manufactured by Felten and Guilleaume are in use, but only in special circumstances. In Munich, Nuremberg, and Landshut there is some underground work, consisting partly of iron pipes,

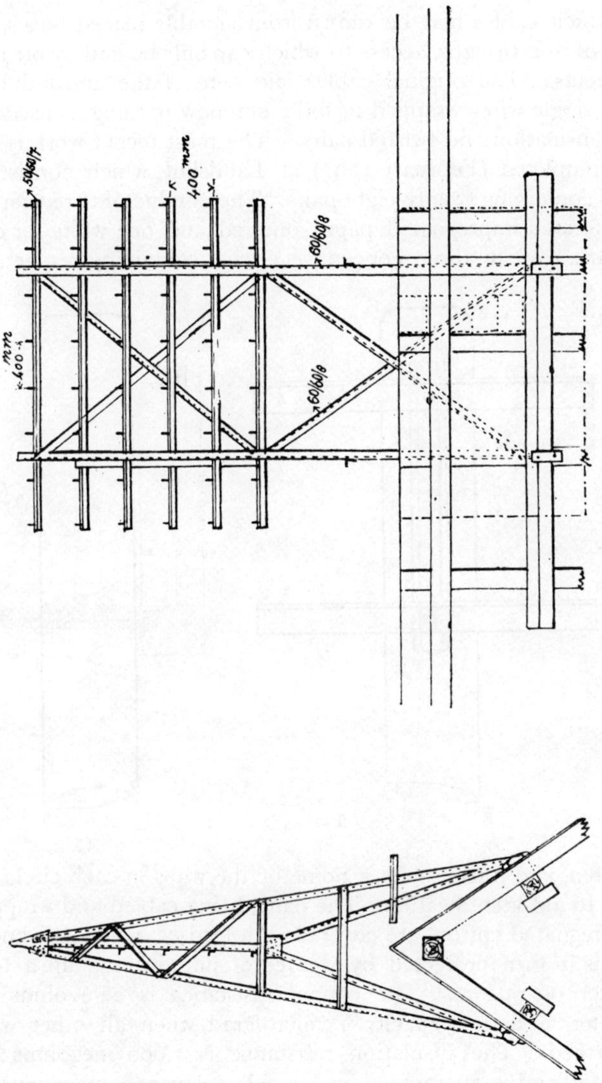

Fig. 14

into which cables may be drawn from suitably placed boxes, and partly of iron troughs, access to which can only be had by breaking the streets. The original cables laid were of the anti-induction type—single wires wrapped in foil; but now nothing is used but paper insulation and twisted pairs. The most recent work is that just completed (February 1895) at Landshut, which consists of cables containing twenty-eight pairs. The conductors are wrapped in perforated impregnated paper, one red and one white for each pair, arranged in three concentric circles containing respectively

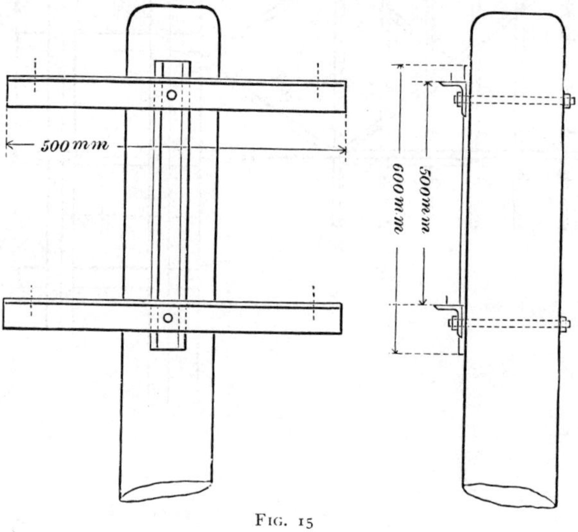

FIG. 15

three, ten, and fifteen pairs. Some of the wires in each circle are tinned to aid identification. The pairs, being cabled and wrapped in impregnated cotton, are covered with lead of 2 mm. thickness, which is in turn protected by a layer of jute, making up a total diameter of 33 mm. The copper resistance is 22·6 ohms per kilometer, and the capacity ·1 microfarad when all other wires are earthed. The insulation resistance is 2,000 megohms per kilometer, and is guaranteed not to fall below 500 megohms for two years. This cable, which has been supplied by Franz Clouth, of Cologne-Nippes, is reported to give every satisfaction. Some of the Munich and Nuremberg cables are from the factory of Messrs. Felten & Guilleaume.

OUTSIDE WORK (TRUNK)

The Bavarian trunks are for the most part constructed of silicium bronze of 3 to 4 mm. diameter, according to the distance to be covered. Wherever practicable, one loop only is run on a route of poles, the wires being arranged in a vertical plane and crossed at long intervals. This plan is found to secure a sufficiently silent line. An iron frame to attach to poles, with the object of facilitating the running and accurate spacing of trunk wires, is shown at fig. 15. It is used on the Munich–Berlin line within Bavarian territory, and on other routes.

PAYMENT OF WORKMEN

Foremen receive from 60*l.* 3*s.* to 120*l.* per annum, according to length of service. When working away from their homes they have 1*s.* 3*d.* per day extra, and when obliged to sleep away, 3*s.* Skilled wiremen are paid from 3*s.* 8*d.* to 4*s.* per day, with no extras; and labourers from 2*s.* 5*d.* to 3*s.* 2*d.* The working day averages ten hours, less meals.

PAYMENT OF OPERATORS

These are all youths. They receive about 2*s.* a day for a duty which varies from six to eight hours.

III. BELGIUM

HISTORY AND PRESENT POSITION

THE original exchange systems in Belgium were established by the International Bell Telephone Company under concessions, granted September 22, 1883, from the Government. Subsequently these were acquired by the Compagnie Belge du Téléphone Bell, which, until the transfer to the Government at the end of 1892, operated in Brussels, Antwerp, Ghent, Charleroy, Verviers, and La Louvière. Liége was worked by a separate company, the Compagnie Liégois du Téléphone Bell, while several of the smaller towns were granted to individuals. Thus a Mr. J. Ryf, a Swiss from Zürich, established exchanges in Louvain, Namur, and Mechlin ; and a M. Cahen in Mons and Courtray. All these, with the exception of Namur, Mechlin (Mr. Ryf), and Courtray (M. Cahen), the concessions for which do not expire until during the current year, have now come into the possession of the Belgian Government. The State had itself established exchanges in four different areas—Ostend-Bruges, Termonde-St. Nicholas-Alost, Hasselt-Landen, and Tournay. After 1895 the State will possess a monopoly of telephonic as well as of telegraphic communication within the kingdom, and intends to preserve it. All the exchanges constructed by companies were on the single wire and earth return system, while all those of the State were on the metallic circuit plan. All exchanges were built, and still consist of, overhead wires. The State, recognising the inadequacy of single wires for the general purposes of a telephone system, intends to gradually convert the whole of the exchanges taken over from the companies to double wires, and, furthermore, to place all main routes of wires in towns underground.

Belgium is at present, for telephonic purposes, divided into seventeen areas, each having one or more towns for a nucleus, and comprising together all the chief centres of commercial activity. The areas have not been apportioned arbitrarily, but with due regard to the business relations and exigencies of the several districts. The shapes and superficial measurements of the areas differ widely, as the requirements and convenience of the telephoning public have in each case been the paramount consideration, and the idea has been to avoid the creation of vexatious barriers between neighbouring towns and villages. A considerable portion of the country is still left unallotted, that is to say, is not included in any of the areas; but this portion is mostly agricultural, or of such small industrial development that no great demand for telephonic communication has as yet arisen within it. The Government, however, is prepared to inaugurate new areas, and provide trunk communication with the old ones, on receiving sufficient evidence of a demand. In the meantime, persons outside the areas are connected to the nearest exchange on payment of an extra subscription proportionate to the length of the line required. Once connected, they partake of all the privileges of subscribers located within the area, both as regards local and trunk services.

SERVICES RENDERED TO THE PUBLIC

1. **Intercourse between the subscribers and public telephone stations of the same area.**—A subscriber paying the prescribed annual rental for connection to his exchange is entitled to free communication with all other subscribers within the area in which that exchange is situated. In estimating the reasonableness of the Belgian rates, some of which superficially appear considerably dearer than those of Switzerland, Sweden, Würtemberg, and some other countries, it must be borne in mind that they apply not to a single town, but to a considerable district, which often comprises two or more towns of notable size. Thus the Brussels area measures, roughly, fifteen miles from west to east, and eight miles

from south to north ; the Termonde–St. Nicholas–Alost area, thirteen miles from east to west, and thirty miles from south to north ; and the Ostend–Bruges, twenty-seven miles from west to east, and twelve miles from south to north. In the Ostend–Bruges area, a three-year subscriber located within one kilometer of the Ostend exchange is entitled for his payment of 6*l.* per annum to speak without restriction or extra charge to Bruges, thirteen miles ; to Blankenberghe, eleven miles ; to Heyst, sixteen miles ; to Nieuport, ten miles. And a Nieuport subscriber can speak to Heyst, twenty-six miles, for his 6*l.* per annum. These distances are measured direct ; as the wires go by the railways they are usually greater. In the Termonde–St. Nicholas–Alost area the distances available for 6*l.* per annum are even longer.

When a subscriber removes to new premises within the same telephonic area his wire and instrument are shifted gratis. He is held responsible for the safety of his apparatus under all circumstances ; if it is destroyed by fire, or otherwise, he must pay its full value to the State. The Government has the right to suspend any part of or all the telephonic communication at its discretion, in which case the subscribers cannot claim any refund of subscription.

The burning question of the use of telephones by non-subscribers has been settled liberally in Belgium by formal permission being given to subscribers to allow strangers to use their instruments provided no payment or other consideration is received. Hotel, restaurant and club telephones are free to all and sundry.

It is rather singular that in spite of the immense traffic at the port of Antwerp no ships are fitted with telephones for the purpose of enabling them to use the exchange when in harbour, as is often done in Sweden and sometimes in Great Britain.

2. **Internal trunk line communication.**—The seventeen areas are already connected by trunk lines, so that practically all the merchants and manufacturers in the kingdom are within hailing and talking distance of each other. The rule is that a subscriber in any area may call up and talk to a client in any other area for five minutes for one franc (9·6*d.*). Nothing could be simpler, and nothing could be more effective. If five minutes does not prove sufficient the conversation may be extended to ten minutes

for an additional half charge. Expert users of the telephone can easily talk at the rate of 100 words per minute, so that a conversation of 1,000 words can be got through in ten minutes. The greatest distance that can be talked over at present is 156 miles, from Nieuport-Bains in the Ostend-Bruges, to Spa in the Verviers area.

3. **International trunk line communication.**—At present the only international connection is with France, but an agreement has been signed with the Dutch Government for a line to Rotterdam and Amsterdam. An understanding has twice been arrived at with the German Government, but as often cancelled by the Berlin authorities prior to actual signature. A line from Brussels to London is also in contemplation. Experimental talking has been carried on between the two cities *viâ* Paris and Calais. The French frontier is crossed at five different points: by the direct Brussels-Paris lines; by a line from Charleroy to Maubeuge; by a line from Mons to Valenciennes; by a line from Tournay to Lille; and by one from Courtray to Lille. The Brussels-Paris line has been a great success, there being now three circuits between the two capitals. From noon till 3 P.M. all lines are engaged without intermission, twenty-six connections per hour being got through, on an average, on each; this could not of course be done if each connection occupied its maximum time of three minutes. The receipts are consequently 26 × 3 × 3 francs = 234 francs (9*l.* 7*s.* 2*d.*) per hour during the busy time. The telegraph traffic between the Brussels and Paris Bourses, formerly very considerable, has been practically killed by the telephone, yet the telegraph receipts as a whole continue to grow. It is a curious fact, as illustrating forcibly the superiority of the telephone for certain purposes, that during the total interruptions of the Brussels-Paris telephone lines which have twice or thrice occurred, the stockbrokers have not reverted temporarily to the telegraph, formerly in incessant use between the two bourses, but have waited for the re-establishment of the talking facilities.

4. **Telephoning of telegrams.**—Subscribers may telephone telegrams to the telegraph offices and receive telegrams by telephone. In the latter case a copy of the message telephoned is mailed, postage paid, to the addressee by the next delivery. No charge is made for this service (the State regarding the telephone system as

a feeder to the telegraph, and as such to be encouraged, not despised), which is very largely taken advantage of. Its value as a feeder may be estimated from the fact that the number of special clerks engaged at the telegraph offices receiving and transmitting telegrams through the telephone exchanges during the day is nine at Antwerp, eight at Brussels, five at Ghent, three at Liége, two each at Namur and Charleroy, and one each at Mons and Tournay. The growth of the traffic has been continuous and rapid. During the month of August 1894, 45,646 telegrams were received from and 39,637 forwarded to subscribers throughout Belgium. Of this total of 85,283, Antwerp had 24,556 ; Brussels 14,081 ; Ghent 7,273 ; Liége 6,790 ; and Charleroy 5,710. The telephone is thus made to bring the telegraph to the merchant's desk and to the family fireside, rendering the employment of messengers to take despatches to perhaps distant telegraph offices, and others to bring them from the telegraph offices to the addressees, unnecessary. Under such circumstances it is natural to expect that telegrams will be more freely sent, and experience shows that it is so. The State also saves considerably in cost of delivery. For instance, in August 1894 no less than 39,637 journeys were saved to the telegraph messengers, or at the rate of 475,644, nearly half a million, per annum. This means that the staff of boys, wear and tear of boots and uniforms, &c., may be greatly economised, while the deliveries themselves are markedly accelerated. The clerks employed at the telegraph offices are competent to receive and telephone messages in French, Flemish, English, German, and Dutch. To avoid mistakes between words and letters of similar sound, each subscriber is furnished with a printed table showing the letters of the alphabet numbered from 1 to 26. A doubtful word is spelt, and a doubtful letter referred to by its number in the table. The French numerals 'six' and 'dix' are liable to be confounded by some speakers. When this is the case, 'six' is dubbed F and 'dix' J. In Flemish a similar uncertainty is apt to arise between one and two, which are then referred to as A and B. With these precautions (which are likewise adopted with modifications suitable to the language in most continental countries) mistakes occur but rarely, and the service grows continually in popularity.

5. **Public telephone stations.**—Several of these conveniences exist in every town, mostly at the post offices, railway stations, and bourses. The principal ones are open all night. There are ten in Brussels, eight in Antwerp, five in Ghent, four in Liége, &c. They may be used for all classes of communications admitted by the regulations. To facilitate payments at the public stations a series of adhesive stamps, similar to those introduced by the author in this country in 1884, of the values of ·25, ·30, ·50, ·90, 1·00, and 3·00 francs, have been issued. The 25-centime stamp is shown in fig. 16.

Fig. 16

6. **Call notices to non-subscribers.**—In connection with the public stations a service of call notices (*Avis téléphoniques*) is in operation, which enables a person to summon to a distant public telephone station any non-subscriber with whom he wishes to speak. He does this by telegraph (at specially reduced rates), specifying the place and time of the requested attendance.

7. **Railway station service.**—This provides for the switching on of subscribers (or of non-subscribers at public telephone stations) to any railway station in the area to enable official information as to the movements of goods or trains to be obtained. An extra subscription is charged for this service, payment of which also confers the right on a subscriber, or on his agents or friends, to use the railway station telephone, even when there is no public station there, for the purpose of communicating through the exchange.

TARIFFS

1. **Rates for exchange communication within an area.**—There is considerable divergence in the local rates. For the most part the system still consists of exchanges taken over from the different companies and individual concessionaries, whose practice was by no means uniform, and whose rates were, as a rule, higher than those imposed by the State in the areas—such as Ostend–Bruges, Termonde–St. Nicholas–Alost, and Tournay—

which it had itself initiated. But in taking over the concessionary single-wire systems in January 1893 the State determined to leave the rates unaltered, at least in the meantime, as—in view of the great expense involved in the determination to ultimately convert everything to metallic circuit and to abolish overhead wires in the centres of towns—it was felt that a reduction, until some further experience had been gained, was unadvisable.

TARIFF IN AREAS EXPLOITED BY THE STATE

All double wires

1*s.* = 1·25 francs Radius	3-year contracts Per annum	1-year contracts Per annum	½-yearly contracts extending over 3 consecutive years Per half-year
	£ *s.* *d.*	£ *s.* *d.*	£ *s.* *d.*
Within 1 kilometer . .	6 0 0	6 16 0	4 0 0
,, 1½ kilometers . .	6 9 6	7 10 0	4 8 5
,, 2 ,, . .	6 19 0	8 4 0	4 16 10
,, 2½ ,, . .	7 11 2	8 18 0	5 5 2
,, 3 ,, . .	8 3 2	9 12 0	5 13 7
Each additional ½ kilometer	0 14 0	0 14 0	0 8 5

TARIFFS IN AREAS TAKEN OVER BY THE STATE

(BEING THE ORIGINAL TARIFFS CONTINUED IN FORCE)

Principally single wires

—	Antwerp and Brussels Per ann.	Charleroy Per ann.	Ghent, Verviers, and La Louvière Per ann.	Mons Per ann.	Louvain Per ann.	Liége Per ann.
	£ *s.*	£ *s.*	£ *s.*	£ *s.*	£ *s.*	£ *s.*
Within 1½ kilometers . .	—	—	—	—	—	7 0
Within 3 kilometers . .	10 0	9 0	8 0	6 0	5 0	9 0
Each additional kilometer .	2 0	2 0	2 0	1 8	1 4	2 0
Extra instrument . .	2 0	2 0	2 0	2 0	2 0	2 0
Extra bell	0 6	0 6	0 6	0 8	0 6	0 8
2-way switch . . .	—	—	—	—	0 4	—
2-way switch with indicator	0 4	0 4	0 4	0 4	0 4	0 8
3-way switch with two indicators . . .	0 8	0 8	0 8	0 8	0 8	0 16

The half-yearly contracts are intended to meet the requirements of subscribers who, like the hotelkeepers in Ostend, desire communication during part of the year only. In view of the fact that metallic circuits are given everywhere and that the subscription covers, not a single town, but an area of many square miles, often comprising several towns, these rates are unquestionably liberal.

Half-yearly subscriptions, taken for three consecutive years, are charged :

	£	s.	d.
In Mons area	3	12	0
In Charleroy and La Louvière areas	4	16	0

In all other areas taken over by the State, three-fifths of the annual rate.

In Antwerp, Brussels, Charleroy, Ghent, Verviers, and La Louvière a reduction of 2*l.* is made when more than one line is subscribed for by the same person or firm ; in Louvain and Mons the reduction is only 10 per cent.

Subscribers may have double wires instead of single, to enable them to obtain better communication over the long-distance trunks,[1] for 50 per cent. above the normal rate. Thus a metallic circuit would cost in Liége 10*l.* 10*s.* if one and a half kilometers long, and 13*l.* 10*s.* if three kilometers long ; in Brussels and Antwerp, 15*l.* for three kilometers or any shorter distance ; while in Louvain the same length of double line would cost only 7*l.* 10*s.*, just the half ; and in Mons only 9*l.*

It will be seen that the retention of the old tariffs leads to considerable want of uniformity in practice, the subscription, for instance, in Charleroy being higher than in the far more important towns of Liége, Ghent, and Verviers. But having applied still lower rates in areas quite as extensive, and, moreover, provided therein double wires without extra charge, the State can scarcely hope to maintain the old companies' tariffs permanently. The subscriptions in every case cover the supply and maintenance of wires and apparatus.

2. **Rates for internal trunk lines.**—Here the liberality of the Belgian Government becomes conspicuous, for there is only one trunk rate throughout the country. A subscriber in one area can

[1] The number of subscribers taking advantage of this arrangement was, in December 1894, in Brussels, 91 ; Antwerp, 37 ; Liége, 13 ; Ghent, 5 ; Verviers, 9 ; Mons, 13 ; and Charleroy, 1.

speak to another located in any other for five minutes for 1 franc (9·6*d.*), and for ten minutes for 1·50 francs (14·2*d.*).

Ten minutes is the longest time permitted to a pair of talking subscribers should others be wanting the line. Otherwise, there is no limit, but the charge is based on the assumption that a new communication is commenced every ten minutes. Thus the charge for 15 minutes is 1·50 + 1 = 2·50 francs; for 20 minutes, 1·50 + 1 + ·50 = 3 francs; for 25 minutes, 3 + 1 = 4 francs, &c.

Subscribers using the trunks have to deposit in advance the estimated value of their monthly traffic. On these deposits interest at the rate of 3 per cent. per annum is allowed. This rule applies equally to the telegram service.

Instead of paying per conversation, it is open to subscribers (or to non-subscribers making use of public telephone stations) to pay monthly in advance for the right to occupy a specified trunk line for any length of time, but not less than ten minutes, every day. The rates are:—

	Per month £	*s.*	*d.*
10 minutes or less per day	1	8	0
10 ,, to 15 minutes per day	2	2	0
15 ,, 20 ,, ,,	2	16	0
20 ,, 25 ,, ,,	3	8	0
25 ,, 30 ,, ,,	4	0	0
Each additional 5 minutes ,,	0	12	0

It is not possible for one man to retain possession of a line to the exclusion of others, since he must split his time into two or more periods of ten minutes if the wire is wanted. The subscriber is not obliged to use up all the daily time for which he has paid at once, nor with the same correspondent, but unutilised time cannot be carried forward to the next day.

No distinction is made in the trunk line charges between subscribers and strangers. The former may use his own instrument or a public station at his option; a stranger is restricted to the public station.

In the event of a trunk line being interrupted for more than twenty-four hours, the subscriber is reimbursed one-thirtieth of his monthly rate for each succeeding twenty-four hours.

In all cases the time unit of five minutes is reduced to three between the Brussels and Antwerp Bourses during business hours.

Formerly, internal trunk rates were doubled from 9 P.M. till 7 A.M., but this was found to kill traffic during the hours when it was most wanted, and there is now no distinction between night and day.

3. **Rates for international trunk lines.**—Communication is now established between most of the Belgian areas and Paris, as well as to Lille, Arras, Dunkirk, Douay, Cambray, Roubaix, Tourcoing, Valenciennes, Maubeuge, and other towns in the north-east of France. Subscribers with double wires naturally possess a great advantage when using this service, and the State accordingly recommends their use, but the single wires in Brussels and elsewhere are put through by means of translators at the subscribers' risk. As in the case of the internal Belgian trunks, each conversation may be paid for separately, or a specified line may be engaged for a stated number of minutes each day. As telephone trunk lines are usually very much occupied during business hours, and very much the reverse during the evening, night, and early morning, an attempt has been made, and with some success, to distribute the traffic better by granting reduced rates, approaching half-price, between the hours of 9 P.M. and 7 A.M. The result has been that the lines are kept as constantly busy during the day as of yore, while the night traffic has sensibly increased. The simplicity of the Belgian uniform rate between areas is replaced in the French communications by a tariff regulated by distance, which is as follows:—

Distance	Day 7 A.M. till 9 P.M.	Night 9 P.M. till 7 A.M.
	d.	*d.*
Up to 50 kilometers	14·4	8·6
50 ,, 150 ,, ...	19·2	11·5
150 ,, 250 ,, ...	23·1	14·4
250 ,, 350 ,, ...	28·8	17·3
Each additional 100 kilometers or fraction thereof	4·8	2·88

The time unit is five minutes, with no reduction for extended talks; but to Paris the time unit is only three minutes, the charge being 28·8*d.* during Bourse hours at Brussels, Antwerp, and Paris (11.51 A.M. till 3.1 P.M.). On Sundays the time unit is uniformly five minutes.

The monthly subscription tariff is as follows :—

FOR TEN MINUTES' DAILY OCCUPANCY OF A SPECIFIED LINE

	£	s.	d.	
Up to 50 kilometers	1	16	0	per mensem
50 ,, 150 ,,	2	8	0	,,
150 ,, 250 ,,	3	0	0	,,
250 ,, 350 ,,	3	12	0	,,
350 ,, 450 ,,	4	4	0	,,
450 ,, 550 ,,	4	16	0	,,

Longer periods of occupancy are charged proportionally by increments of five minutes.

In case of interruption, monthly subscribers are reimbursed one-thirtieth of the subscription for each twenty-four hours after the first. Monthly subscribers may not occupy the Paris line during Bourse hours.

The rate agreed upon for the Belgo-Dutch trunk line under construction is 28·8*d.* between Brussels or Antwerp, and Rotterdam or Amsterdam.

4. **Rates for telephoning of telegrams.**—Every telephone exchange is connected to the nearest telegraph office for the despatch and receipt of telegrams. No charge is made for the service, but subscribers availing themselves of it have to deposit the estimated value of a month's traffic, on which deposits interest at the rate of 3 per cent. per annum is allowed. Copies of the telegrams dictated by subscribers through the telephone are furnished, if desired, to the senders at ·96*d.* each, which is also the charge for a formal receipt.

Copies of telegrams telephoned to subscribers are sent on by the next post free. If desired, copies may also be delivered by special messenger at a cost of 2·4*d.* each.

5. **Rates applicable at public telephone stations.**—Non-subscribers are charged 2·4*d.* for five minutes' talk within the area in which the public station is situated. The time unit is, however, reduced to three minutes between Brussels and Antwerp during Bourse hours. The distances comprised within the areas are, as

already stated, very considerable, so that from twenty to thirty miles may be talked over for 2·4*d.*

The rates for internal and international trunk talks are the same as those from subscribers' offices, already given.

Subscribers, on producing cards of identity and signing their names, may use the public stations free within the limits of their subscriptions. Beyond such limits, or if they do not produce cards, they pay exactly like non-subscribers.

A non-transferable public station card is supplied gratuitously to each subscriber, who is also entitled to a second one in favour of a partner, employee, or member of his family. If more than two cards are required an annual charge of 16*s.* is made for the third, and of 8*s.* for each additional one. Each card must bear the signature of the person to whom it is issued, and when using a public telephone station he must sign a sheet kept there for the purpose. The attendant must see that the signatures correspond. These regulations are identical with those introduced by the author in Scotland in 1884 in connection with the issue of telephone stamps.

Monthly public station cards for local use only are also issued to non-subscribers at a charge of 4*s.*

Automatic slot boxes for checking payments are not used.

6. **Charges for call messages.**—The charge for a telegram to a non-subscriber requesting his attendance at a specified public station at a certain time, is 2·4*d.* within an area, and 3·36*d.* without.

7. **Rates for the railway station service.**—For this a supplementary subscription of 4*l.* per annum, or 2*l.* 8*s.* per half-year, has to be paid.

WAY-LEAVES

The State has no right to place telephone poles or fixtures on lands or buildings without the consent of the proprietors. Under the Telegraphs Law (No. 593) of June 11, 1883, proprietors and tenants may not refuse to allow unattached wires to hang over their lands and buildings, but they are entitled to compensation for their presence. No work of any kind must be done over or under private property without the previous consent of the proprietor, and tenant if there is one. With respect to poles, standards, and other attachments, the absolute right to refuse exists and is often exercised. Actually, the State pays frequently as much as ·48*d.* to ·96*d.* and 1·44*d.* per wire per annum for their standards. Sometimes a free connection is asked and given in consideration of a standard. In 1893 the way-leaves paid for standards alone throughout Belgium was 40,000 francs (1,600*l.*). On one occasion a proprietor in Brussels consented to the erection of a standard conditionally on its colour and that of the insulators attached to it harmonising with his building. To meet his ideas of harmony the State had to go to the expense of having the necessary number of porcelain insulators of a peculiar tint specially manufactured. The Government has the right to erect poles and wires along railways which, like the Grand Central Belge, are still in the hands of companies, but only on payment of a way-leave to be agreed upon.

SWITCHING ARRANGEMENTS

These are not noteworthy for any speciality in design or arrangement. In the larger centres—Brussels, Antwerp, Verviers, &c.—one or other of the older forms of Western Electric single-wire, double-cord, series, multiple switch-board is employed; in the smaller, Gilliland and Western Electric 'standard' non-multiples. The leading idea everywhere has been to concentrate as much as possible in one switch-room in each town. Thus in Brussels and Antwerp, the two largest cities, there is practically but one switch-room, the outlying ones (Vilvorde, Hal and Nivelles in Brussels, and Boom in Antwerp) being of quite insignificant size and several kilometers away. As a rule, each operator manages 100 local subscribers' lines. Trunk line switching is effected at a separate table upon which the local lines are not multipled. At Brussels thirty-eight trunks are shared by four girls during the busiest time, and the three Paris circuits are looked after by one operator. Fig. 17 is a plan of the trunk board at Brussels, which, with a few modifications of detail, is also used in the other towns excepting Mons and Namur. The Van Rysselberghe system compels the adoption of a few special features, such as the phonic call and alarm. W1 W2 are the two wires of a metallic circuit coming from the condensers of a Van Rysselberghe telegraph line. W1 leads through the jack J1 and thence through the coil P1 of the phonic call to the plug F1 and, by its base contact, through the secondary coil T1 of the translator to earth. W2 goes

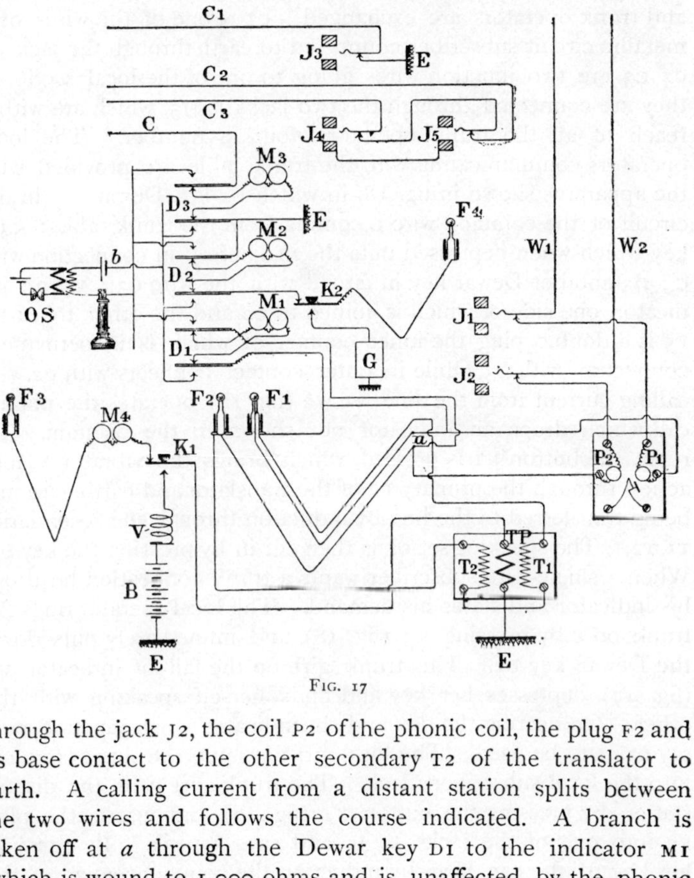

FIG. 17

through the jack J2, the coil P2 of the phonic coil, the plug F2 and its base contact to the other secondary T2 of the translator to earth. A calling current from a distant station splits between the two wires and follows the course indicated. A branch is taken off at *a* through the Dewar key D1 to the indicator M1 (which is wound to 1,000 ohms and is unaffected by the phonic call currents) and the base contact strip of the plug F1. Depressing the key D1 cuts the indicator out and the operator's set OS in. F3 is a plug connected to earth through the indicator M4, the calling key K1, the vibrator V, and the battery B. This is for calling on the Van Rysselberghe circuits ; for use on ordinary lines there is another plug F4 and key K2 which brings a magneto generator G into play. C is a wire common to all the sections of the local multiple, by which all communications between the local and trunk operators are exchanged. C1 is one of the wires of a metallic circuit subscriber connected to earth through the jack J3. C2 C3 are two junction wires going to one of the local sections ; they are connected through the two jacks J4 J5, which are within reach of all the trunk operators, four in number. The local operators communicating with the trunk table are provided with the apparatus shown in fig. 18, in which D2 is a Dewar key in the circuit of the common wire C coming from the trunk table ; K3 a key which when depressed puts the magneto G in connection with C ; D3 another Dewar key in circuit with the wire C3 ; M5 an indicator, one side of which is joined to C2 and the other to earth. F5 is a double plug the inner contact of which is in permanent connection with C3, while its outer contact connects with C2. A calling current from the line W1 W2 (fig. 17) operates the phonic call which drops an indicator not shown in the diagram. To reply, the button K1 is pressed, which brings the vibrator V into action through the primary TP of the translator and earth, currents being transferred to the line by induction through the secondaries T1 T2. The speaking set OS is then cut in by pressing the key D1. When a single-wire subscriber wants a trunk connection he drops his indicator and states his demand. The local operator rings the trunk on C by pressing K3 (fig. 18), and immediately puts down the Dewar key D2. The trunk girl, on the fall of indicator M3 (fig. 17), depresses her key and finds herself speaking with the other. On hearing the demand she indicates which junction wire, say C3, is to be used. The local girl then puts the plug F5 (fig. 18) into the local subscriber's jack. The trunk girl calls the distant station by pressing the button K1 (fig. 17), and inserts the plug F3 into one of the jacks J4 J5 and says 'speak.' By pressing her key D1 she can hear by induction the commencement of the talk. When finished, each subscriber rings off and the indicator

M4 falls. The presence of these ring-off indicators (one at each end of the line), which, as well as the phonic call coils, have to be talked through, is a bad feature of the system. The plug F5 (fig. 18) is in connection with the wire C2 by its second contact, and through a 1,000-ohm indicator M5 to earth. Therefore when F5 is in a trunk jack the test line of the local subscriber is in connection with C2, now insulated, and through the indicator M5 to

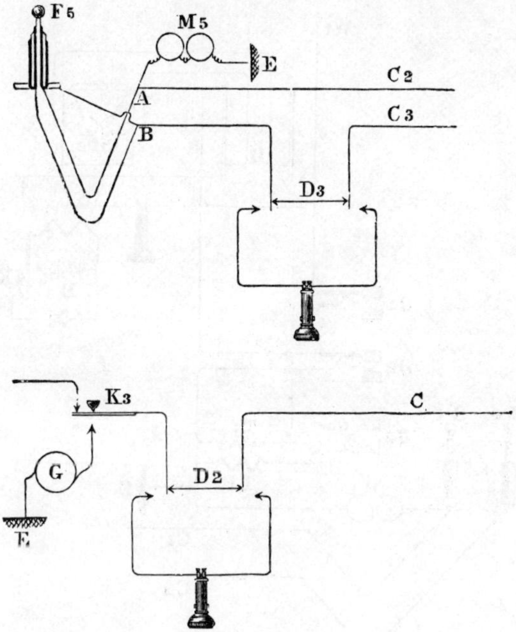

Fig. 18

earth. When the calling subscriber has a double line, the connection when established comprises (if the operators have agreed to use junction wire C3) F5, C3, to jack J3 (fig. 17), plug F2, jacks J4 J5, plug F1, coils P1 P2 of phonic call, jacks J1 J2 and W1 W2. In this case the indicator M1, which is in shunt with the phonic call coils, acts as ring-off. The test is managed as before. The translator is cut out, but the phonic call coils still have to be spoken through. When two trunks are connected the phonic call and the indicator of one of them are cut out, leaving the remaining indicator to act as ring-off.

At Mons and Namur a more simple arrangement, devised by M. Delville, is in use, the plan of which is shown in fig. 19. The wires of the trunk W1 W2 come to the Dewar key D and the jacks J1 J2. The spring of J1 is connected to the contact *a* of jack J3,

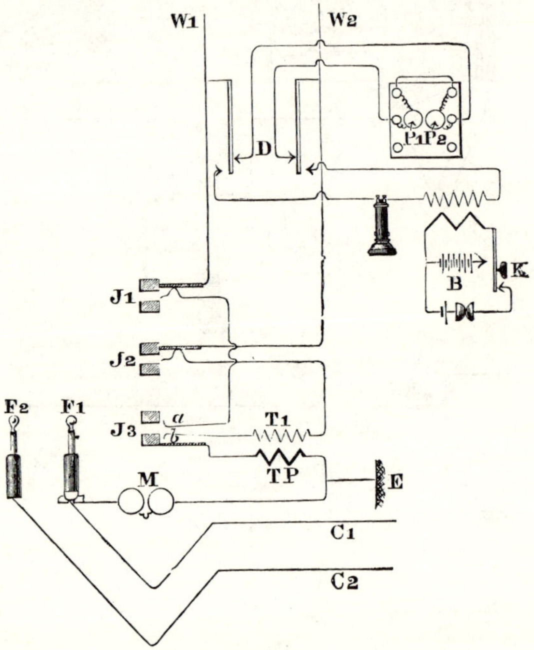

FIG. 19

while that of J2 is joined to one end of the translator secondary T1. The other end of the secondary goes to the contact *b* of J3, which is normally insulated from *a*. The frame of J3 is connected to earth through the primary TP of the translator. When D is up, the phonic call P1 P2 is in circuit with the line; when depressed, the operator's phone and the secondary of its transmitter induction coil are cut in. The primary circuit of this coil is closed through the microphone, transmitter battery, and the top stop of a Morse key K. When this is depressed it closes the circuit of the ringing battery B, and by making and breaking contact impulses are sent into line by induction to the secondary coil sufficiently powerful to start the phonic relay at the distant station. The wires (one pair of which is shown at C1 C2) coming from the local table end in plugs F1 F2, of which F1 is an ordinary single plug, while F2 has a metallic tip insulated from the piece which is in connection with the cord. This tip brings *a* and *b* into contact when inserted in J3, and so closes the translator secondary circuit. F1 rests on a metallic earth strip when out of use. On the phonic relay indicating a call, the operator depresses D and is then enabled both to ring and speak to trunk. If the connection demanded is with a single-wire subscriber the insertion (after the necessary communication with the local operator) of F2 in J3 completes it, since by this movement the translator secondary is brought into use through the contacts *a* and *b*, while the junction wire C2 utilised for the connection finds circuit through the main contact of F2, the socket of J3, primary TP of translator, and earth. The phonic relay P1 P2 remains in shunt (key D being up) across the loop and serves as ring-off. Two metallic circuits are joined direct by inserting a double-conductor cord terminating in two single plugs at each end in the jacks J1 J2 and the corresponding jacks of the second metallic circuit. The indicator M between the earth stop of F1 and earth serves for calling from the local to the trunk operator; the latter has also a battery push or generator for calling the former. There can be no doubt of the superiority of this plan over that in use at Brussels, since there are no coils to speak through, while the contacts are fewer and the arrangements simpler in every way. M. Delville evidently understands that in telephony, as in most things, the shortest road with nothing to jump over is the best.

Several patterns of phonic relay are used. One of the best that designed by M. Sieur, is shown in fig. 20. It consists of two coils P having soft-iron cores polarised by the permanent magnet M. A soft-iron diaphragm D placed in front of, and close to, the cores, is furnished with a platinum disc *d*, on which rests the light metal hammer H provided with an adjustable counterpoise A, by which the pressure of H on *d* can be varied. Normally the battery B is short-circuited through H and *d*; but when the intermittent calling current from line traverses the coils P, the diaphragm D is vibrated and momentarily casts off the hammer H, breaking the circuit or greatly increasing its resistance, whereupon the battery current traverses the coils of the ordinary indicator I and brings down its shutter. The work is severe on the battery, which is almost continuously on short circuit; but as the Van Rysselberghe system necessitates signalling through two sets of condensers in

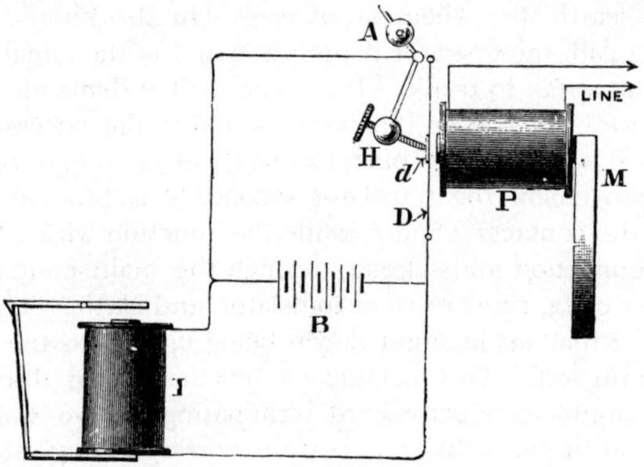

FIG. 20

series, something delicate and at the same time certain in its action is a necessity.

The translators employed are of the form designed by Van Rysselberghe (fig. 21), and consist of two induction coils fixed at right angles on a base-board. Each coil has a core of split soft iron tube, a primary of 80 ohms, and a secondary of 300 ohms, resistance, the two coils being usually joined in series. The ratio of the primary to the secondary, 1 : 3·75, is practically that which the author found to be best when experimenting with the original translator, but the actual resistances are very much greater. The subscriber's single wire is brought to the terminal S and earth to S1, while the trunk wires are connected to T1 and T2. The remaining terminals are joined by a short piece of wire.

There are no specialities in cross-connecting, but the lightning-guard boards at Bruges, Tournay, and elsewhere are on a plan designed by Mr. H. Frenay, Engineer to the Belgian Telephone Administration. They comprise a long earth strip separated from plates, to which the line wires are connected, either by paraffined paper or an air space. Beneath the earth strip every line passes through a testing jack the upper spring of which is elongated forward and curved upwards. Above the row of jacks and normally clear of them, extends a long metal cylinder turning on an eccentric axis, which is in permanent connection with the earth. One turn of a crank suffices to bring the cylinder against the elongated springs, so putting every line to earth instantaneously. Sometimes the crank is placed in a switch-room on the ground floor and connected with the cylinder in an attic by means of a long spindle, an arrangement which enables an operator, on the approach of a storm, to ground all the wires without outside assistance. The erection of a magnificent new telephone building is proceeding at Brussels, and Mr. Frenay is at present occupied in settling the details of the new switch-room and other arrangements. Whatever plans may be decided upon, it may safely be left to the Belgian technicians to provide their Administration with an installation that will rank second to none. The connections at Brussels average eight per subscriber per day ; Antwerp is understood to be busier, but records of the ordinary calls are not kept.

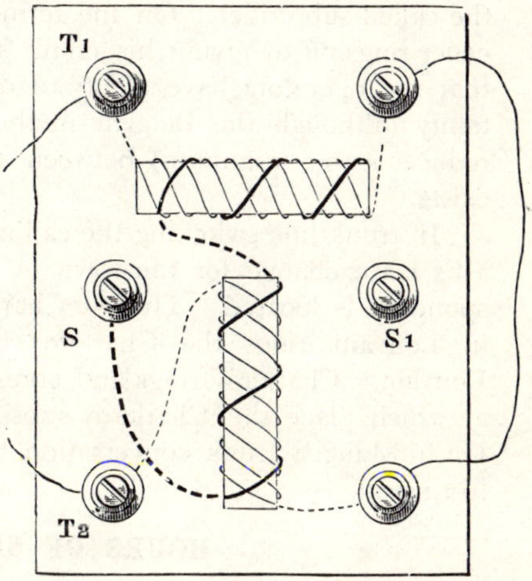

FIG. 21

In Brussels, Antwerp, and Verviers, subscribers are asked for by their list numbers only ; in Ghent, Liége, and elsewhere, by their names *and* addresses, although furnished with numbers in the list. On. receiving a call from a subscriber the operator always says 'I hear No. ——,' mentioning the list number of the caller, who thereupon gives the number (or name and address) of the person he wants, which the operator repeats. The caller then hangs up his phone and awaits a ring from the exchange, which in Brussels, &c., signifies that his correspondent is there. In Ghent, &c., the operator both rings and speaks to notify the establishment of a connection. In all cases, the switch girl rings the called subscriber. On the termination of a conversation the caller rings off by giving his crank several turns. It will be seen that the operators have plenty to do, and that the usual uncertainty (although the Belgian method of using the instruments reduces it to a minimum) between a ring-through and a ring-off exists.

In trunk-line switching the calling subscriber in the first place asks his exchange for the town in which his prospective correspondent is located. Thus, a Charleroy subscriber wanting one in Louvain, rings the Charleroy operator and says 'give me Louvain.' Charleroy rings and connects Louvain, to the operator at which place the Charleroy subscriber gives his order direct. On finishing a trunk conversation both subscribers are expected to ring off.

HOURS OF SERVICE

The service is continuous day and night at all the principal towns; the smaller ones are open from 5 A.M. till 11 P.M.; 7 A.M. till 11 P.M., and in some cases 7 A.M. till 9 P.M.

SUBSCRIBERS' INSTRUMENTS

The sets now fitted consist of magneto with base-board and battery-box; a Hunnings, or 'solid-back' transmitter modified somewhat from the original American design; and double-pole receiver. The magnetos are provided with a lightning-guard consisting of two metal plates separated by paraffined paper; in some cases this is combined with a point discharger, and mounted on a separate base-board fixed above the instrument. Test-plates or lightning-guards at the point of entry into a building are not employed, the outside lead of guttapercha-covered and braided wire being soldered direct to the inside lead of cotton-covered wire. The instruments are by different makers, but appear to be of uniformly good quality. Many instruments have a second receiver attached.

OUTSIDE WORK (LOCAL)

The wire used for local work is of bronze, 1·4 mm. gauge, 30 per cent. conductivity, and 114 kilogrammes breaking strain, the insulators being small double-shed in white porcelain. The insulator groove often contains a thick india-rubber ring, and sometimes also a strip of lead, with the object of stopping vibration. For junction lines the gauge is 1·6 and 40 per cent. conductivity: these are always metallic circuits. All wires are at present overhead, although extensive underground work is in contemplation in connection with the new exchange in Brussels. Aerial cables are not employed in the capital, but there are a few short lengths at Antwerp, Blankenberghe, and other provincial centres. All joints are soldered. The standard and pole work is exceedingly good in Belgium, both as regards design and execution, and constitutes the most striking feature of all to an English eye. The standards are built of angle and bar iron riveted together, and generally consist of uprights with widely-spread struts on both sides, the uprights and struts being rigidly connected by cross pieces. The whole is bolted to an iron base-plate or wooden platform made to suit the contour of the roof. The base-plate is generally separated by thick layers of felt sandwiched between thin leaden sheets from the rafters on which it rests, with the view of intercepting the vibrations from the wires. The whole forms such a rigid structure that stays are generally dispensed with, even on angles. In the event of a storm or fire suddenly destroying a bed of wires and throwing a heavy strain on one side of a standard, there is no danger of its yielding and allowing the damage to spread beyond the particular space involved. In fig. 22 are given front and side views, with details, of a standard, carrying

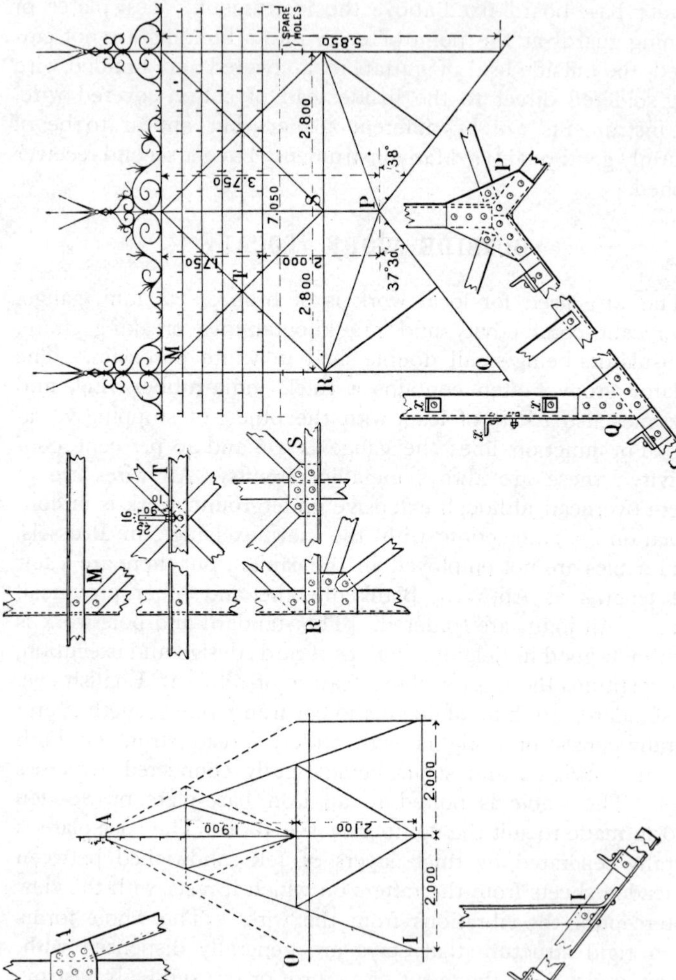

FIG. 22.—Dimensions in meters.

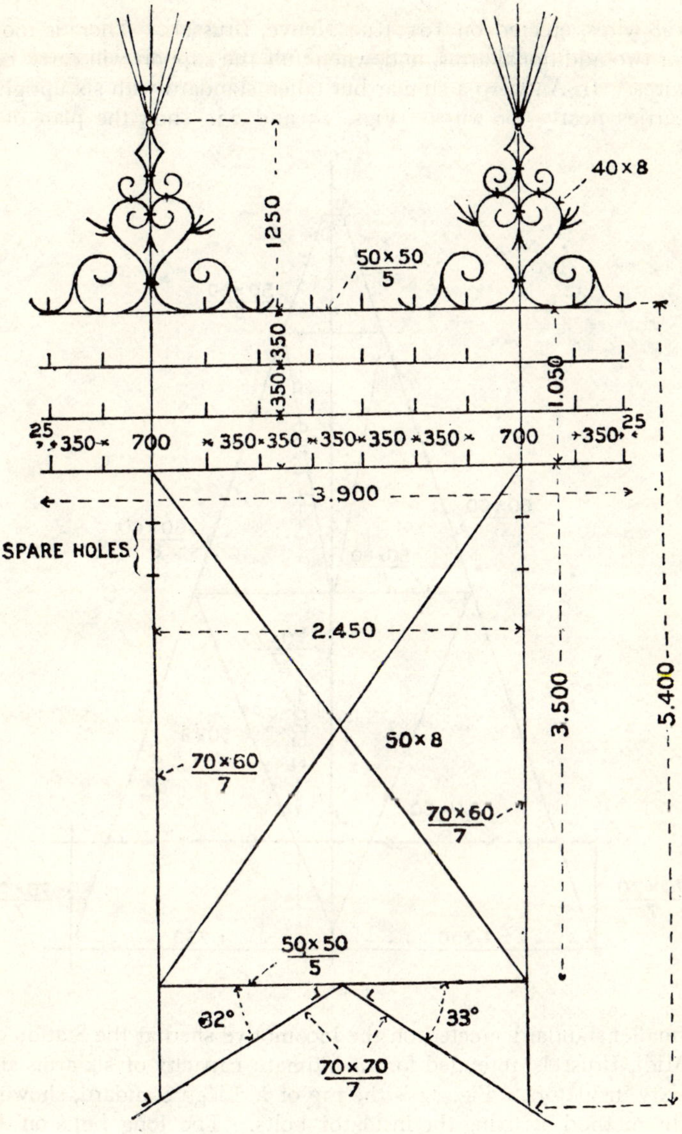

FIG. 23.—Dimensions in meters.

108 wires, erected on 101 Rue Neuve, Brussels. There is room for two additional arms, and when full the support will carry 144 wires. In Antwerp a similar but taller standard with six uprights carries nearly 600 wires. Figs. 23 and 23A show the plan of a

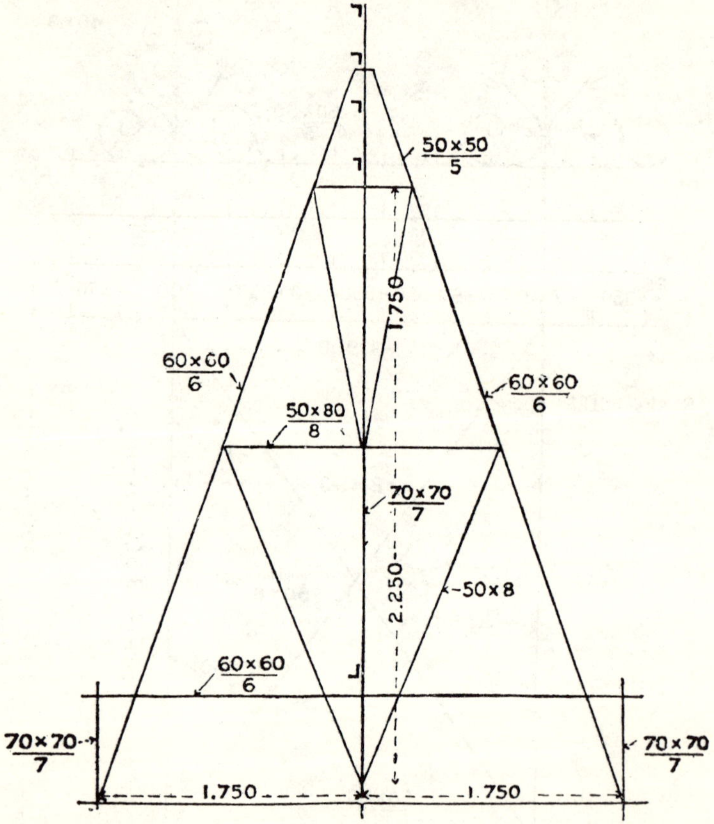

FIG. 23A

smaller standard, erected on the locomotive shed at the Station du Midi, Brussels, intended for an ultimate capacity of six arms and sixty insulators. Fig. 24 is the top of a Liége standard, showing the method of fixing the insulator bolts. The long bolts on the upper arm are for large double-shed insulators carrying trunk wires. The top of a bolt is wrapped tightly with tow and the insulator cup forced down upon it: the resulting fixture appearing everything that can be desired. The ornamental finials *e* are in galvanised iron. Each upright of a standard is connected to earth by an iron wire of 5 mm. diameter. With earthed single wires this is a somewhat superfluous precaution, but as metallic circuits multiply its utility will increase. All zinc, lead, and other metal about the roof is put in connection with the ground wire. No accidents from lightning are recorded in the ten years of telephonic experience in Belgium, although there have been violent storms during that period and many buildings have been struck. Single standards, except when intended to carry three or four wires only, are usually of lattice construction and

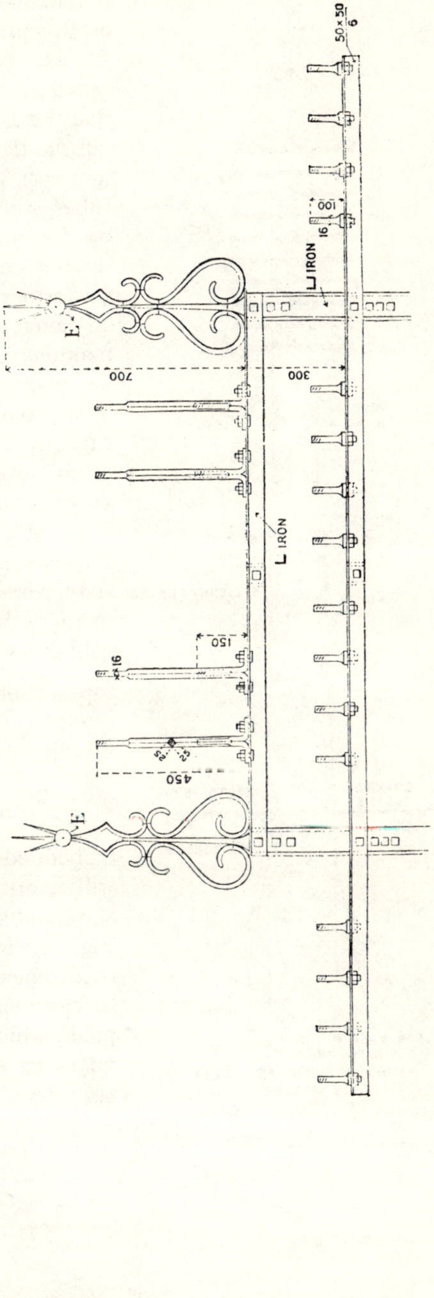

FIG. 24.—Dimensions in meters.

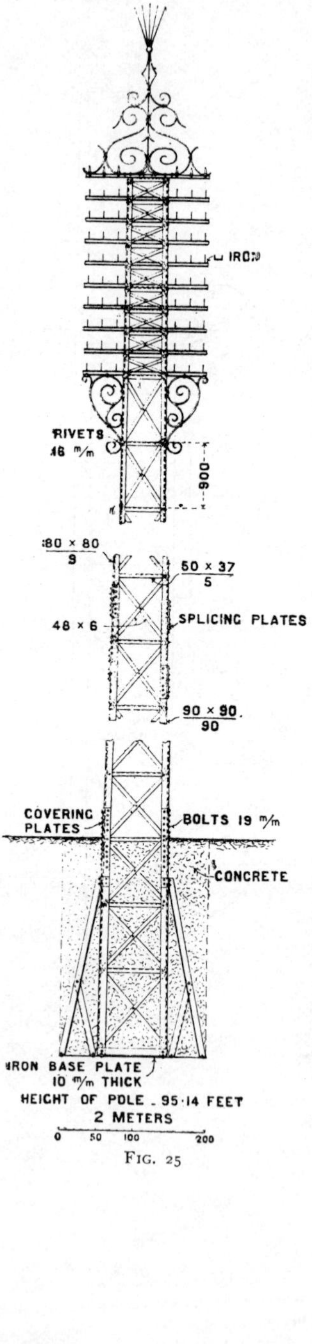

FIG. 25

practically resemble the top portion of the iron ground pole shown in fig. 25, the junction with the roof being as in fig. 22. It will be seen that the Belgian standards are both substantial and handsome. Their first cost is doubtlessly higher than tubes stuck into sockets and held up by wire ropes, but then they do not collapse under the various misfortunes to which standards in all countries are subject, and their maintenance (unless they have to be bodily shifted) is a bagatelle. The ground poles, when of small capacity, are usually of wood; when designed to carry many wires, or when located where appearance is an object, usually of iron lattice. Fig. 25 shows a ground pole carrying sixty wires, typical of the practice in Liége, with its details. It is built with two splices, the angle-iron of the top section being $\frac{70 \times 70}{9}$, of the middle $\frac{80 \times 80}{9}$, and of the bottom $\frac{90 \times 90}{9}$ millimeters. The foot is embedded in concrete. Such a pole will stand on a sharp angle without stays and without visible deflection. Fig. 26 shows the plan, with details, of a somewhat similar pole designed to carry eight ten-wire arms. Such poles, which are common along the quays at Antwerp, at Namur, and elsewhere, are 62 feet high and

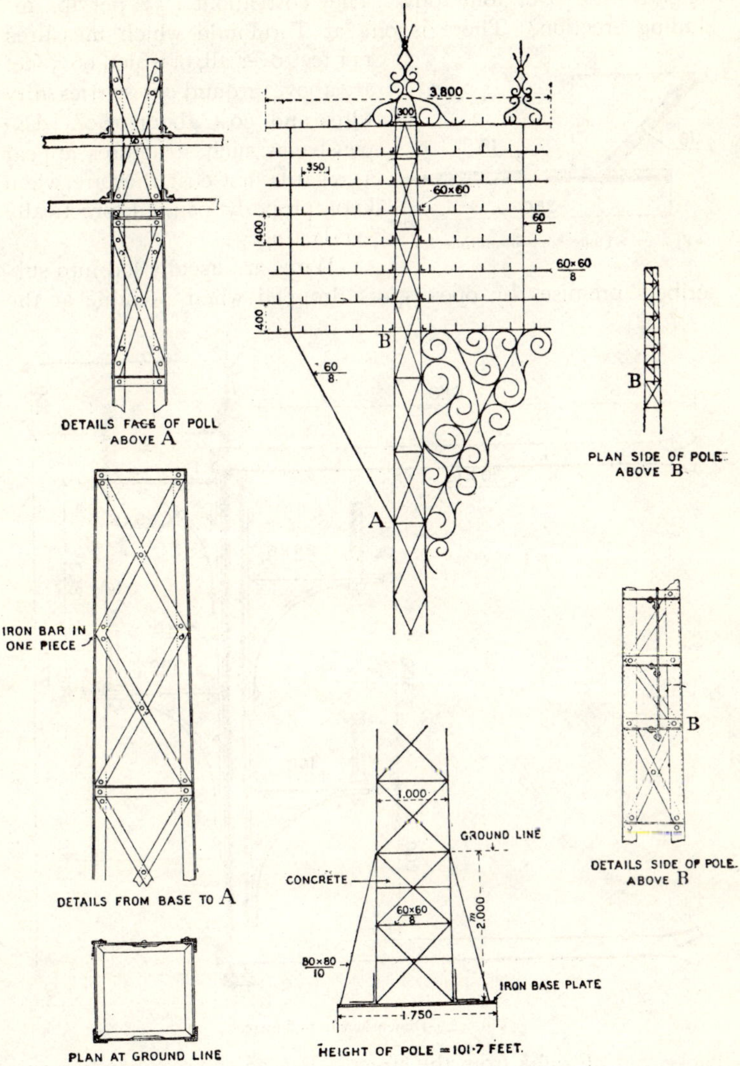

FIG. 26.—Dimensions in meters.

weigh a little over four tons. They cost about 1·5*d.* per lb., including erection. There is one at Termonde which measures 111 feet over all, of which 96·5 feet are above ground: it carries fifty wires and cost about 160*l.* Expensive as such structures appear as regards first cost, they are, when kept properly painted, practically everlasting.

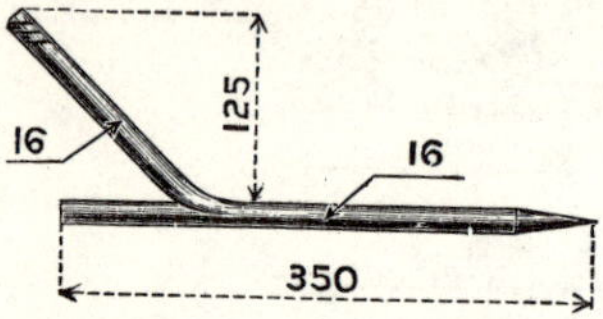

FIG. 27—Dimensions in millimeters

Wires are usually led into subscribers' premises by open spurs dropped when possible at the

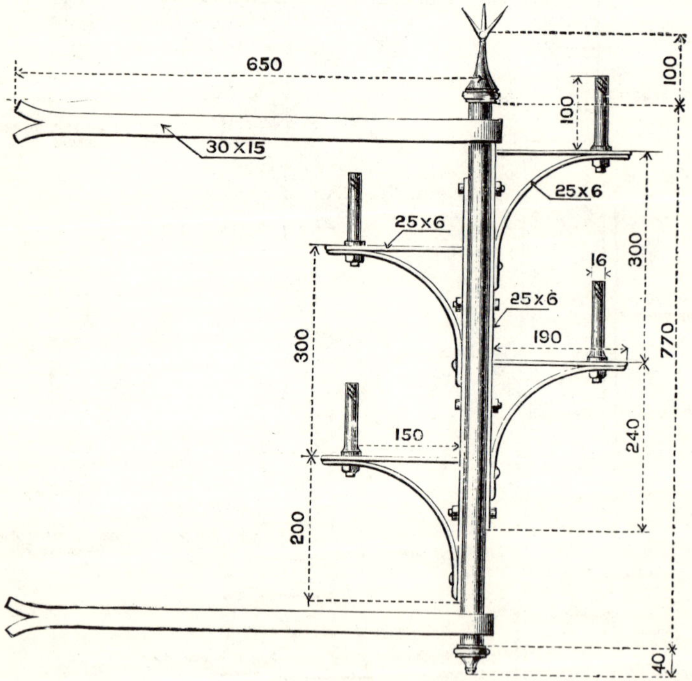

FIG. 28.—Dimensions in millimeters.

backs, out of sight from the street. Fig. 27 shows a handy insulator spike for this purpose. The insulator, which is fixed with tow, stands at a convenient angle for receiving the drop wire. Fig. 28 shows a neat bracket standard, useful for running a few wires along walls or houses. There is a twelve-wire route of this nature along the Fossé-aux-Loups, Brussels.

With one exception, that of Louvain, none of the exchange fixtures in Belgium offer novel points. The Brussels central standard, soon to be superseded, has 3,000 wires already attached, with space for 400 more. It is the original American erection, square, with wooden uprights and arms. At Antwerp the fixture is built of angle iron in the same way essentially as the ordinary standards. At Antwerp the site of the central station was not too wisely chosen, being adjacent to the great cathedral, which blocks it entirely on one side. As a consequence, very heavy routes have to be crowded on the old houses on either side of the cathedral. An elevation of the handsome octagonal tower of the new combined telegraph and telephone office at Louvain is given in fig. 29. Belgium has always been celebrated for its steeples ; now here is a new variation of that architectural embellishment which, in time to come, may share with the ecclesiastical variety the admiration of antiquaries. The accommodation provided for wires is far in excess of present requirements at Louvain, but then the end is not yet. In Antwerp some immense lattice iron arches were erected astraddle of some of the principal streets for the purpose of supporting the conductors in connection with the projected travelling balloon at the 1894 Exhibition. These have since been acquired by the State for use as telephone wire supports. As a general rule the outside work in Belgium is so well designed and so thoroughly well executed that it is difficult to suggest where there is room for improvement.

OUTSIDE WORK (TRUNK)

As far as supports are concerned there is nothing special about the Belgian trunk work. The poles, away from the towns, are generally creosoted wooden ones, sharpened at the tops and without roofs. For the most part they are carried along the railways, but where exposed to stone-throwing the insulators, which for trunk work are large double-sheds, are of brown or slate-coloured

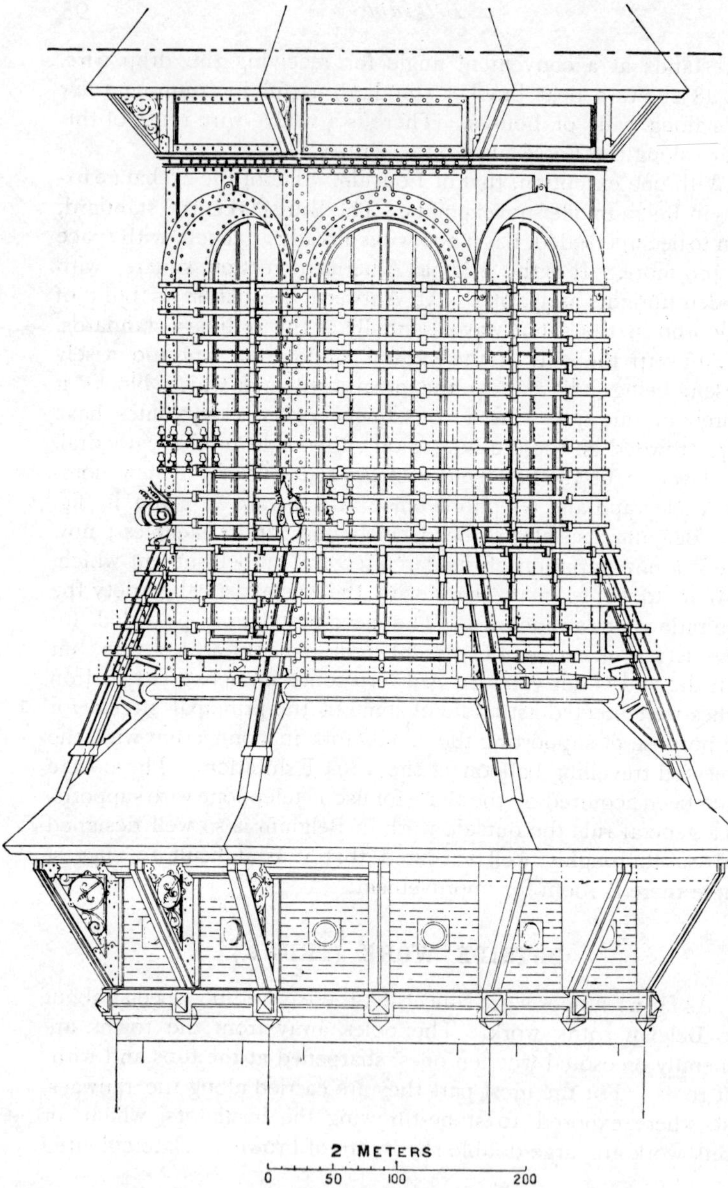

FIG. 29

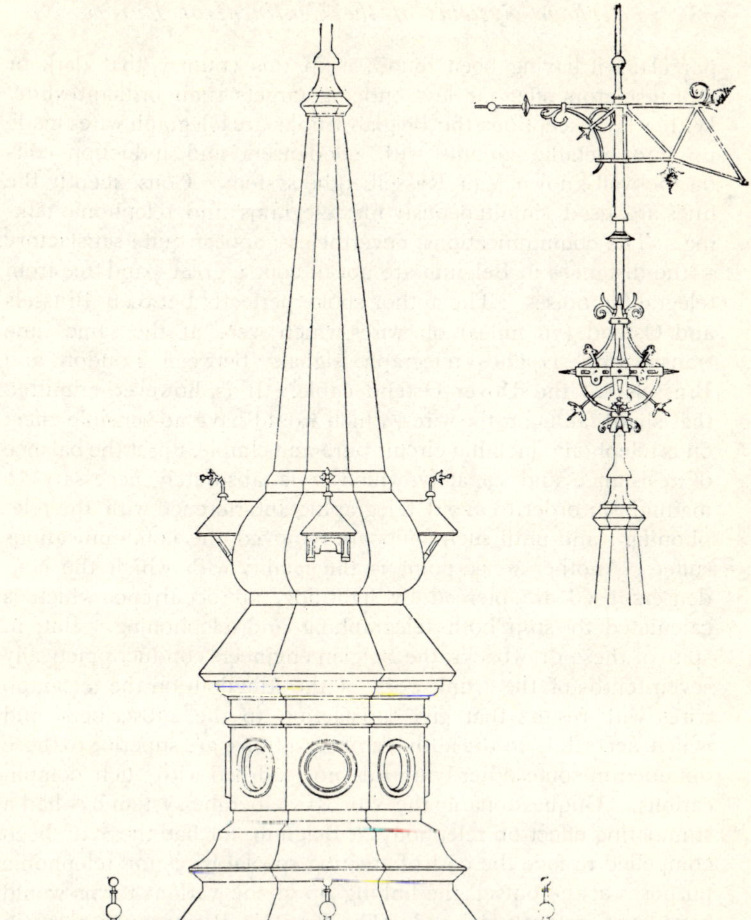

Fig. 29

porcelain, it having been found, as in this country, that dark or dull insulators offer far less enticing targets than brilliant white. With a few exceptions the Belgian trunks are telegraph wires made up into metallic circuits with condensers and induction coils on the well-known Van Rysselberghe system. Consequently the lines are used simultaneously for telegrams and telephonic talking. The communications, nevertheless, appear quite satisfactory —the distances in Belgium are not of course great—and free from telegraphic noises. The author spoke perfectly between Brussels and Ostend (76 miles) on wires which were at the same time transmitting Hughes telegraph signals between London and Brussels *viâ* the Dover–Ostend cable. It is, however, admitted that slight faults on the wires, which would have no sensible effect on a telephonic metallic circuit pure and simple, upset the balance of resistance and capacity which it is absolutely necessary to maintain in order to avoid telegraphic interference with the telephoning; and until such faults are removed the communications suffer. Another weak point is the facility with which the condensers used are pierced by lightning, an occurrence which is calculated to stop both telegraphing and telephoning. But, in spite of these drawbacks, the Belgian engineers conduct practically seven-tenths of the trunk work of the kingdom on the telegraph wires with results that give satisfaction to the subscribers, and which, according to the author's observations, are superior to those obtained in some other countries not saddled with such complications. Unquestionably the Van Rysselberghe system has had a stimulating effect on telephony in Belgium, for had the State been compelled to face the cost of erecting special wires for telephonic purposes at the outset, the linking up of the various towns would have been seriously delayed. The Brussels–Paris trunks, three in number, are exclusively telephonic and are composed on the Belgian side of 3 mm. bronze wire of 95 per cent. conductivity. The wires are revolved on the Moseley - Bottomley system adopted by the British Post Office. The revolutions on the Brussels–Paris lines are in Belgium made by aid of the fixtures shown in fig. 30, the former effecting the vertical and the latter the horizontal changes. Such fixtures require the tops of poles; consequently their definitive adoption would limit the number of trunks to that of existing pole lines. As a fact, the three Brussels–Paris trunks follow different routes for a considerable portion of the way. If placed alongside each other on a cross-arm and crossed every kilometer at the same places, all three would have been got on the same poles and been equally effective.

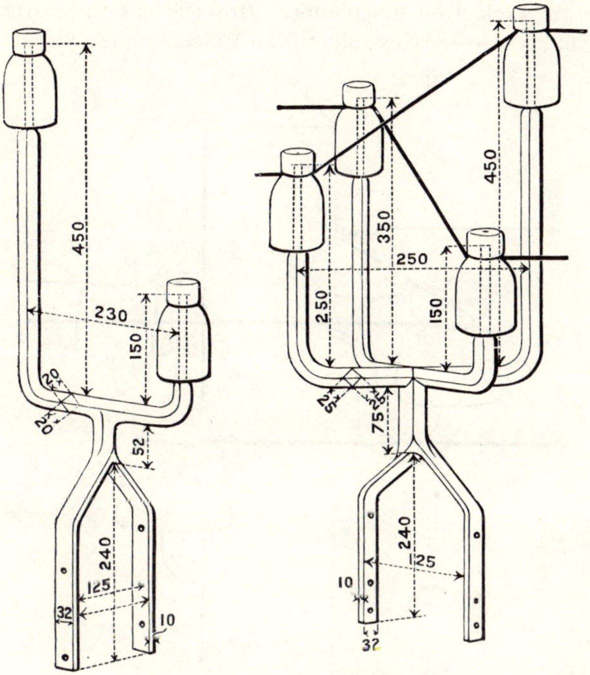

FIG. 30.—Dimensions in millimeters.

As it is, the speaking between Brussels and Paris is practically perfect.

The same twisting plan was originally employed on other trunks, but has since been discontinued, the simple horizontal crossing introduced by the author on the Dundee–Arbroath trunk line in 1884 being found equally effective. The non-use of cross-arms on telegraph poles in Belgium, as in other continental countries, brings about curious complications whenever the simplicity of the normal line has to be departed from. Fig. 31 shows the crossing, devised by M. Saboo, most in favour in Belgium, and at the same time gives a good idea of a Belgian pole and insulators.

There are eight trunks between Brussels and Antwerp, four of which are exclusively telephonic. Brussels has three trunks to Liége and two to Verviers, all on Van Rysselberghe's plan. Of the

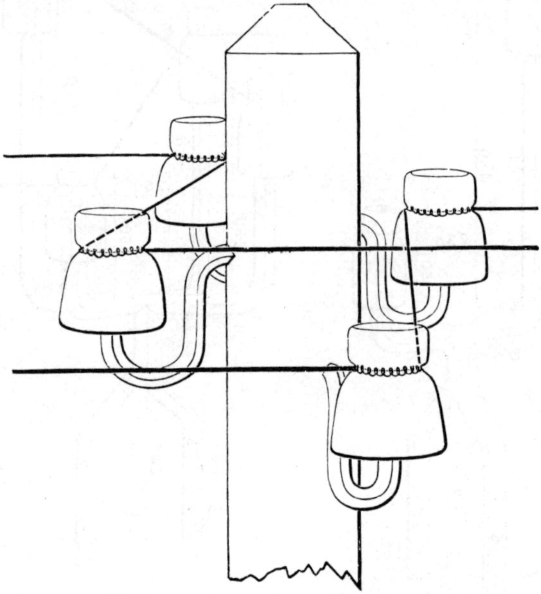

FIG. 31

three trunks between Brussels and Ghent only one is exclusively telephonic. The wire used for the Belgian trunks (excepting the Brussels-Paris) is 2 mm. bronze of 95 per cent. conductivity.

PAYMENT OF WORKMEN

Foremen receive from *6l.* *8s.* to *8l.* per month, with an allowance of 1*s.* *8d.* per day when working away from home.

Workmen commence as lads at 1*s.* 10*d.* per day ; when competent to go on the roofs, they get 2*s.* per day, afterwards rising to 2*s.* 2*d.* and 2*s.* 4*d.* Assistant foremen get 2*s.* 7*d.* per day. All workmen are allowed 10*d.* per day when engaged away from home. In the summer the men are supplied with cocoa, and in winter with brandy, a pint to every ten men daily, gratis. They are expected to make grog of the brandy, which, with the cocoa, is supplied partly with the object of preventing the men drinking unboiled water of bad quality. The hours are ten per day, less one hour for dinner and half an hour for breakfast, making a working day of eight and a half hours. Carpenters, masons, plumbers, and other skilled workmen incidentally required receive 2*s.* 8*d.* per day.

PAYMENT OF OPERATORS

Girls are taken on at eighteen years of age and commence with 32*s.* per month, rising gradually to a maximum of 68*s.* per month. On entry they have to pass an examination in common subjects. The daily duty is from seven to eight hours. Night duty is performed by men.

STATISTICS

The continuous growth of the telephonic telegram traffic is illustrated by the following figures :—

Year	Number of telegrams telephoned throughout Belgium	Year	Number of telegrams telephoned throughout Belgium
1887	469,823	1891	873,266
1888	587,383	1892	900,933
1889	691,098	1893	946,168
1890	800,269	1894	1,023,396 *

* Estimated from August traffic.

The 1894 traffic means receipts for the Telegraph Department amounting to at least 20,000*l.*, which the officials consider more than balances any loss through trunk-line competition. As the telegraph revenue continues to increase year by year, this view is no doubt correct.

At the end of 1894 the particulars of the areas, exchanges, and subscribers stood as follows :—

Name of Area	Names of Exchanges	Number of subscribers	Total subscribers per area
Le Littoral	Ostend	119	269
	Bruges	114	
	Blankenberghe	18	
	Heyst	6	
	Middelkerke	3	
	Nieuport	9	
Termonde—St. Nicholas—Alost	Termonde	22	64
	Alost	21	
	St. Nicholas	14	
	Lokeren	7	
Tournaisis	Tournay	102	133
	Peruwelz	31	
La Hesbaye	Landen	17	113
	Waremme	13	
	St. Trond	25	
	Tirlemont	22	
	Hannut	14	
	Hasselt	22	
Brussels	Brussels	2,474	2,506
	Hal	15	
	Vilvorde	17	
Antwerp	Antwerp	1,832	1,840
	Boom	8	
Verviers	Verviers	649	659
	Spa	10	
Louvain	Louvain	129	129
Liége	Liége	1,073	1,073
Charleroy	Charleroy	328	328
Ghent	Ghent	865	865
La Louvière	La Louvière	51	51
Mons	Mons	400	400
Total for State	31 exchanges	8,430	—
Still in hands of concessionaries at December 31, 1894			
Courtray—Roulers	Courtray	74	—
Mechlin	Mechlin	55	—
Namur	Namur	198	—
Total for Kingdom	—	8,757	—

The growth of the internal trunk traffic has been as follows :—

	1887	1888	1889	1890	1891	1892	1893	1894
Number of conversations .	46,720	53,621	61,575	80,120	108,459	131,189	150,436	
Receipts in francs	49,489	56,344	65,172	88,399	125,415	156,818	187,259	

The lines carrying this traffic numbered and measured at December 31, 1893 :—Sixty-four metallic circuits, each made up of two telegraph wires, measuring in total length 8,408 kilometers, and worked by Van Rysselberghe's apparatus ; eleven exclusively telephonic metallic circuits, measuring 1,124 kilometers of wire.

The actual receipts by the State for the telegraph and telephone services respectively for the five years 1889-93 were as follows :—

Year	Telegraph receipts (francs)	Telephone receipts (francs)
1889	3,463,267	136,359
1890	3,614,930	181,612
1891	3,721,805	242,971
1892	3,650,146	306,503
1893	3,684,068	1,845,010[1]

It will be observed that in 1891, the year of the greatest development of the telephonic trunk lines, the number of trunk talks increasing from 80,120 in 1890 to 108,459 in 1891, the telegraph revenue was better than ever before. During 1892, however, in face of 131,189 trunk talks and a trunk revenue of 156,818 francs, it dropped 71,659 francs. This reads a large sum in francs, but reduced to English money it means only some 2,866*l.*, a small matter for a State department, which was partly made up by the increase of 1,256*l.* in the telephone trunk receipts. In 1893 the telegraph had recovered to within 1,509*l.* of its 1891 figure, in face of an increase in the telephone trunk revenue of 1,217*l.* over 1892. In 1893 the telegraph receipts had decreased 1,509*l.*, and the telephone trunk receipts increased 2,473*l.* over 1891, while the telegraph had resumed its upward course. It must be concluded therefore that the new service had in 1893 drawn 187,259 francs (7,490*l.*) from the pockets of the Belgian people without sensibly affecting the old one.

1 The State acquired most of the companies' systems at the beginning of 1893.

IV. BOSNIA-HERZOGOVINA

No telephone exchanges exist. So far, the telephone has been employed exclusively for military purposes.

V. BULGARIA

TELEPHONE exchange work is a Government monoply. All lines are, and are to be in future, metallic circuits. The development attained at date of writing (February 1895) is but modest, the total number of subscribers in the country being only 151, of which the capital, Sofia, possesses 90. The total length of local lines is $47\frac{1}{2}$ kilometers; of trunk lines actually working, 160 kilometers; and of trunk lines under construction, 330 kilometers. For these lines silicium bronze of 3 mm. diameter is being used, while the local connections are run with wire of the same kind, but of 2 mm. diameter only. Sofia has also telephonic communication with Philippopolis by means of the Van Rysselberghe system fitted to ordinary telegraph wires. Switch-boards with indicators and cords are used in the three exchanges open: these have been supplied by Messrs. Jenisch Böhmer, Berlin; M. Hipp, Neuchâtel (Switzerland); and Deckert and Homolka, Vienna. It is obvious that the Bulgarians are making a good beginning with metallic circuits and bronze wire everywhere, and may be cordially wished success.

SERVICES RENDERED AND TARIFFS

1. **Local exchange communication.**—For a subscriber located within the town limits:—

First year	*8l.*
Subsequent years	*6l*

These rates are inclusive of installation, maintenance, and all charges.

2. **Internal trunk line communication.**—Time unit, five minutes. Charge, any distance, 9·6*d.* Express or urgent conversations are admitted at triple rates.

3. **Public telephone stations.**—Time unit, five minutes.

LOCAL TALKS

Subscribers	free
Non-subscribers	4·8*d.*

Express talks, triple fee.

TRUNK TALKS

Subscribers and non-subscribers	9·6*d.*

Express talks, triple fee.

VI. DENMARK

HISTORY AND PRESENT POSITION

In Denmark, as in Holland and Norway, and at first in Sweden, telephonic development has been left in the hands of concessionary companies and individuals, and to an even greater extent, for it is only within the last two years that the Government Telegraph Department has taken any part, directly or otherwise, in telephone exchange work. The plan has been for municipalities and other local authorities to grant licences for the areas under their control, and exchanges have been thereupon established, usually with locally-subscribed capital. This system, open as it doubtless is to the reproach of want of uniformity and homogeneity, has had, wherever brought into use, a most beneficial effect in stimulating telephonic development and in bringing the new mode of communication within the reach of the masses. It has placed Holland and the three Scandinavian countries telephonically far in advance of Great Britain, where the alternative of doing without telephones at all is apparently preferred to allowing the people any opportunity of acting for themselves, or of breaking loose from the fetters forged, in the name of public policy, by the Post Office. In Denmark, as a consequence, a country not much larger than some of our English counties, there exist and flourish—that is to say, are worked at a profit—some sixty-six telephone exchanges, which means that not only every town, but almost every townlet and village in the country, possesses one. Copenhagen, the capital, a city with a population rather exceeding that of Islington, boasts (November 1894) of 4,510 instruments in connection with its exchange, and outside Copenhagen, in the same small island of Zealand, there are 900 more. How many are there in Islington? Possibly 100, although that is extremely doubtful.

The International Bell Telephone Company commenced work in Copenhagen in 1880, and held the ground without competition until the sale in 1882 of the system to a local association called the Copenhagen Telephone Company, which, under the able management of Mr. E. B. Petersen, has not only preserved the monopoly, but has extended its system until the highly-creditable development mentioned above has been reached. The absence of competition has prevented the low rates enjoyed by the subscribers in Stockholm and Christiania being attained, and the handsome figure, redolent of telephonic clover, of 8*l.* 6*s.* 8*d.* per annum is still maintained in the Danish metropolis. It is not, therefore, surprising to learn that the company maintains a dividend of about 7 per cent. on its capital of 112,000*l.*, a capital which has not only sufficed to construct the Copenhagen exchange, but to cover Zealand with trunk lines too. And it must be clearly understood that the rate of 8*l.* 6*s.* 8*d.* covers not only communication within Copenhagen itself, but with every subscriber in the island of Zealand, whether a member of the Copenhagen company or any other. As Zealand measures some eighty miles from north to south and sixty miles from east to west, and contains some 900 subscribers outside the limits of Copenhagen, the liberality of this arrangement is beyond question.

The geographical character of Denmark has not favoured the erection of long-distance trunks. Within the three chief divisions, Zealand, Funen, and Jutland, the country has long been well telephoned, the local companies being left to construct what trunks they chose free from Government interference. It was not until the question of joining up the three divisions and of making a connection with Sweden, works necessitating the use of submarine cables, came to the front, that the State bestirred itself. The Royal Telegraph Department then announced that it would itself undertake the construction and maintenance of these through main lines; and accordingly it has recently established communication with Sweden by utilising an old telegraph cable, and opened a line to Funen, which is to be extended as soon as practicable to Jutland, and eventually thence to Hamburg. For the purposes of these trunks the Government has established a small switch-room at the Central Telegraph Office in Copenhagen, but, with the exception of one public station, there are no other connections to it, the company's subscribers supplying the necessary customers. To enable them to make the best use of the trunks, those subscribers who are willing to pay 2*l.* 15*s.* 7*d.* down, a first and last payment, are being supplied with metallic circuits. The company, being gifted with the faculty of rightly interpreting the signs of the times, intend to gradually convert the whole of its system to double wires, and all new work and alterations are designed accordingly, especially its grand new central station at Copenhagen, which is being fitted throughout for metallic circuits. Underground work has already been undertaken in Copenhagen on an extensive scale, and much more, with paper insulation and twisted pairs, is in contemplation. Altogether, Denmark may be complimented on being a practical, advancing, and exemplary member of the telephonic family, and one which may be safely trusted to look after its own interests, both technically and financially.

Although comparatively high rates prevail in the capital, the provincial towns enjoy subscriptions which range from 1*l.* 18*s.* 8*d.* to 4*l.* 8*s.* 11*d.* per annum. As in Norway, the subscribers sometimes supply or pay for their instruments, but in the majority of cases the subscription is an inclusive one. There are some fifty independent companies in Denmark, all, or nearly so, having rules which differ in some or other respect from those of their neighbours. An exhaustive account of these small concerns would be equally tedious and unprofitable, but, thanks to the courtesy of the managers of some of them, the author is enabled to present herewith a tabulated statement in which their chief characteristics are set down. It will be noticed that the subscriptions rule higher than in Norway, but that, on the other hand, the members are seldom called upon for any supplementary payments, while the distances over which they are entitled to speak are often considerable. The full accounts for 1893 of the Aarhus Telephone Company, which afford an insight into the methods prevalent in Denmark generally, are printed at the end of this section.

STATISTICS OF SOME PROVINCIAL DANISH EXCHANGES

Town	Population	Date of opening	By whom owned	Number of switch-rooms	Number of subscribers	Do subscribers pay for their instruments?	Do subscribers pay for their lines?	Amount of annual subscription
Aalborg . .	19,500	April, 1884	Syndicate of three members	1 Central, 7 Branch	394	No	No	3*l.* 12*s.* 2*d.*
Aarhus (*a*) . .	33,000	July, 1883	Company	1 Central, 9 Branch	500	No	No	4*l.* 3*s.* 3*d.* for town ; 5*l.* 16*s.* 7*d.* whole district
Esbjerg .	1,529	Sept., 1885	Syndicate of four members	3, Esbjerg, Ribe, Fano	220	No	Yes	Esbjerg, 2*l.* 15*s.* 7*d.* ; Ribe, Fano, 3*l.* 17*s.* 8½*d.*
Fredericia .	10,042	Dec., 1889	Company	1	116	No	No	3*l.* 6*s.* 8*d.*
Frederikshavn .	2,891	Jan., 1885	Hjorring County Telephone Company	1 Central, 7 Branch	218	No	No	Town, 2*l.* 15*s.* 7*d.* ; whole county, 5*l.* 11*s.* 1*d.*
Hillerod . .	—	Jan., 1889	Mr. T. Schäffer	1 Central, 5 Branch	102	No	In towns, no ; in country, yes	3*l.* 6*s.* 8*d.* locally ; 8*l.* 6*s.* 8*d.* whole of Zealand
Horsens . .	12,654	Jan., 1885	Company	1	200	No	No	3*l.* 6*s.* 8*d.*
Korsör .	—	Jan., 1895	Company	1	96	No	No	1*l.* 18*s.* 8*d.*
Middelfart .	—	July, 1885	Company	1 Central, 2 Branch	40	No	Yes	3*l.* 17*s.* 8½*d.*
Odense . .	30,268	March, 1884	Company	1 Central, 12 Branch	500	No	No	Town, 4*l.* 8*s.* 11*d.* ; suburbs, 5*l.* 11*s.* 1*d.*
Randers (*b*) .	16,617	Nov., 1883	Company	1 Central, 14 Branch	345	No	No	4*l.* 3*s.* 3*d.*
Ringkjöbing .	—	Oct., 1890	Company	7	112	No	No	2*l.* 15*s.* 7*d.*
Skive . . .	—	Feb., 1888	Co-operative Company	1 Central, 1 Branch	103	No	No	3*l.* 12*s.* 2*d.*
Sorö and Ringsted	—	1887	Mr. Charles Heidemann	2	113	No	No	3*l.* 12*s.* 2*d.*

(*a*) See full accounts for 1893 at end of this section. (*b*) Multiple switch-board at Central. Town wires, 1·5 mm. bronze ; country wires, 2 mm. steel.

Distance of radius to which subscription applies	Hours of daily service	Total expenditure on system	Amount of annual revenue	Annual amount of working expenses and maintenance	Annual profit	How is profit disposed of?	Number of connections per month	Description of instrument used by subscribers
32 kilometers	14 hours summer; 13 winter	£ (*c*)	£ 1,420	(*c*)	12 per cent. on the subscribed capital	(*c*)	100,000	Magnetos
About 20 kilometers	Day and night	8,241	2,307	Not yet ascertained for 1894	Not yet ascertained for 1894	Reserve fund and shareholders	—	Magnetos
About 9 kilometers	7 till 9 summer; 8 till 9 winter	3,309	658	£ 275	£ 383	Shareholders	30,000	Battery calls
15 miles to Veile; 4 miles to Middelfart, through a submarine cable	8 till 8	1,318	302	152	150	6 per cent. on capital to shareholders; rest to reserve	—	Magnetos
Towns and county respectively	7 till 9 summer; 8 till 9 winter	3,571	879	549	330	Half to shareholders, half to reserve	—	Magnetos
About 20 kilometers	(*d*)	2,200	594	(*d*)	(*d*)	(*d*)	(*d*)	Magnetos
Town and vicinity	Not stated	2,947	659	385	274	Half to shareholders, half to reserve	15,000	Magnetos
Town and vicinity	12 hours	(*e*)	(*e*)	(*e*)	(*e*)	To be divided between shareholders, subscribers, and reserve fund, according to fixed rules	—	Battery calls
Not stated; submarine cable to Fredericia	8 till 8	1,099	176	99	77	Divided between shareholders, reserve, and employees' fund	4,500	Battery calls
30 kilometers	8 till 10	5,200	1,757	879	878	⅓ to shareholders; ⅓ to reserve; ⅓ to employees' fund	—	Magnetos
About 20 kilometers	Day and night	5,824	988	450	538	6 per cent. to shareholders; rest to reserve	—	Ericsson's magnetos
22 Danish miles	14 hours	1,819	385	186	199	6 per cent. to shareholders; rest to reserve	9,000	Battery calls
8 miles	13 hours	1,540	439	181	258	Profit is used for new works; paid-up capital is only 384*l.*	—	—
Whole Sorö country	8 till 8	(*f*)	395	275	120	—	1,500	Magnetos

(*c*) Not given. (*d*) Not stated. (*e*) New company. (*f*) Not properly known: present owner having bought a part of the system subsequent to its construction.

SERVICES RENDERED BY THE COPENHAGEN TELEPHONE COMPANY

1. **Intercourse between the subscribers and public telephone stations for the same town or district.**—A Copenhagen subscriber is entitled to free communication with every other subscriber in the island of Zealand, even when the exchange of which this latter is a member belongs to another company. This means that Korsör (63 miles), Elsinore (34 miles), Slagelse (56 miles), Naestved (57 miles), Praestö (53 miles), Kiöge (25 miles), are all covered by the Copenhagen subscription. Conversely, however, subscribers in these and other Zealand towns (most of which are in the hands of local concessionaries) must pay extra for the right to originate communication with Copenhagen. Thus in Korsör, Elsinore, Roskilde, Kiöge, Sorö, Slagelse, &c., there are three tariffs in operation : (1) for local town communication only ; (2) for communication within the limits of the same county (there are five counties in Zealand) ; and (3) for communication with the capital. The Copenhagen subscribers are at present, pending the completion of the new central station, scattered amongst twelve switch-rooms, four of which are in the town and eight in the suburbs. The number of junction wires which connect these last to the main offices is insufficient to always ensure getting through without waiting, in consideration of which the subscribers whose wires go to the suburban switch-rooms are charged only 5*l.* 11*s.* 1*d.* per annum. Apart from having to wait their turn for the junction wires, their privileges are on a par with those of the city subscribers. The liberal policy of the Copenhagen Company receives another demonstration in its treatment of subscribers changing offices or residences, whose telephones are shifted gratis.

2. **Internal trunk communication.**—This practically extends from Copenhagen to every town and village in Zealand and in the island of Funen. The exchanges in Jutland and in Laland are in communication with each other locally. Funen is connected to Zealand by a cable, twelve miles long, across the Great Belt, between Korsör and Nyborg, which cable touches in passing at the famous island and lighthouse of Sprogö. It is an old tele-

graph cable. As the Copenhagen local subscription covers the use of the Zealand inter-town wires, the company has no trunk revenue if certain express fees and charges for inserting provincial subscribers' names in the Copenhagen list be excepted. As in Norway, the cost of constructing and maintaining the trunks is apportioned between the companies using them. The Government has not interfered in any way, and has even granted way-leave facilities freely when required. The Swedish and Norwegian practice of booking talks over the trunks in advance is not permitted in Denmark.

3. **International trunk communication.**—The lines intended for this purpose are constructed and maintained by the Government. Communication is at present limited to Sweden, with the southern portion of which Denmark has necessarily extensive commercial relations. The distance being short (10½ miles) two wires of an old four-wire telegraph cable, touching at the island of Hveen, have been utilised with sufficiently satisfactory results, communication between Copenhagen and Stockholm (375 miles) being good enough for all purposes. That there is a fair demand for the Swedish connection is evidenced by the fact that 100 Copenhagen subscribers have already paid *2l.* 15*s.* 7*d.* and had their lines converted to metallic circuits in order that they may use it. The company has, with the same object, also provided eight of the public stations with double wires. The long-distance trunk connections are made through three metallic circuit junction lines which join the telephone central to the State telegraph office. It is the intention to follow up the connection of Zealand with Funen (completed) and Jutland (constructing) by a line to Hamburg, which will be made up as follows :—

	Miles
Copenhagen to Korsör	63
Korsör to Nyborg (cable)	12
Nyborg to Strib	45
Strib to Fredericia (cable)	2
Fredericia to Hamburg	155
	277

4. **Telephoning of telegrams.**—There are two distinct forms of this service. Firstly, telegrams can be forwarded and received

by the subscribers by means of wires connecting the telephone exchange with the Government telegraph office. In this case subscribers availing themselves of the facility have to enter into an agreement direct with the State authorities regarding the payment of the charges accruing on their traffic. Secondly, the telephone company undertakes the duty, for those who desire it, of writing down messages dictated by their subscribers and sending them by messenger to the nearest telegraph office, where they are handed in and paid for, the charges being afterwards collected from the senders. Similarly, subscribers can order the telegraph people to deliver telegrams addressed to them at a telephone station, whence they are telephoned. The second plan obviates any formal agreement with the State, although it is necessarily less rapid. It is noteworthy that the company exacts no deposits from its subscribers to cover telegram and trunk charges and yet suffers no loss, an experience which agrees with that of the author in Scotland during 1885–1890. A simple undertaking to pay was then found sufficient, and in no single instance led to loss. In Copenhagen accounts for these extra charges are rendered monthly, but are collectable oftener at the company's discretion. Copies of telegrams telephoned to subscribers are afterwards delivered by messenger in the usual way.

5. **Telephoning of messages for local delivery.**—With this telephonogram service the Copenhagen Company scores a good point. It amounts, in effect, to a twopenny ten-word telegram rate for the city and a 3·3*d.* rate for the suburbs. The State telegraph department, although legally invested with a telegraphic monopoly, has not interfered, and is apparently content to let the company provide the citizens with a cheaper service than the department itself sees its way to. The company accepts written messages addressed to non-subscribers at all its offices, switch-rooms, and public stations, and transmits them by telephone to the nearest points, delivering them thence by messenger. The subscribers can likewise call the head office and telephone such messages. The only restriction is in the matter of language, Danish being obligatory, as the mass of the employees understands no other. Still, it is to be apprehended that the Londoners or Glaswegians would not absolutely refuse to use a twopenny

telegram service even though restricted to English. The traffic in these twopenny telephonograms amounted in 1892 to 40,266; in 1893 to 44,249; and in 1894 to 47,069.

6. **Public telephone stations.**—These are numerous, and are available for local and trunk talks, and for the transmission of both long-distance and local telegrams. The company sells books containing ten tickets, each of which entitles the presenter to a free local talk at a public office, or from the premises of any subscriber who may allow his instrument to be used. Such a subscriber, on sending the tickets he collects to the telephone office, is paid ·66*d.* on each by way of remunerating him for his trouble. A subscriber may go in regularly for the public station business by paying an additional subscription of 2*l.* 15*s.* 7*d.* per annum, in which case the company supplies him with a signboard and allows him to keep all he can manage to take. There is another arrangement, by which a person occupying suitable premises pays only 2*l.* 4*s.* 5*d.* by way of annual subscription, and is charged by the company 2*d.* for each talk had from his instrument. On talks had by strangers he collects 2*d.* and pays over to the company only 1·33*d.*. Automatic slot boxes (Schäffer's patent) are used in about fifty public stations and give satisfaction.

7. **Messenger service.**—As in some other countries, non-subscribers are called to public stations to converse with subscribers who want them. Nothing is charged for the service. The company's messengers do not run ordinary errands or carry parcels, there being a separate organisation (Adam & Co.) in Copenhagen for this purpose.

TARIFFS

1. **Rates for communication within Copenhagen and Zealand:**

	Per annum. £	s.	d
One instrument on a direct line to central exchange	8	6	8
For a second connection	6	13	4
One instrument on a direct line to a suburban exchange	5	11	1
Extra instruments	1	7	10

There is also an elaborate tariff for several instruments on the same line. Contracts are for one year only. Subscriptions are payable quarterly in advance. The difference of rate between the central and suburban exchanges is due to the small number of junction wires employed, which necessitates occasional waiting for connections by the suburban subscribers.

The tariff covers connection in any part of the town or suburbs, and includes the right to originate communication with any telephone subscriber in Zealand.

2. **Rates for Zealand and Funen trunk communications.**—As the local rates cover the Zealand trunks, the company has no trunk revenue except that derived from the express fees, there being a rule that any subscriber who wants immediate connection may speak out of his turn on payment of 4·6*d.* As the provincial subscribers are entitled to be called up from Copenhagen, it is important for them to have their names in the Copenhagen Company's list, although they may themselves be members of a local exchange owned by another association. For this service the Copenhagen Company charges 11*s.* $1\frac{1}{2}$*d.* per annum per insertion. To these two sources of income must be added the fees —4·6*d.* per five minutes—payable by strangers at public telephone stations for talks to Zealand towns. The tariff for trunk talks to Funen and Jutland has been fixed at 1*s.* $1\frac{1}{4}$*d.* and 1*s.* 8*d.* per three minutes respectively, these charges going to the State. At date of writing (February 1895) the Jutland line had not been completed.

3. **Rates for international trunk communication:**

	s.	*d.*
To Malmö	1	8
,, Stockholm	2	$2\frac{1}{3}$
,, Gothenburg	2	$2\frac{1}{2}$

Time unit, 3 minutes.

4. **Rates for the telephoning of telegrams:**

When the message is telephoned direct between the subscriber's office and the State telegraph office, in either direction, per message	2·6*d.*
When dictated to the company's office for handing to the State, or a message is received by the company from the State to be telephoned, per word	·133*d.*

5. **Rates for written messages accepted, telephoned, and delivered by the company.**—Within the limits of Copenhagen: a first charge of ·66*d.*, with ·133*d.* per word; minimum charge, 1·99*d.*

Within the suburbs: a first charge of ·66*d.*, with ·266*d.* per word; minimum charge, 3·3*d.*

A town telephonogram containing ten words therefore costs ·66 + ·133 × 10 = 1·99*d.*

And a suburban, ·66 + ·266 × 10 = 3·32*d.*

Subscribers may telephone similar messages from their own offices at the same rates. Accounts for these are rendered monthly; no deposits.

Were it not for the Spanish rate for a corresponding service in Madrid, &c. (Spanish section, p. 329), of 1·92*d.* for *twenty* words, the country of the Dane would have been fairly entitled to a record in this matter.

6. **Rates levied at public telephone stations:**

	£	s.	d.
Five minutes' local talk	0	0	2
Five minutes' talk with any town in Zealand connected by trunk	0	0	4·6
Books containing ten 2*d.* tickets are sold for	0	1	1½
Annual rate, covering free use of all public stations for local talks	2	4	5

The police are entitled to use the public stations gratis.

Subscribers are allowed to use a number of the public stations without charge. These free stations include the Bourse, where there are eight sound-proof compartments containing instruments, and the Custom House, where there are three instruments. At these last two stations messengers are kept who fetch (without charge) non-subscribers wanted by subscribers to the instruments. From eight of the public stations the international line to Sweden may be used at the usual rates (p. 116).

7. **Messenger service.**—This is performed by the company gratis.

WAY-LEAVES

The author is not aware whether Denmark is one of those fabled regions, about which partisans wax eloquent whenever anybody complains of high rates, wherein way-leave grantors are supposed to cease from troubling and monopolists enjoy halcyonian rest. If so, he is sorry to dispel the illusion once more. None of the Danish companies possess any way-leave rights other than they bargain and arrange for. In Copenhagen especially (and this city is certainly one of the worst on the Continent in this respect) overhouse way-leaves are difficult, and in some quarters even impossible, to procure. For a standard of any size a free telephone has generally to be given. So thorny grew the company's path that, at a very early date, it obtained a concession from the municipality permitting the laying of wires under the streets, a privilege for which 388*l.* per annum is at present paid, a tribute which is liable to be revised—i.e. increased—every five years. The country authorities have, however, been easy-going in respect to the roads, since permissions to erect the trunk line poles have generally been accorded at reasonable rates. The Government, too, although owning the railways and telegraphs, has not played the dog-in-the-manger, and has lent the companies a helping hand where difficulties, otherwise insurmountable, have presented themselves.

SWITCHING ARRANGEMENTS

At the present central station the switch-board is an ordinary Western Electric single-wire double-cord series multiple ; at the branches Gilliland boards are still employed. The test, lightning-guard and cross-connecting boards are neatly arranged round the interior walls of small rooms or cupolas. The number of connections asked for by each subscriber daily averages eleven, is frequently twelve, and sometimes as high as fourteen. The operators attend to from 50 to 100 lines each. Called subscribers are rung by the operators. For this purpose a magneto generator, driven by an electro-motor supplied with current from the municipal lighting mains, is employed. But the present arrangements are to vanish in a few months, as soon as the company's new building is ready. In May 1893 Messrs. Ericsson & Co., of Stockholm, delivered a sample single-cord, parallel-jack board, manufactured to the designs of Mr. J. L. W. V. Jensen, the Copenhagen Telephone Company's chief engineer, which was put into use for the trunk and other metallic circuits converging at the present central station, and being found entirely satisfactory, an order was placed with Messrs. Ericsson for a complete installation on the same plan for (ultimate capacity) 10,200 subscribers' metallic circuits and 480 trunks and junctions for the new central station. The board, which is equipped at present for 6,240 lines only, has been delivered, and is only waiting the completion of the switch-room. It presents several new features, and will be clearly understood with the help of fig. 32. The main idea has been to keep only one indicator in shunt across the metallic loop when two subscribers are coupled, and this has been effected by the combined aid of the jacks, the plugs, and of the special relays *sr*. l_1 and l_2 are the subscribers' two lines through the multiple system; *t* the test wire. j^{I} j^{III} show jacks at different boards, j^{II} at the subscriber's own board. *j* shows a jack with a plug inserted, causing the line springs s_1 and s_2 to make contact with the head and tube of the plug respectively, while the test spring *ts* is insulated from the jack and thrown into connection with the testing battery, which, in the manner explained below, cuts out the subscriber's drop. *if* is the intermediate field. *sd* is the subscriber's drop, also acting as a ring-off drop, having a very high self-induction. *sr* is the subscriber's relay, which cuts out the *sd* when a plug is inserted in one of the jacks j^{I} j^{II}. The relay and drop are shown separate; if preferred, they may be combined. Although the armature of *sr* is shown inserted in one of the branching wires to the drop, and thus leaves the drop coils connected to one side only of the loop when a connection is on, it could as easily have been placed midways if the wire on the drop-magnet had been wound in two halves. Experience shows, however, there is no advantage in doing so, because the exceedingly small capacity of the drop does not perceptibly disturb the balance of the metallic circuit. *sp* is the subscriber's plug with flexible cord ; p_1 and p_2 are the head and tube of the plug respectively. *pc* is the subscriber's plug contact, cutting out the relay *sr*, when the plug *sp* is removed, and at the same moment joining the battery to the test wire *t*, so causing the subscriber to test 'busy.' *sk* is the line key. When this is pressed down, the operator's telephone apparatus is cut in between l_1 and l_2. *sk* and *sp* are for convenience placed in close proximity. The operator's apparatus consists of *kp* and *kl*, keys for speaking to plug or to line respectively. These keys are not used under normal conditions, and only when it is necessary to speak to one side, insulating at the same time the other side. *cp* and *cl* are calling keys for effecting the ringing to plug or to line respectively. When one is depressed, it is at the same time possible to speak to the other side. Under normal conditions only *cp* is used. *g* is the generator for the ringing current. *i* and *mb* the microphone and telephone combined into a microtelephone set, suspended and balanced by a counterpoise. The apparatus is connected by a flexible cord to a four-way plug and jack (only shown in the figure by dots), so that a new microtelephone set may be immediately inserted. *mc* is the microphone battery contact, and *tc* the test contact, giving a road to earth through a self-induction coil (not shown in the diagram) for the test current when this has passed the telephone. These contacts, *mc* and *tc*, may be left out if another four-way plug, connected to a second microtelephone set without such contacts, or to a head telephone and microphone, be inserted. The connecting wire between *tc* and ground might then be connected to a wire in the telephone between the magnet spools, the test current going in this manner through only one of the coils, which is quite sufficient for the testing.

The mode of operating is as follows : When *sd* falls, *sk* is depressed and *sp* lifted by practically the same movement with the right hand, while the number of the wanted subscriber is received through the telephone. After testing by touching the jack of the line called for with the plug head and pressing at the same time *tc* with the left hand, the plug is inserted in the jack and *cp* is depressed for calling. When the connection is through, *sk* is released and *sd* replaced. When *sd* falls again, in response to

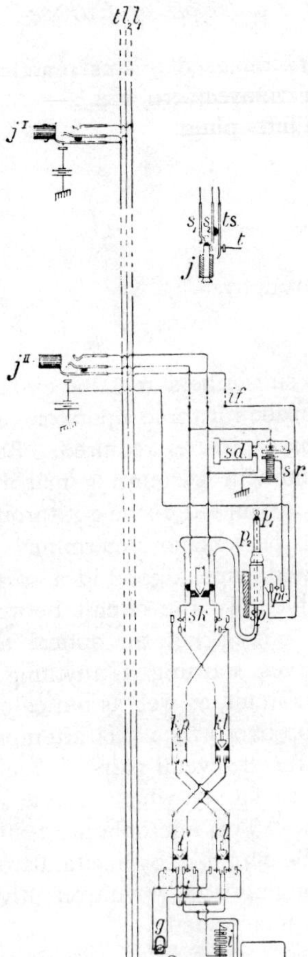

tll
j
ts
t
if.
sd.
sr.
sk
sp
pr
kl
g
i
mb.
m
mc
tc.
te.

FIG. 32

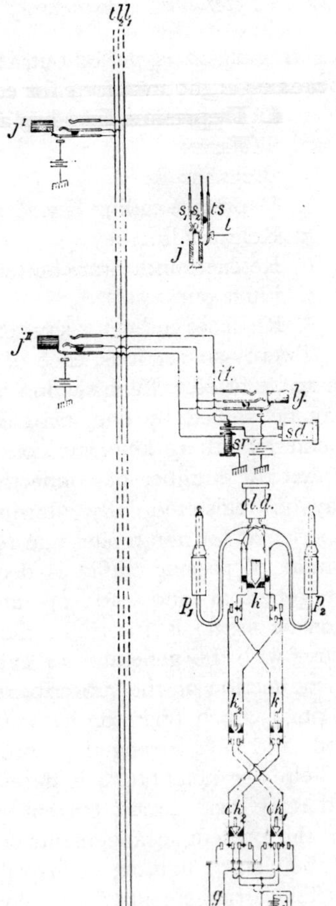

tll
j
ts
l
if.
lj.
sd.
sr
k
i
g
mb
m
mc.
tc

FIG. 33

the ring-off, *sp* is pulled out, and *sd* replaced. This practically means eight motions for each connection, viz. :—

1. Depresses line key and lifts plug.
2. Tests.
3. Plugs in.
4. Depresses calling key.
5. Releases line key.
6. Replaces indicator shutter (through).
7. Plugs out.
8. Replaces indicator shutter.

Twenty-six sections, each of 240 subscribers' metallic circuits, are to be fitted. Each section has space for three operators, and may be served by one, two, or three girls as required. Each operator's calling key *cp* is connected with a counting machine, so that the number of connections attempted to be got through may be registered. By allowing a percentage determined by experience for non-replies and repeated rings, a good idea of the volume of passing traffic is deducible. In the circuit between the generator and each operator's calling key an optical and acoustic signal is inserted which gives warning if anything is wrong with the generator or calling circuit, as well as notice of a disconnection on the subscriber's loop over which it is attempted to ring. Each operator has within reach several pairs of double cords and keys, arranged according to fig. 33, which enable her to help her neighbours if necessary. The microphone, testing and relay local circuit current will be supplied by accumulators, and the present arrangements for ringing from generators driven off the electric lighting mains will be maintained.

The arrangements for the local and trunk inter-switching—a very important matter in Copenhagen—have not yet been finally matured.

Fig. 33 shows Mr. Jensen's adaptation of his idea to a double-cord parallel multiple board. *lj* is the local jack ; *cl.d* the ring-off drop ; p_1 and p_2 the plugs ; *k* the line switch ; k_1 and k_2, ck_1 and ck_2, keys for speaking and ringing to either side. Normally, when p_1 is used as answering plug, only *k* and ck_2 are brought into play.

Mr. Jensen has further modified his system to act with self-restoring drops. Fig. 34 shows the alterations made on the fig. 33 arrangement in order to bring this about. *sd* is the self-restoring drop and cutting-out relay combined in one piece, while *cl.d* is the self-restoring ring-off drop without a relay.

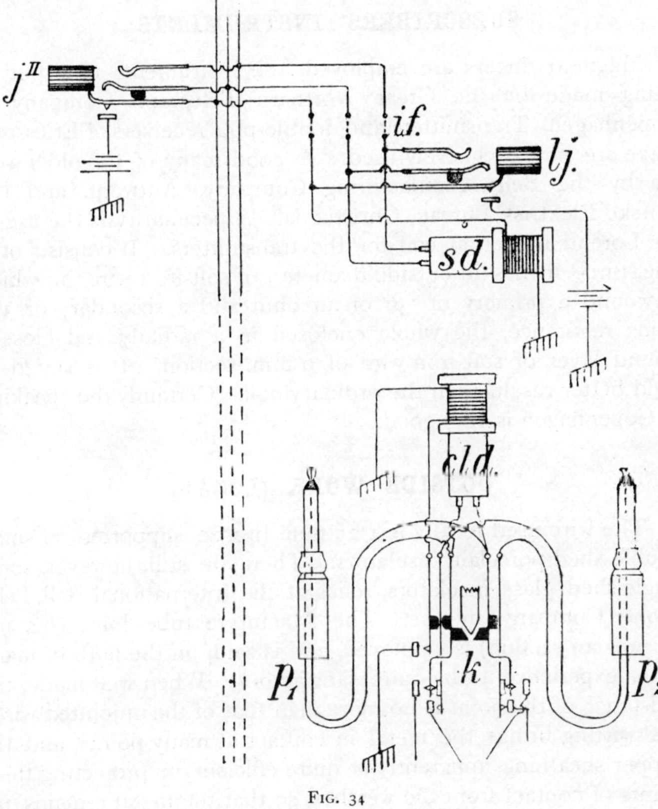

FIG. 34

Pending the introduction of the metallic circuits, the subscribers are connected to the Zealand trunks through translators of the author's construction, manufactured by Messrs. L. M. Ericsson, of Stockholm.

HOURS OF SERVICE

The Copenhagen central station is open day and night ; the suburban ones from 6 or 8 A.M. till 8 or 10 P.M., which are also about the hours of the provincial exchanges.

SUBSCRIBERS' INSTRUMENTS

Magneto ringers are employed, the instruments now put in being made by the Great Northern Telegraph Company of Copenhagen. Transmitters and double-pole receivers of Ericsson's make are now exclusively used. A good many of the older sets are by the Bell Manufacturing Company, Antwerp, and the Norske Elektrisk Bureau, Christiania. A peculiarity is the use of the Lorentz induction coil for the transmitters. It consists of a ring, three inches in outside diameter, of soft-iron wire, on which is wound a primary of ·36 of an ohm and a secondary of 360 ohms resistance, the whole enclosed in a radially and closely-wound layer of soft-iron wire of 9 mm. section. It is stated to yield better results than the ordinary coil. Certainly the speaking in Copenhagen is very good.

OUTSIDE WORK (LOCAL)

The wire used locally is 1·25 mm. bronze, supported on small double-shed porcelain insulators. There are still, however, some single-shed glass insulators, relics of the International Bell Telephone Company, in use. The Macintyre tube joint (fig. 99, Norwegian section) is employed, and is said, on the faith of many years' experience, to be quite satisfactory. When well made, the resistance of this joint is no more than that of the unjointed wire ; the twisting brings the metal in contact at many points, and the copper sheathing apparently is quite efficient in protecting these points of contact from the weather, so that the metal remains uncorroded and even bright after prolonged exposure of the joint. Mechanically, the joint is stronger than the wire. Solder could not produce better results than these, and the elimination of the soldering bolt in any form is a decided gain. Of course the

FIG. 35

joint to be effective must be well made, but so must soldered ones. There are no single standards in Copenhagen, all having two or more uprights. They are built of channel and angle iron; are well stayed, and generally strong and well constructed. Fig. 35 shows a typical Danish standard with its details. All house-top fixtures are protected from lightning by a conductor and special earth-plate. The pole routes are substantially built, and many of the ground poles erected within the city limits are of highly ornamental design. In this respect it is strange how far the Danes, in common with most continental peoples, are in advance of us. In Great Britain the mere mention of a telegraph pole conjures up visions of something offensive, both to the eye and the nose; in many cities on the Continent, on the contrary, such a structure evokes no disagreeable feeling because, by means of a graceful outline and regularly-renewed paint, it is made to harmonise with its surroundings. It appears, when so treated, to drop into its natural place, and nobody thinks of objecting to it any more than to a lamp-post. To find anything more obtrusively ugly than a British telegraph pole, it is necessary to view a French railway telegraph or cross the Atlantic to the dominions of Uncle Sam. The present central station fixture is the original wooden one of American design. It will be replaced on the new building by an iron tower with attachments for 4,000 wires. An important feature of the Copenhagen system is the underground work. By virtue of its agreement with the municipality, for which it pays 388*l.* per annum, the company is allowed, under supervision, to open the streets and put down conduits and cables. The original conduits consist of cement troughs of rectangular section, covered with an arched lid which fits, and is cemented, into grooves formed along the tops of the trough walls. The custom has been, when additions or repairs are necessary, to open the ground, remove the lid section by section, lay in the cable, replace the lid, and make good the ground. This plan, although it permits of the cables being laid neatly in the trough without friction or chafing, necessitates long lengths of open trench and frequent disturbance of the streets. On these grounds the municipal authorities have objected, and in future the conduits will be permanently buried, and the cables drawn in. The conduits now being laid have an ultimate capacity of 8,000 metallic circuits, and consist partly of cement blocks, with ducts for the cables, and partly of small iron tubes stacked together, the object being in each case to provide a separate channel for each cable, an object which cannot be too strongly commended. The cables, which in the centre of the town convey nearly one-third of the total number of subscribers, have hitherto been chiefly of the 'anti-induction' type, i.e. the single wires are insulated with india-rubber and sheathed with metal foil joined to earth; but in connection with the new exchange the cables will be all paper-insulated, with conductors of ·8 mm. copper, and a capacity of ·05 microfarad per kilometer, the wires being laid up in twisted pairs. There are a few aerial cables, each containing fifty-two twisted pairs of copper conductors, ·8 mm. copper, insulated with paper, capacity ·05 microfarad per kilometer, protected by lead, and hung from stranded steel suspenders.

OUTSIDE WORK (TRUNK)

The trunk lines which radiate from Copenhagen to every town and village of Zealand are mostly metallic circuits built of 2 mm. hard-drawn copper, the wires being crossed at intervals to counteract induction. The poles are wood, and the insulators double shed; as a rule, the routes, which follow the country roads, are both substantial and neat. The Government line to Sweden, *viâ* Vedbok, is of 3 mm. high conductivity bronze wire, twisted on the Moseley Bottomley plan. On the Swedish side the construction is with 3 mm. hard copper, the two sections being joined by an old four-line telegraph cable with parallel wires. The Danish section of the projected line to Hamburg is to be of 4 mm. high conductivity bronze with twisted wires, but the twelve-mile submarine section between Zealand and Funen will in this instance also be an old telegraph cable.

PAYMENT OF WORKMEN

The foremen receive 8*l.* 6*s.* 8*d.* per month; skilled wiremen 4*s.* 5*d.*, and labourers 3*s.* 4*d.* per day, hours being from 7 A.M. till

7 P.M. in summer, with one and a half hours for meals; in the winter the men work only from daylight to sunset, but their pay is not reduced.

PAYMENT OF OPERATORS

Girls are taken between the ages of eighteen and twenty-four only. After a month or two of probation and a successful examination in common subjects, they begin with 22*s.* 1*d.* per month, with four hours per day duty. The next step is to 38*s.* 8*d.* per month, with six hours' daily duty. The maximum to an ordinary operator, attained after five years' service, is 55*s.* 7*d.* per month. The day's duty never exceeds six hours. Night and Sunday duty, for which extra payment is given, is performed by the girls. The chief operators, of course, receive better pay still, but it is subject to no rules.

STATISTICS, &c.

ACCOUNTS OF THE AARHUS TELEPHONE COMPANY FOR 1893

1 krone = 1*s.* 1¼*d.* 1*l.* = kr. 18·2

Working Account

Cr.	Kr.	Öre	Dr.	Kr.	Öre
Town subscribers' rentals	24,001	43	Manager's salary	800	00
Suburban ,, ,,	1,448	76	Wages, lady operators	4,906	75
Country ,, ,,	8,470	26	Bookkeeping and audit	400	00
Corporation ,, ,,	1,381	00	Messengers' wages	222	76
Subscriptions for suburban lines	2,639	46	Firing and light	492	60
Talks over suburban lines	810	00	Rates, and repairs to property	157	22
Night talks	334	00	Fire insurance	182	00
Interest on bank balance	29	38	Contribution to the Jutland United Telephone Society	102	50
			Cleaning, travelling expenses, advertisements, printing, books, postage	393	58
			Interest on mortgage	776	25
			Other interest	363	90
			Superintendence at the following branch stations: Hammel, Haselager, Morke, Ronde, Tranbjerg, Vrinders	822	08
			Repairs to town lines	4,129	73
			,, suburban lines	4,312	19
			Reserve for reconstruction of various country lines	1,500	00
			Balance, being net revenue	19,552	73
	Kr. 39,114	29		Kr. 39,114	29

Cr. *Profit and Loss Account* **Dr.**

Cr.		Kr.	Öre
Balance from last year		368	14
Balance from Working Account as above		19,552	73
		Kr. 19,920	87

Dr.	Kr. Öre	Kr.	Öre
Value of the company's telephone system at Jan. 1, 1893	72,298·99		
New lines in 1893	9,599·86		
	Kr. 81,898·85		
10 per cent. written off in accordance with bye-laws		8,189	89
Written off the company's building, standing at Kr. 29,365·86 in the books		1,000	00
Written off furniture and fixtures		60	00
Commission to the manager, 5 per cent. on Kr. 11,302·84		565	14
Directors' fees, 6 per cent. on same amount		678	16
Dividend to shareholders, 5 per cent. on Kr. 60,000 = Kr. 3,000, to which is added Kr. 3,000 under Bye-law 14		6,000	00
Placed to reserve fund under Bye-law 14		3,000	00
Balance to next year		427	68
		Kr. 19,920	87

Balance Sheet

Assets	Kr.	Öre
Construction account	73,708	96
Building ,,	28,365	86
Stores ,,	7,824	14
Aarhus Private Bank	102	97
Sundry debtors	1,977	66
Fixture account	500	00
Cash in hand	629	72
	Kr. 113,109	31

Liabilities	Kr.	Öre
Capital	60,000	00
Mortgages	16,500	00
Loan from Aarhus Private Bank	10,000	00
Sundry creditors	9,148	94
Profit and Loss Account	6,000	00
Reserve fund ,,	9,532	69
Repairs ,,	1,500	00
Balance from Profit and Loss Account to next year	427	68
	Kr. 113,109	31

[Signed] Otto Monstea, Kjen.
Joh. Baume, Springborg.

Aarhus: *December* 31, 1893.

The undersigned, auditor, has examined the books and accounts of the company, and has no remarks to make.

J. H. Frank.

Aarhus: *February* 20, 1894.

Note.—Since going to press, the accounts for 1894 have been received. They show an amount available for dividend of Kr. 7,200; Kr. 2,700 carried to reserve, and Kr. 399·55 to 1895. The value of the system at January 1, 1895, was Kr. 90,828·80.

VII. FINLAND

LIKE the other northern continental countries, the Grand-Duchy of Finland has become the scene of great telephonic activity. There would seem to be something in the Scandinavian blood, to which the inhabitants of the capital and all the more important coast towns mostly belong, which renders the possession of many telephones an essential to their owners' happiness. Wherever two or three Swedes, or Norwegians, or Danes, or Finns of Scandinavian descent, are gathered together, they almost infallibly proceed to immediately establish a church, a school, and a telephone exchange. Whatever else in life that is worth having generally comes after. Thus the inhabitants of Mariehamn in the Aland Islands (the whole group of 300 islands contains only 18,000 souls) support and find uses for a flourishing exchange, while our own islands of Wight, Jersey, Guernsey, Arran, &c., incomparably richer and better peopled, show no sign of consciousness of even the existence of such a facility.

The telephonic development has been conducted on Scandinavian lines—that is to say, by local companies and co-operative societies, which have been formed in every town in the country under concessions from the Finnish Government, which has not dabbled directly in telephones at all. The telegraph lines in Finland belong to the Russian Posts and Telegraphs Department, the only telegraphs owned by the Grand Duchy being those erected along the State railways. The first telephone exchange was opened in Helsingfors in 1882. As a general rule, a member pays for the cost of his line and instrument and for his share of the exchange

Fig. 36

apparatus, and afterwards contributes a modest annual amount to cover the cost of working and maintenance. In the capital, Helsingfors, where, with a population of 64,641, there are, in March 1895, 2,150 subscribers, and also in Abo (population 31,671, subscribers 575) and Wiborg (subscribers 670), there is competition between co-operative societies and companies which work on an inclusive annual subscription. Free intercommunication is, however, allowed between the subscribers to the rival systems. The rates in force in these towns are as follow :—

Town	Co-operative Societies			Companies
	Entrance fee		Annual subscription	Inclusive annual subscription
	Wire	Instrument		
Helsingfors . .	6*l.*	4*l.*	2*l.* 16*s.*	4*l.* to 4*l.* 16*s.*
Abo . . .	6*l.*	4*l.*	2*l.* 8*s.*	4*l.* 16*s.*
Wiborg . . .	8*l.*	—	2*l.*	3*l.* 4*s.*

The co-operative rates in the last two towns may be taken as typical of those prevailing in the remaining thirty-two exchanges of Finland, the most northern of which is Uleaborg.

It will be seen that the telephones in Helsingfors number 3·3 per 100 inhabitants, a proportion which gives it a prominent place amongst the best-telephoned cities of the world. There is no telephoning of telegrams, as the Russian Posts and Telegraphs Department cannot be induced to concur in the necessary linking up with the various companies. The telephonogram service is also wanting. Helsingfors and Wiborg exchanges are always open, and several others can be used at night on payment of a fee to the attendant.

Enterprise is not confined to local exchanges, for a company, bearing a name which means, being translated, the Southern Finland Interurban Telephone Company, acting under a Government concession, has connected by metallic circuit trunk lines all the coast towns from Wiborg to Abo, nine in number, and spread over a distance of 400 kilometers, the actual length of the circuits

Fig. 37

used being 900 kilometers. The company's charges are, the time unit being five minutes,

0 to 100 kilometers	·19 pennis per kilometer
100 ,, 200 ,, ...	·18 ,, ,,
Exceeding 200 ,, ...	·17 ,, ,,

Thus a talk between Helsingfors and Borgā, a distance of 59 kilometers (36·6 miles), costs 59 × ·19 = 11·2 pennis; and one between Helsingfors and Wiborg, 300 kilometers (186·4 miles), 300 × ·17 = 51 pennis. As ten pennis make one penny, it follows that 36½ miles can be spoken over for 1·12*d.*, and 186½ miles (practically London to Manchester) for 5·1*d.* This is even slightly cheaper than in Sweden. All the other towns, with the exception of seven of the most northerly ones and Mariehamn in the Aland Islands, have been connected to the capital and to the Interurban Company's lines by other concessionaries, so that Finland is actually covered with an almost perfect network of telephone trunk wires which bring the shores of the Baltic into instantaneous communication with those of Lake Ladoga, and the far-off interior with both. The Finnish trunk lines extend to the Russian frontier and to within a few miles of St. Petersburg, but the establishment of communication with the Russian systems has not yet been permitted.

Although the trunks are double, the subscribers' wires are single, so that translators must be used when connecting them together. The town wires are usually of 2·2 mm. galvanised steel, as bronze is reported to be too liable to be affected by the formation of frost, which frequently proceeds with great rapidity and adheres to and breaks down the wire by sheer weight. Some bimetallic wire—steel coated with copper—of 1·8 and 2 mm. diameter, is also being tried experimentally. The trunk lines are partly of copper and partly of the same bimetallic wire of 2·2 mm. gauge. The insulators have a bolt right through fastened by a nut at the top, like a single shackle bell used as an upright. The standards are built up of angle iron, and closely resemble the Russian fixture shown in figs. 110 and 110 A (Russian section). Fig. 37 is a view of the exchange fixture at Helsingfors.

The subscribers' instruments are all of the magneto type, the

Finnish engineers having been wise enough to eschew galvanic batteries for ringing purposes. Originally the instruments were of American manufacture, but latterly the market has been monopolised by Messrs. Ericsson & Co., of Stockholm, and by one or two Christiania firms.

The exchange of the Helsingfors Telephone Company, of which a view is given in fig. 36, is fitted with a Western Electric multiple for 1,400 lines; the other switch-boards are non-multiple and of varied design and manufacture.

VIII. FRANCE

HISTORY AND PRESENT POSITION

UNLIKE some other countries, France was prompt, on the appearance of the telephone, to determine how to treat the intruder. By the law of 1837, confirmed by that of 1851, the monopoly of telegraphic communication rested with the State, and the French authorities had little difficulty in pronouncing the telephone a telegraph. But it was a new-fangled one, nevertheless; and who was to be at the trouble, risk, and expense of proving its suitability for the sphere claimed for it by its introducers, and of sampling the public taste and estimate of the commercial and social value of the innovation? Soon the Government decided that that was eminently the function of the sponsors themselves, so as early as 1879 three five-year concessions, comprising between them the whole of Paris, were granted. But the town council naturally took exception to the arrangement, and brought pressure to bear on the concessionaries to force a fusion, so that Paris might be worked as a whole, and not split into, possibly hostile, camps. Thereupon the concessionaries, very wisely, determined to join hands, a resolution which led to the formation on December 10, 1880, of the afterwards powerful association, the Société Générale des Téléphones. The Société found that it had to a certain extent to dance in fetters, since the State claimed a royalty of 10 per cent. on the gross receipts, and stipulated that the Department of the Posts and Telegraphs should construct and maintain the company's system, so far as the outside wiring was concerned, at prices which might appear fair and reasonable to that department. Moreover, the State claimed a general control, including the right to fix the charges, and reserved power to buy the system at the value of the material employed on the termination of the five-year concession. The exchange rate approved of was, for Paris 24*l.* per annum, and for the provinces 16*l.* And so the quest for the telephonic chestnuts was embarked upon, the position at the start being that the company was willing to risk its money and hoped to gain experience, while the State was willing to risk nothing—but still hoped for experience. Not content with Paris, the company soon undertook the concessions for Marseilles, Bordeaux, Lyons, Havre, Rouen, Lille, Nantes, and several other leading towns, while it was not till 1883 that the Department of Posts and Telegraphs timidly took its maiden telephonic dip by opening exchanges at Tourcoing, Roubaix, and Rheims. The plan adopted in these three towns was to make the subscribers pay for their lines and instruments in consideration of a reduced annual subscription. Paris was opened on September 30, 1879, and it is here necessary and just to award to our neighbours the credit of being the first to recognise the merits of the metallic circuit (first pointed out by Hughes) for practical exchange work by constructing Paris on that system. It is probably true, since its provincial exchanges were made single-wire, that the company was driven to metallic circuits in Paris by the necessity it was under of going for the most part underground by means of cables laid in the sewers (in which position, in those days, before the 'anti-induction' type of cable was known, the overhearing between single wires would have been intolerable); but nevertheless it remains a fact—and a most important and creditable one it is—that the first double-wire exchange was opened and systematically developed in France.

The Paris exchange soon acquired respectable proportions, but those in the provinces hung fire, and even in Lyons and Marseilles the increase was remarkably slow, doubtless due in a large measure to the high rate of 16*l.* This rate, too, like the Parisian one, was exclusive of the subscribers' transmitters and receivers, which, strangely enough, it was decreed that they should buy themselves. The intention of the State in authorising this system is believed to have been a desire to obtain the most perfect type of instrument possible by encouraging competition between manufacturers; but the only concessionary, the Société Générale des Téléphones, was also primarily a maker of instruments, and owner of some of the most important patents connected with them. It worked out, therefore, that practically the Society sold its own telephones to the subscribers, and thus made a manufacturer's profit first, and collected a liberal subscription to cover the exchange service afterwards. The first concessions expired in September 1884, at which time the State possessed exchanges in six of the smaller provincial towns—Roubaix, Rheims, Tourcoing, Troyes, St.-Quentin, and Halluin. The experience gained in these places was not considered sufficient to justify the taking over of the concessionary's systems by the State, and a prolongation of the licence for another five years was accordingly granted. The rates of subscription were not altered, but permission was given to open public telephone stations, to connect the exchanges with telegraph offices for the despatch and delivery of subscribers' telegrams, and to establish communication between town and town by means of trunk lines constructed by the State, which also again reserved to itself the erection and maintenance of all outside wires, the Society's staff being confined strictly to work in the exchanges and subscribers' premises. It was ordained that the Society's employees should be all of French nationality, and subject to the oath of secrecy imposed on all servants of the Posts and Telegraphs Department. The original royalty payable to the State was continued at 10 per cent. of the gross receipts, with a minimum of 40*l.* per annum for each provincial exchange opened. During this second term of five years the Paris exchange increased rapidly, those in the provinces very slowly; a few internal trunk lines of inconsiderable length were erected, and the first metallic circuit between Paris and Brussels put into use. Early in the second term—in 1886—a step was nearly taken which would have totally changed the history of French telephony. The Minister of Posts and Telegraphs signed a concession for thirty-five years, giving a telephonic monopoly to a new company, with a capital of 1,000,000*l.* sterling, which was to acquire not only the business of the Société Générale, but also the exchanges already opened by the State. At the end of the thirty-five years the company's system was to lapse to the

State without payment. But the House of Deputies would not endorse the project, which was accordingly shelved.

In the autumn of 1889 the second term came to an end, and the State, which had opened some twenty-five additional provincial exchanges since 1884, decided to assume possession of the concessionary's system in the terms of the licence. This it did on September 1, 1889, eight days before the concession had expired, but not without friction. The Société Générale des Téléphones had conceived the impression that the Government did not intend to treat it fairly, and not unnaturally objected to give up possession before its concession had expired. It asked that the amount to be given for the property should be at least fixed, if not paid, before possession was yielded; pointed out that the leases of the various switch-rooms belonged to it, and that there was nothing in the concession compelling it to part with leases or anything beyond the plant and instruments. This ingenious contention—that the Société had sold the kernel but not the shell, and that the State must take the former, if it wanted it, without touching the latter—was, however, treated with scant consideration, for on the date named—a Sunday—a State engineer, attended by a commissary of police, took possession of each of the Société's exchanges, in spite of protests by the officers in charge, who declared they submitted only to main force. At each switch-room a sheriff's officer was in attendance, who served writs on the Government engineers as soon as they had taken possession, in which damages for breaches of concession were claimed and protests against confiscation set forth. It was stated that the Government had appointed their own arbitrators to fix the amount due to the Société, and had refused to admit any representative of the latter, while the Press expressed a conviction that the haste to take possession was due to the Government's anxiety to have the telephone system under its control during the approaching general election. Whether this was so or not, is not material; the Cromwellian coup was successful, and thenceforward the French telephones belonged to the State. Since then the atmosphere of the law courts has been heavy with rumours of claims and counter-claims, in which millions figure as freely as do units in the transactions of ordinary mortals. At the date of writing (January 1895) the judges have not succeeded in evolving order from the chaos arising from the circumstance that the Société claims over fifteen millions, while the arbitrators award ten millions, and the State is only willing to pay five.

The first act of the Government was to reduce the rates of subscription, a process for which there was certainly plenty of room. The Parisian tariff came down from 24*l.* to 16*l.*, and the provincial from 16*l.* to 8*l.*, with the reservation, however, in the latter case that the subscriber should not only find his own transmitter and receiver, but contribute 15 francs (12*s.*) per 100 meters of single wire towards the cost of his line; that is to say, practically pay its entire cost and to spare. Further, in towns possessing any considerable amount of underground work the provincial subscription was to be 12*l.* It cannot, therefore, be contended for a moment that telephone rates are low in France. They were very high during the reign of the company (but with the State's connivance, since it reserved power in the concessions to fix rates), and the reductions and alterations made since do not put the French subscribers on such good terms as those of most other continental countries. For instance, the French provincial subscriber finds the capital for his line and instrument, and yet pays some 10*s.* per annum in subscriptions more than his German competitor, whose line and instrument are found for him, and who gets off, everything included, for 7*l.* 10*s.* per annum. It is true that the Frenchman generally gets a metallic circuit, but so do the Swedes and Belgians, and at a much lower charge. Even some of the British provincial subscribers have easier terms than the French; and this fact of universal dearness may perhaps account for the slow progress made by the telephone everywhere in France outside Paris. Not even the great towns of Le Havre, Marseilles, Lyons, and Bordeaux yet count, after some fourteen years' development, more than from 1,000 to 1,200 subscribers each, and, compared with many in Germany, Switzerland, Belgium, &c., &c., rank as third- and fourth-rate centres. They are beaten even by provincial Italy (Milan) and provincial Spain (Barcelona), so that there is evidently something in the French Government policy that fails to commend it to the multitude. Would-be subscribers may possibly be deterred not only by high rates, but by their complexity, and by the multiplicity of the rules which regulate exchange connections. The French *bourgeois* is a cautious individual who likes to understand exactly what he is undertaking, and it is quite comprehensible how even a business man possessing no previous knowledge of the subject may be fogged into indifference on the threshold of his investigations. There is nothing like simplicity both for fostering and administering business. The French machine would move more freely if it had fewer wheels, for additional wheels mean friction, and friction expense.

SERVICES RENDERED TO THE PUBLIC

1. **Intercourse between the subscribers and public telephone stations of the same town.**—The local rates apply without modification, whatever the lengths of the lines may be, sometimes within the *octroi* limits, sometimes within the free telegram delivery radius, and sometimes within the boundaries of a commune or parish. Occasionally even several neighbouring communes are banded together and treated as a local area. The subscribers fall under numerous categories, which will be detailed under the heading *Tariffs*. Briefly, it may be said that the French regulations are marked by a decided lack of liberality towards the public. The acknowledged idea is to make subscribers find the capital for their own lines, besides buying their own instruments, either in the form of a slump payment at the rate of 12*s*. per 100 meters of single line, or in that of an *amortissement* or half-yearly payment in excess of the tariff until the cost of the line has been paid off. This system is carried out everywhere except at Paris and Lyons, where the cost of the line is considered to be included in the subscription. The cost of overcoming any exceptional difficulties in construction must also be borne by the subscriber. That individual, besides buying his transmitter and receiver, has to find any extra bells, indicators, or switches he may require, and to pay the State 15 per cent. on their value annually for maintenance, with a minimum of 4*s*. Thus, 4*s*. per annum may be charged for maintaining a trembling bell, value 5*s*. or 6*s*., which is, moreover, the subscriber's own property. New exchanges are not taken in

hand, too, unless the local chamber of commerce, town council, or a syndicate of persons interested advances the necessary money to the State without interest, these advances being refunded out of the subscriptions when collected, or by mortgaging the subscriptions. Subscribers changing premises have to bear the cost of shifting the lines and instruments. When a subscriber is located outside the local area he has to pay, besides an initial charge of 4*s.* per 100 meters of single wire, an extra subscription of 24*s.* per annum per kilometer if his line is underground, and of 12*s.* if overhead, in addition to paying the railway or other fares of the inspectors who look after his apparatus. The subscription for clubs and public establishments is increased 50 per cent. Under such circumstances as these, it is not surprising to read in the Finance Reports that the provincial exchanges are worked at a large profit; but the meagre proportions attained by them show that the State regulations operate to the restraint of trade.

2. **Intercommunication between a town and its suburbs.**—Subscribers connected to suburban or branch switch-rooms in the neighbourhood of a town are not on the same footing as those located in the town itself. A town subscriber's rate includes the right to call up the suburbs, but the member of a suburban exchange can only originate communication with the town by paying 4·8*d.* per five minutes, unless he likes, instead of paying the local suburban subscription, to pay the town rate plus 8*s.* per annum for each kilometer of single wire separating the two exchanges. The policy of discriminating against suburban subscribers is a most unwise one; it reacts on the town itself by deterring shopkeepers and other candidates for suburban custom from joining, and puts a brake on the whole machine. Branch switch-rooms subject to this differential treatment are known as *annexes*. St.-Denis, near Paris, is an *annexe*. The distance is five and a half kilometers, equal (as all junction lines are metallic circuits) to eleven kilometers of single wire. The local rate is 8*l.*, which gives communication only with those subscribers who are attached to St.-Denis switch-room. To be free to call up Paris and the other suburbs the rate becomes 20*l.* 8*s.*, that is to say, the Paris subscription, 16*l.*, + 11 kilometers of single wire × 8*s.* St.-Germain is worse off still, having to pay 16*l.* + 22 kilometers of single wire × 8*s.* = 24*l.* 16*s.*

3. **Internal trunk line communication.**—The French internal trunk service has recently experienced a wide extension. Some of the lines date from 1885, when the system was commenced by the connection of Paris to Rouen, Le Havre, Lille, and Rheims. In 1888 Lyons and Marseilles were added, and now there are but few of the leading provincial towns without communication with Paris. No fewer than fifty-four long-distance trunks meet (January 1895) at the Paris Central Station in the Rue Gutenberg. The rates are based on distance, being 4·8*d.* per 100 kilometers, and so considerably cheaper than those proposed by the British Post Office. Thus the rate from Paris to Marseilles (560 miles) is 3*s.* 7*d.* for five minutes, while for a similar distance the British would be 7*s.* 6*d.* for three minutes—a vast difference. The French have, too, reduced rates during the night, and a system of monthly subscriptions which secures a specific line to the subscriber every night at a cost of less than one half of the normal tariff. Unquestionably the French trunk line policy is more liberal and better adapted to actual requirements than their local. In Algeria, which telephonically is also administered by the French Posts and Telegraphs Department, there is a trunk line between Oran and Sidi-Bel-Abbas. The number of trunk communications in France is certainly very large, but the officials scout the idea that the trunk service has injured the telegraph revenue.

4. **International trunk line communication.**—At the present time this is opened to England, between Paris and London ; to Belgium, between Paris and most of the towns in the north-east of France, and Brussels and the chief Belgian cities ; to Switzerland, *viâ* Besançon, and from St.-Julien to Geneva ; and to Monaco, from Nice and Mentone. A trunk to Madrid is spoken of, but nothing has yet been heard of lines to Italy or to Germany.

The receipts of the Anglo-French trunks are pooled, and divided between the two Governments in the proportion of eleven-twentieths to France and nine-twentieths to England. Similarly, France receives three-fifths of the total receipts derived from the Franco-Belgian intercourse.

5. **Telephoning of telegrams.**—This is the one matter in which the French have shown a commendable liberality, for, as a rule, they charge nothing for the telephono-telegraphic service.

Were it not that they make exceptions in the cases of Paris and Lyons, the two most active telegraphic centres in France, where the subscribers who want their telegrams telephoned have to pay an additional subscription of 2*l.* per annum, one would have imagined that the necessity of compensating for the draining effects of the trunk lines on the telegraphic system by encouraging the telephone as a feeder had been duly recognised. Outside Paris and Lyons the only obligation imposed on the subscriber is a deposit to cover the value of his telegrams; but everywhere the language used must be French, and no message must exceed fifty words in length. In Paris, copies of telegrams telephoned to subscribers are posted; elsewhere, delivered by messenger.

6. **Telephoning of messages for local delivery.**—Subscribers from their own instruments, and non-subscribers from the public stations, between the hours of 7 A.M. (8 A.M. in winter) and 9 P.M., may telephone messages in French to the telegraph office to be written out and delivered by messenger to addresses in the same town. The charge is not by word, as in most other countries, but by the time occupied in taking down the message, the rate being 4·8*d.* per five minutes or fraction thereof. The service is consequently dearer than elsewhere, at least for short messages. Matter for mailing, as letters and post-cards, cannot be telephoned.

7. **Public telephone stations.**—There are some 350 of these in France, generally situated at post and telegraph offices. Subscribers may use them for local talks without charge on producing a card of identity bearing a photograph of the person to whom it is issued. Payments are managed exclusively by the aid of telephone tickets, which are on sale at the public stations and elsewhere. The right to use the public stations for local talks may be acquired, if desired, by a non-subscriber for an annual payment, which varies with the town. Messages for local delivery may be telephoned from these stations, but long-distance telegrams cannot be sent.

8. **Municipal telephone stations.**—Towns or communes desiring telegraphic or telephonic communication which the State is not willing to undertake may demand a connection to the nearest telegraph or telephone office on advancing the money necessary to defray the cost of the installation. This is fixed at

10*l.* per kilometer of line as a maximum, and 12*l.* for supplying and fitting the instrument. The local post-office is generally used as the station, and the employee in charge is repaid for the extra work involved by an allowance of 1·44*d.* on each message forwarded, and ·96*d.* on each received. The advance is gradually repaid, without interest, out of the proceeds of a surcharge of 2·4*d.* on each telegraphic or telephonic message transmitted, which surcharge ceases as soon as the cost of the line has been wiped out. In the middle of 1894 there were but ten municipal telephone stations in operation, and these appear, for the most part, to be essentially telegraph offices with telephones in lieu of the ordinary apparatus. The results achieved by the Swiss parochial stations, which these to some extent resemble, are certainly not attained.

9. **Special exchanges, or connection of groups of subscribers to an existing trunk line.**—When several persons located near the route of a trunk line wish to avail themselves of telephonic communication they are formed into a 'special exchange.' Each subscriber has to pay 2*l.* per annum, in addition to the cost of his line, which may, at his desire, be spread over several years, but this entitles him to nothing except actual connection to the system and to be rung up by anybody who may want him. If he originates a conversation, even with his next neighbour, he must pay at the rate of 4·8*d.* per five minutes' talk. All such special exchanges are connected to the trunk line which passes near, so that communication to and fro over it is available to the subscribers on payment of the trunk rates. This system has its analogue in Switzerland, but there the subscribers may talk freely locally, and only have to pay when the trunk line is brought into requisition.

TARIFFS

1. **Rates for local exchange communication.**—**Paris.**—Within Paris proper the annual rate is 16*l.*, the subscriber finding his own transmitter and receiver and any extra bell or switch that may be required, but paying nothing towards the cost of his line.

An extra instrument in the same building costs 2*l.* per annum.

A second instrument, not in the same building but on the same line, can be attached for 6*l*. 8*s*. per annum. This second instrument may, by agreement with the original subscriber and permission of the State, be used by a person unconnected with the original subscriber.

If any special difficulties are encountered in installing a line, the subscriber has to pay the actual cost of overcoming them, plus 5 per cent. The subscription covers maintenance of line and apparatus, including the transmitter and receiver supplied by the subscriber, but not of any extra bell, indicator, battery, or switch-board. These have to be furnished by the subscriber at his own expense, but he is not allowed to maintain them. That is done by the State at an annual charge of 15 per cent. on their value, with a minimum charge of 4*s*.

A subscriber whose line extends beyond the free limits must pay extra at the rate of 24*s*. per annum per kilometer of additional length.

The foregoing payments entitle a subscriber to speak all over Paris and with the suburbs.

Clubs, and establishments where the public have admittance to the instrument, pay 24*l*. per annum.

Lyons.—Owing to the amount of underground work in this city, the rate is dearer than in Marseilles or Bordeaux. Within the limits of the Lyons telegram free delivery the rate is 12*l*. per annum, the subscriber supplying his transmitter and receiver and any extra apparatus, but paying nothing towards the cost of his line. An extra instrument in the same building is charged 2*l*. per annum. A second instrument on the same line, but not in the same building, costs 4*l*. 16*s*. per annum ; this second instrument may, by arrangement, be used by a person not connected in business with the original subscriber. The cost of overcoming any special difficulties in constructing a line must be defrayed by the subscriber.

The subscription covers maintenance of line and all apparatus except extras required and supplied by the subscriber. The State maintains these too, but at an annual charge of 15 per cent. on their original value, no charge being less than 4*s*.

A subscriber whose line extends beyond the free limits must pay extra at the rate of 24*s*. or 12*s*. (according to whether his line is underground or aerial) per kilometer per annum.

Clubs, and establishments where the public have admittance to the instrument, pay 18*l.* per annum.

All other towns with a population of over 25,000.—The rate within the free limits is 8*l.* per annum; beyond the limits, 12*s.* per kilometer additional is exigible. Subscribers supply their own transmitter, receiver, extra bells, &c., and pay for their line at the rate of 12*s.* per hundred meters of single wire, equal to about 10*l.* 4*s.* for single and 20*l.* 8*s.* for metallic circuit per mile. If the line extends beyond the limits and requires a special route of poles, the cost per 100 meters of single wire is increased to 16*s.* Clubs and public establishments pay 12*l.* per annum.

In all other respects the rates are the same as those charged at Lyons.

Towns with a population of less than 25,000.—The rate within the free limits is 6*l.* per annum for ordinary subscribers, and 9*l.* for clubs and public establishments. In all other respects the rates and regulations are the same as in the larger towns, Paris and Lyons excepted.

General.—Rates are everywhere reduced 50 per cent. for Government and 25 per cent. for municipal connections.

Agreements are for one year dating from January 1 or July 1 after connection. Subsequently they are subject to three months' notice on either side.

It would seem that some subscribers join for the purpose of using the trunks only. In such a case only half the usual local subscription is charged.

When there are several switch-rooms in the same town, a subscriber joined to one who has frequent communication with a subscriber joined to another may arrange to retain the use of one of the junction lines between the two switch-rooms, and to be left through permanently to his correspondent (unless a special disconnection signal is given), for an extra annual subscription of 6*l.* in Paris and Lyons, and of 1*l.* 10*s.* elsewhere, per kilometer of junction line involved.

Subscribers located outside the free limits of a town have to pay the fares and expenses of the inspectors who attend to their instruments.

The State reserves the right to disconnect any subscriber at any time without notice. In such a case the proportion of subscription paid in advance for the unexpired period is refunded.

Subscribers are held responsible for all apparatus belonging to the State placed on their premises.

In some towns, busy during a season only, half-yearly subscriptions at half rates are admitted for the whole or part of the subscribers. In this case the subscriber must pay for his line in one sum when the first six-monthly subscription becomes due.

Subscriptions are payable half-yearly in advance at the telephone office, but will be collected at the subscriber's on payment of 2·4*d.*

When a subscriber's line becomes interrupted for more than fifteen days he is entitled to a proportionate refund of his subscription.

2. **Rates for suburban connections.**—The rates set forth in the preceding paragraph cover the right to originate communication with subscribers connected to suburban switch-rooms, but such subscribers are on a different footing, as they cannot call up the town subscribers without incurring extra charges.

The local suburban rates follow the provincial according to whether the population is over or below 25,000. Thus the rate (the cost of their lines being defrayed by the subscribers) at Versailles (population 51,000) is 8*l.*; at St.-Denis (population 50,000), 8*l.*; at St.-Germain-en-Laye (population 14,000), 6*l.*; which rates secure communication within the respective towns only. A St.-Germain subscriber calling up a client in Paris, Versailles, or St.-Denis must pay 4·8*d.* per five minutes' talk, and he will not be connected at all unless he has made a deposit in advance to cover such charges. Alternatively, he can make himself free of Paris and all its suburbs by paying, instead of his local subscription, the Paris one, plus 8*s.* for each kilometer separating his local switch-room from the Paris central. As already pointed out, this means 24*l.* 16*s.* per annum for a St.-Germain subscriber, a heavy impost for a suburban tradesman or residenter. The same system applies throughout France wherever suburban exchanges or *annexes* exist.

3. **Rates for internal trunk communication.**—The time unit for internal trunk talks is five minutes. The duration of a con-

versation between the same persons must not exceed ten minutes if others are waiting. The tariff is simple—50 centimes, =4·8*d*., per 100 kilometers or fraction thereof, measured by the actual length of the line. This is very high compared with the German universal rate of 1*s*., and very low compared with the proposed rates of the British Post Office.

Between the hours of 9 P.M. and 7 A.M. in summer and 8 A.M. in winter, the rate is reduced to 2·88*d*. per 100 kilometers.

A particular trunk line may be engaged for any length of time daily by paying in advance a monthly subscription based on the unit rate of 1·92*d*. per 100 kilometers per five minutes. Thus a Parisian subscriber holding a five-minute talk with Lyons (600 kilometers) every evening would pay 1·92*d*. $\times 6 \times 30 =$ 1*l*. 8*s*. 9½*d*. per month. Such talks are limited to the night hours.

4. **Rates for international trunk communication.**—To England: Time unit, three minutes. Charge 8*s*. Only two consecutive periods of three minutes allowed between the same correspondents if others want the line.

To Switzerland: see Swiss section, p. 385.

To Belgium: see Belgian section, p. 75.

5. **Rates for telephoning of telegrams.**—In all centres except Paris and Lyons this service is free. In those towns, owing to a large portion of the system being underground and consequently more expensive, subscribers using the telegram facilities have to pay an additional subscription of 2*l*. per annum. The telegram charges are deducted from deposits made in advance.

6. **Rates for messages telephoned for local delivery.**—The charge is 4·8*d*. per five minutes or fraction thereof occupied in transmitting the message, irrespective of its length.

7. **Rates levied at public telephone stations.**—The time unit for local and internal trunk talks is five minutes. A local talk costs 4·8*d*. in Paris and 2·4*d*. in the provinces. Annual subscriptions are accepted for the local use of all the public stations in a town at the following rates: In Paris, 3*l*. 4*s*.; in Lyons, 2*l*. 8*s*.; elsewhere, 1*l*. 12*s*.

The trunk rates are the same as from subscribers' offices. Messages for local delivery may be telephoned from the public stations at the same rates as from subscribers' offices. Payments at public stations must be made in telephone tickets. These tickets are perforated, and on presentation one half is retained by the attendant, and the other is stamped and given to the user as a receipt. Subscribers use the public stations free on producing a photographic card of identity. Long-distance telegrams are not accepted.

8. **Charges levied at municipal telephone stations.**—These are used also as telegraph stations. All transactions under both the telephone and telegraph tariffs are subject to a surcharge of 2·4*d.* until the cost of installing the station and its connecting line has been wiped out.

9. **Rates for special exchanges or connection of groups of subscribers to an existing trunk line.**—Each subscriber pays the cost of his line, in addition to finding his transmitter and receiver, and 2*l.* annually. Local conversations originated by him are charged 4·8*d.* per five minutes, and trunk conversations according to the tariff.

WAY-LEAVES

Subscribers are bound by their agreements to obtain the consent of their landlords to the fixing of their wires and instruments, and to bear the cost of all dilapidations caused by the installing or eventual removal. Although this is made part of the contract with each subscriber, the State claims the right to erect standards without charge on any building that lies on a route of wires, provided it is not surrounded by a boundary wall; similarly, to erect poles on any unenclosed ground, private or otherwise. A fence or hedge does not constitute an enclosure—a regular wall is alone competent to turn aside State telegraph or telephone wires. In executing work on private property the State is only responsible for dilapidations brought about. This is the only instance which the author has been able to find in Europe of compulsory way-leave powers being vested in the State, and it is at least singular that it should occur in the Republic of France, where private rights are theoretically more inviolable than in other and more autocratic countries. The influence on rates, which partisans in the United Kingdom so freely ascribe to compulsory way-leave powers, is shown by this example to be practically non-existent. Such powers exist only in France, and what do we find? That the French telephone rates are the lowest in Europe? Not at all. On the contrary, with the sole exception, and that only a partial one, of Russia, the French rates (bearing in mind that the subscribers have in the first place to pay for their wires and instruments) are the dearest.

SWITCHING ARRANGEMENTS

Transition is the present state of these. In Paris there are ten switch-rooms within the fortifications, which serve (January 1895) about 12,500 subscribers. The chief room is at the Rue Gutenberg, and is situated in a special building of ample proportions, made fire-proof throughout. The basement contains the access to the sewers, in which, with the exception of a few junction routes between the switch-rooms, practically all the Parisian telephone lines are laid. The cable wires, after being opened out, pass through test and cross-connection boards, and are carried to the switching department upstairs. Here, in a lofty, well-ventilated and well-lighted room, is a Western Electric Company's double-cord, series multiple table for 6,000 metallic circuits, of which some 5,500 are already connected, together with a junction line section of 1,000 lines and a long-distance trunk switching section communicating with the trunk-line switch-rooms on another floor, where are located fifty-four long-distance and ninety-four suburban and short-distance trunks divided between twenty tables. At the Rue Gutenberg more than half the connections asked for have to be got through over trunks or junctions. There is nothing special in the construction of the table, the test arrangement only being slightly modified to permit of the use of single- instead of double-wound receivers for the operators. There are three girls to each table of 240 subscribers. A subscriber requiring a trunk notifies his operator, who has a service jack to each of the long-distance operators. As each long-distance girl has only one indicator from all the local ones, the service jacks being multipled along the board, the local operator, before calling, must test. The trunk wanted being free, the long-distance operator asks the trunk switching section of the local board, on which all the subscribers' lines are multipled, for the calling subscriber. The junction lines are divided into 500 outgoing and 500 in-coming, the subscribers' lines being multipled also on the junction tables. When a calling subscriber wants a client on another switch-room, the local operator advises the junction girl, who obtains the connection from the other switch-room and completes it through the caller's repeat jack. When a demand comes from another switch-room the junction operator can, of course, satisfy it herself. Junction lines must not be occupied longer than ten minutes for one connection if other subscribers are waiting. At the expiry of that time the talkers

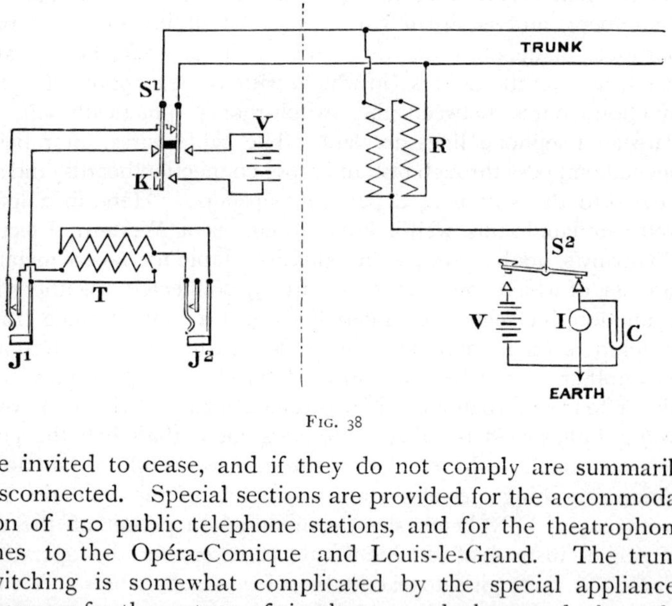

Fig. 38

are invited to cease, and if they do not comply are summarily disconnected. Special sections are provided for the accommodation of 150 public telephone stations, and for the theatrophone lines to the Opéra-Comique and Louis-le-Grand. The trunk switching is somewhat complicated by the special appliances necessary for the systems of simultaneous telephony and telegraphy so much used in France. Three systems are employed—Van Rysselberghe's, Cailho's, and Picard's. The first is too well known to require description. The second is a modification of the plan generally associated in this country with the name of Mr. Frank Jacob, although M. Cailho is understood to claim that he described the system in the 'Annales Télégraphiques' prior to the date of Mr. Jacob's patent. The arrangement used in France is shown in fig. 38, in which S[1] S[2] respectively represent the telephone and telegraph stations. At S[1], K is a calling key, V a calling battery, J[1] J[2] jacks for the loop and single line switching, and T a translator. At S[2], R is a double-wound bobbin of small resistance and high self-induction, in derivation with the two wires of the metallic circuit trunk line, and connected so that currents passing through the equally-wound coils oppose and kill each other. The other terminals are joined to the telegraph instrument I and the adjustable condenser C. M. Cailho, whose plan, it will be seen, differs only from Mr. Jacob's in the character and connection of the resistances, states that the thick wire and opposite winding

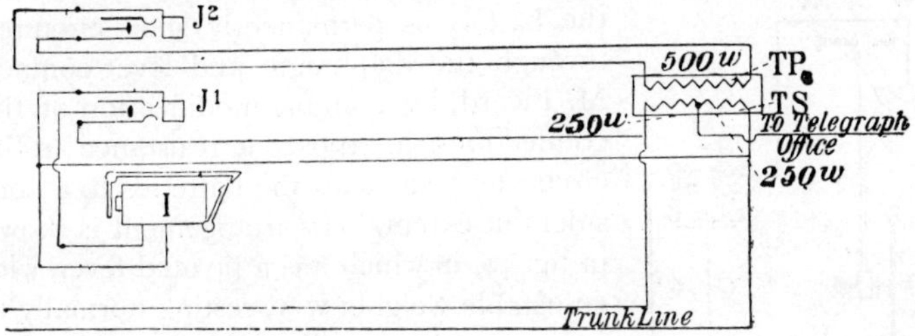

FIG. 39

allow the telegraphic currents to pass uninfluenced by resistance and self-induction, while the bobbin acts as a choke coil for the telephonic currents. The calling battery V may be too weak to operate the indicator at the distant end directly, as it is found that the return or extra current from the double-wound bobbin at the further station which follows the lifting of the key K, is always strong enough to actuate the drop.

M. Picard's plan depends on the use of a differential translator as indicated in fig. 39, which is a plan of the connections used at the Rue Gutenberg. The currents arriving from the telegraph office split between the equal branches of the translator secondary TS and produce no effect on the primary TP; they also neglect the double-wound indicator I. The calling is done by inserting a battery plug in the jack J[1], while subscribers are connected through J² ; consequently, the talking is done by translation, although the subscribers' lines are double. At Paris the translator primary could be dispensed with, the secondary replaced by a pair of balanced resistances on Jacob's plan, and the talking done direct through the jack J¹ ; but this would not be so at single-wire centres.

Of the three systems the Cailho seems to be preferred, as being the most trustworthy under adverse influences. Of course, both the Cailho and Picard are far simpler than the Van Rysselberghe, but, on the other hand, they furnish only one telegraphic circuit from each telephone trunk, while Van Rysselberghe makes two.

In the Belgian section the phonic call designed by M. Sieur has been described (fig. 20), and it is pointed out that its use involves much waste of chemicals, since the battery is permanently short-circuited through the diaphragm and lever contact. M. Picard, by a simple modification of the connections, interposes a resistance in the circuit and so saves the batteries to a considerable extent. His arrangement is shown in fig. 40, in which A is a pivoted lever, with adjustable weights W W, resting normally in contact with the diaphragm D. M is an electro-magnet with coils of equal resistance, joined in parallel, which oppose each other in respect to the armature ; and V is a battery, the current from which splits between the diaphragm, lever and coil C, and coil C¹. Whilst the diaphragm remains quiescent the two opposing circuits are of equal resistance, and no effect is produced on the armature ; when it vibrates, the intimacy of the contact between it and the lever is destroyed, circuit C becomes of greater resistance than C¹, and the battery, acting through the latter, actuates the armature, which is generally arranged to release a shutter. At night the shutter closes another local circuit and rings a bell.

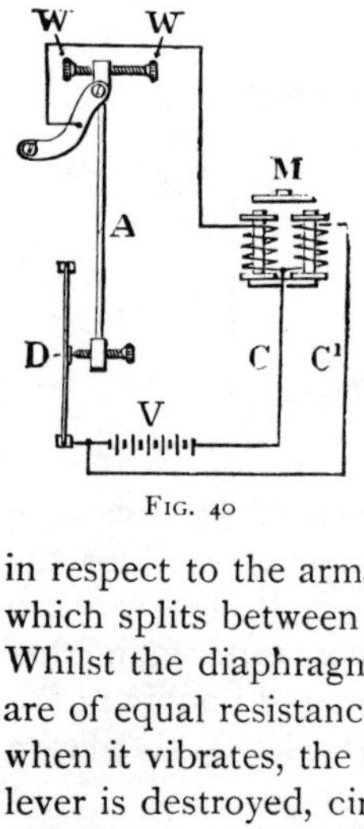

FIG. 40

An excellent plan for the speedy determination of disputes and complaints is in operation in Paris. Apart in a small room, at a switch-board provided with 20 indicators, sits an inspector. To the switch-board are brought two lines from each of the ten switch-rooms in Paris. When a subscriber at any switch-room prefers a complaint, the chief operator puts him through on one line to the special switch-board at the central and gets through herself on the other. The inspector then switches them together, and listens while the subscriber states his grievance and the chief operator makes her defence. If necessary, the working operator responsible for the subscriber's line is allowed to give evidence. Having heard both sides, the inspector delivers judgment and enters the proceedings and result on a form which is sent to headquarters. In this miniature court of justice 90 per cent. of the complaints are settled in about four minutes each, a rate of progress which has not, the author understands, yet been equalled in any court in Britain. In this instance Justice is truly blind, but is provided with particularly long ears. Ultimately, perhaps, there will be no going on circuit—the judges and juries will sit in London, and loud-speaking transmitters and receivers will bring to them and an inquisitive audience the evidence and speeches, and convey to the litigants the verdicts and decisions. There is nothing impossible in this—it could be done to-morrow.

The use of voltaic batteries for ringing is a grave disadvantage; but a worse exists. That is, that subscribers must be asked for by their names and addresses, not by numbers. In Berlin the service suffers from a plethora of numerals; in Paris they have none at all. This was the original way, and for a long time it resulted in no inconvenience, as the operators learned the names and switch-board numbers of new subscribers as they came on; but, now, when the Paris system comprises some 14,000 subscribers and several hundred operators, including necessarily many juniors, and when, moreover, old switch rooms have been closed and their lines concentrated at new ones, it may well be imagined that confusion must result. The subscribers, however, resent any suggestion that they should be numbered—the lists at present contain names, trades, and addresses only—and the Government is weak enough to refrain from making the change. Imagine the long formula that must be spoken by a calling subscriber in order to discriminate between seventeen Rousseaus, fifteen Bertrands, and thirteen Blancs; the mistakes likely to be made by an operator who thinks she knows the number of the person asked for; the delay caused by an operator who knows she doesn't know and accordingly refers to the list!

As an inevitable consequence the service is slow, although the speaking, when once through, is excellent. The service instructions are simple, and probably the best for the circumstances. A caller pushes his button several times and then puts the receiver to his ear and waits for the exchange's reply, which obtaining, he states the name and address of his client—which is repeated by the operator—and again waits with the receiver to his ear until he hears his friend's voice. On receiving a call, a subscriber lifts his phone and speaks without ringing back. The ring-off is a pressure of both buttons. There being no discriminative disconnection signal, subscribers must refrain from touching their buttons during intercourse, a disadvantage, great and grievous as it is, which prevails everywhere on the Continent. The easy-going temperament of the Gaul in telephonic matters is further evidenced by his tolerance of the rule that no new connection must be demanded within half a minute of a ring-off. Fancy a subscriber brought up on the Mann system standing that! He would expect to obtain and get rid of at least two connections in the time. It is but fair to state, however, that the engineers fully recognise the shortcomings of the system, especially in regard to ringing batteries and calling by name, but have to submit to the inevitable, which decrees their continuance. The average number of calls per subscriber in Paris is stated to be 5·5 per day; in the suburbs it rarely exceeds two per day. The traffic to be dealt with is consequently comparatively small. In Paris, all the work being underground, it has not been found necessary to fit lightning protectors at the exchanges, but in the suburbs and provinces this is never omitted. The protector which seems to find the most favour consists simply of a strip of paper, silvered on one side only, 3 mm. wide and 30 mm. long, inserted in the line by means of two metal clips. It is found to invariably fuse and save the coils during a discharge, but it of course never acts without interrupting the communication with the exchange of the line affected, a grave disadvantage.

At Rouen an American multiple with parallel jacks and self-restoring drops, essentially similar to that at Zürich (see Swiss section), has recently been fitted. Another board of this kind has been ordered for Le Havre from M. Aboilard, the Western Electric Company's agent in Paris. M. Portel Vinay, of Paris, is building a multiple for Bordeaux according to the patents of M. Adhémar. It is said to comprise parallel jacks and indicators which are restored in the act of making a connection, while their coils are automatically cut out, leaving only the ring-off drop in circuit. This is the same idea which has been given effect to in Stockholm and Copenhagen. (See Swedish and Danish sections.)

At Marseilles there is a multiple designed jointly by MM. Berthon and Ducousso, of which a promised description has not reached the author in time for inclusion in the present work.

To Lille the Société Générale des Téléphones supplied, some three years back, a multiple on the patent principle of M. Berthon, which presents several points of divergence from ordinary practice. Especially has the inventor aimed at compactness, screening of jacks from dust, and accessibility. The jacks are moulded while the metal is hot in steel dies, so as to insure absolute uniformity. The insulator used is *ivorine*, a composition which, it is said, possesses the good qualities of ebonite, combined with greater toughness and workability. The jacks are only 10 mm. thick, which is one millimeter less than the Stockholm Brunkeberg jacks, to be described in the Swedish section. The rows of jacks are very accurately fitted in their frames, and may be pulled out for repairs from the front almost like tiers of drawers. A diagram of the connections is given in fig. 41, and a plan and front view of the jacks and plugs in fig. 42. In this latter, the two wires of the subscriber's loop, or one wire and earth if the system be single, are joined to the springs v v' through the screws z z'. Normally the springs are in contact with studs a a', from which they are lifted by the nose of an inserted plug. The jack sockets are divided into two halves r r', of which r is joined permanently to the studs and r' to the test wire. The plugs are also in two halves, and shaped to fit into the divided sockets. Referring now to fig. 41, S^1 S^2 are two subscribers joined to the exchange by the metallic circuits L^1 L^2. T T^1 are two operators' sets with calling keys C and C^1, which may, as required, direct the current from the battery P through the plug $F1$ or $F1^2$, and calling keys $C2$

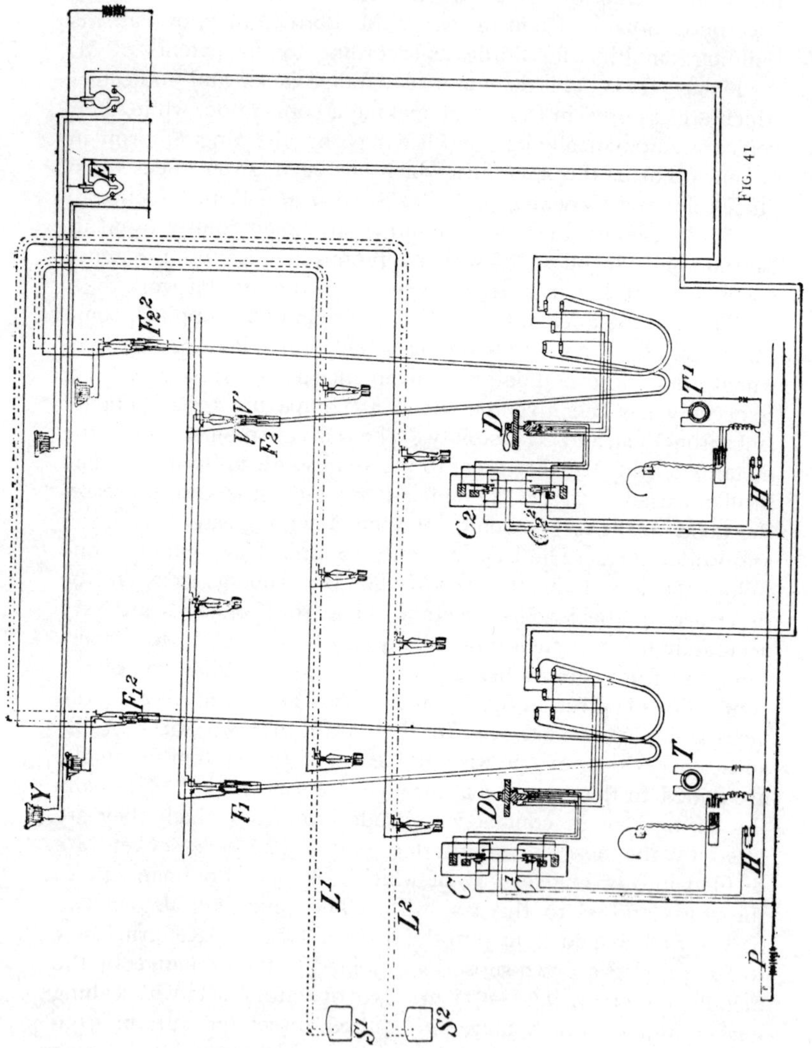

FIG. 41

and $C2^2$ which send it through *F*2 or $F2^2$. A lever switch *D*, when in the upright position, puts the two plugs *F*1 and $F1^2$ in circuit with a relay *E*, which is arranged to close a local circuit through the coils of the ring-off drop *V*. When *D* is turned down, *E* is short-circuited, and the speaking set cut in. Each operator has a test battery *H*. When a line is free, the test wire and the socket

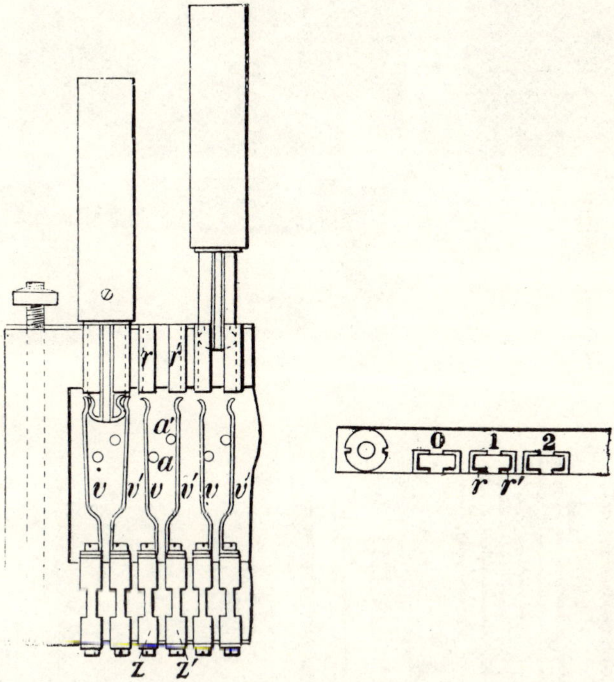

FIG. 42

halves *r* to which it is connected are insulated, and the application of a test plug, as *F*2, to the sockets produces no result, since *H* finds no circuit. But if a connection is on at another section, the socket halves *r* and are in communication through the inserted plug and the line, and a current will now circulate when *F*2 is applied. Figs. 43 and 44 give a front view and end section of the Lille board.

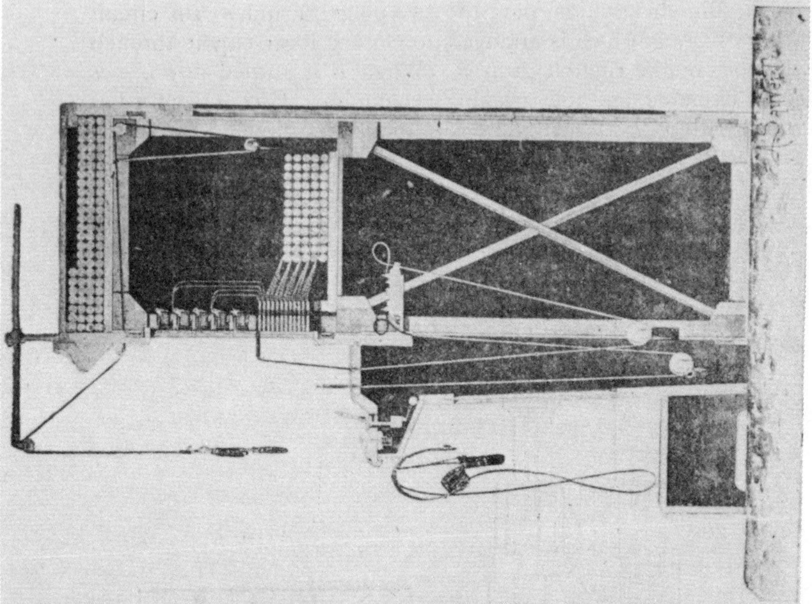

FIG. 44

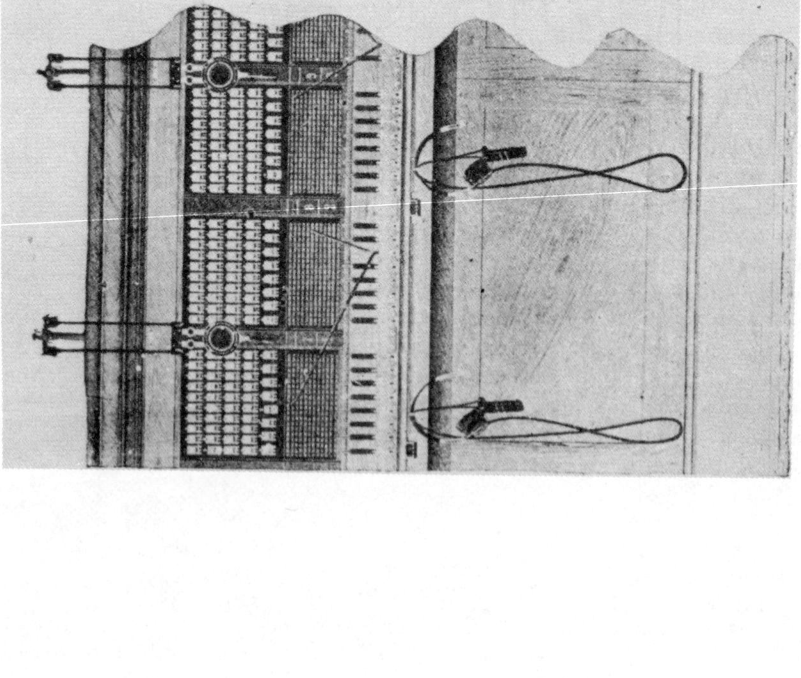

FIG. 43

M. Berthon has likewise devised a novel self-restoring drop, in which the solenoid principle is utilised, perhaps for the first time in telephonic work. The plan has not, however, received a practical application.

HOURS OF SERVICE

These, as a rule, correspond with the hours of telegraphic service, which are continuous in Paris and eleven of the other chief towns, and generally extend from 7 or 8 A.M. till 8 or 9 P.M. in the smaller places. But Aix, St.-Etienne, and Châlons are open till midnight, and Rheims and Pauillac till 10 P.M.

SUBSCRIBERS' INSTRUMENTS

The arrangement by which subscribers were left to purchase their own instruments has produced some curious results. Except within very wide limits the type was not defined until recently, anything that would work in with the existing switching arrangements being at first admitted. The methods of switching practised by the Société Générale des Téléphones required battery and not magneto ringers at the subscribers' offices, so that that system obtained such a hold that it has had perforce to be continued, much to the dissatisfaction of the present engineers, who would change it if they could on account of the great expense of maintaining so many voltaic cells scattered over a large area. There being 14,000 subscribers in Paris and suburbs, each using six Leclanché cells, it follows that there are $14,000 \times 6 = 84,000$ cells to maintain. This would be bad enough if they were collected in one building, but when distributed irregularly over some sixty square miles, the task is recognised as a formidable one. While the Société Générale held the ground, the subscribers' choice of instruments was limited, since it would not allow any but those of its own manufacture to be used ; but this restriction vanished when the State took over the system, and the field was thrown open to all. The wide market thus created gave rise to keen competition between manufacturers and to a great multiplication of types of instruments. Each maker had a type of his own, which he pushed as the best, so that the uninstructed subscribers were greatly exercised as to the respective merits of the Ader, Maiche, Pasquet, Journaux, Dejongh, Crossley, Bréguet, Roulez, Ochorowicz, Bert, D'Arsonval, Mors-Abdank, Mildé, Hunnings, and twenty others, each of which was represented as the only one worthy of attention. The State has the fixing and maintaining of the instruments, although the subscribers buy them, and, after a long time, began to recognise the fact that it had a vast number of cheap and defective instruments on its hands to maintain, and that the operation threatened to become a serious one in respect to cost. So, in 1893, the State issued a specification, intended to secure good workmanship, to be observed by all makers, under pain of having their instruments rejected ; and subscribers were required at the same time to submit the instruments they bought to the telephone authorities to be tested and passed prior to fitting. These regulations have brought about a great improvement in quality, but a vast mass of the older material remains in use, while the diversities of type have not been lessened. The instrument fitters and inspectors have consequently to be familiar with the mechanism and connections of some forty different kinds of apparatus, many of widely diverging patterns. This must lead to delay in removing faults. The hope of the French Government that competition between makers would in time develop an instrument of exceptional merit has scarcely, so far, been realised, since the best transmitters, if not receivers, have originated outside France. Space will not permit of the diversities of design being particularly referred to here, and it must suffice to say that while the battery ringer is universal, and the Ader receiver continues to occupy the position of first favourite, which it won in the days of the Société Générale, the latest tendency in transmitters is towards one or other form of Hunnings. The French instruments now supplied are, as a rule, both well made and tasteful in design and decoration. An ingenious instrument which, although at present employed almost exclusively for private lines, may become more familiar in exchange work later on, when the time for the inevitable change from batteries to magnetos arrives, is the magneto-electric call of M. Roulez, shown in figs. 45 and 46. The soft-iron cores C^1 C^2 of the electro-magnets E^1 E^2 are clamped between the poles of the same name of the powerful permanent magnets

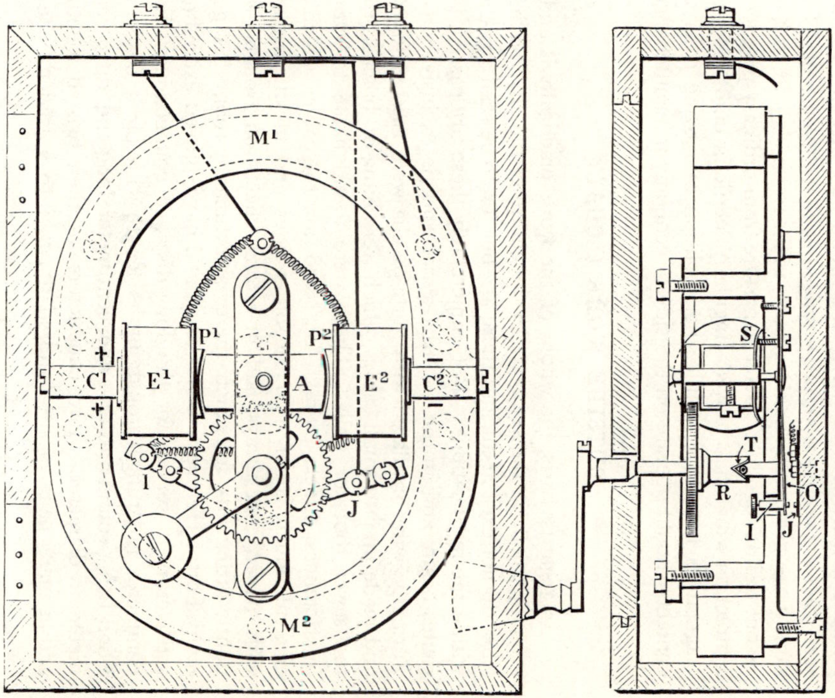

FIG. 45

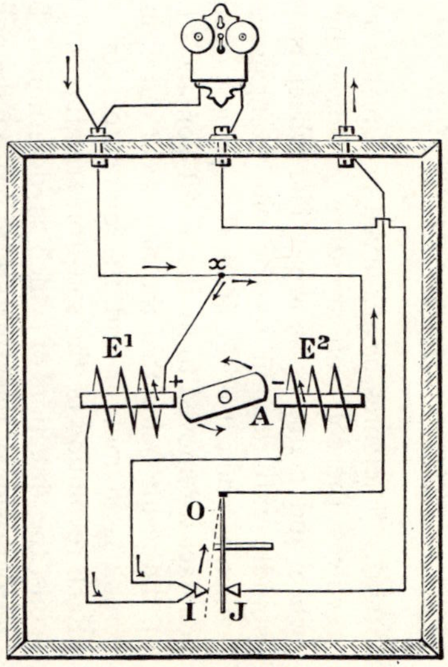

FIG. 46

M^1 M^2. The cores have curved pole-pieces P^1 P^2, between which revolves the soft-iron armature A driven by a pinion, toothed wheel and crank as shown. The wheel is loose on the crank spindle until caught by the pin T engaging with the recessed collar R, when it is revolved, the spring S compressed, and spring contact O removed from the fixed contact I to the fixed contact J, the result being that the bell is cut out from the line when the crank is in motion, and the generator coils when it is at rest. Each revolution of the soft-iron armature induces currents in the coils, the direction of which is determined by the approach or retrogression of the armature to or from the pole-pieces. It will be seen that the connections are arranged so that the currents of the same name generated simultaneously in the two coils join at x, and go out to line together. As a departure from ordinary practice in a direction which has proved singularly sterile in innovation, this magneto is interesting, while the abolition of moving coils and contacts should operate on the side of economy in maintenance.

OUTSIDE WORK (LOCAL)

As regards Paris, the whole of the work practically is underground, chiefly in the sewers. In the centre of the city, overhead wires do not exist at all, and there are but few to be seen anywhere within the fortifications. But immediately these are passed, pole routes begin, and in the suburbs aerial work is exclusively used. Formerly, wires insulated with gutta-percha were twisted in pairs and made up into small cables, which were hung on brackets from the sewer roofs or walls. These were found liable to various interferences, attacks by rats, &c.; and now the cables, which are as a rule much larger than the older ones, and mostly insulated with paper, are always laid in strong sheet-iron troughs with tightly fitting lids, for which there is fortunately room. A good many Fortin-Hermann cables exist, and have proved exceedingly satisfactory in service for the long-distance connections. Each conductor is strung throughout its length with birch beads, the wood being sound and dry, one centimeter long and three centimeters in diameter. Two conductors are then twisted together, and as many pairs as are required drawn into a leaden tube to form a cable. From the size of the beads it will be seen that the Fortin-Hermann system conduces to a very bulky cable ; six pairs, which is the size commonly used in Paris, occupying a space of over an inch. For this reason, and in spite of its electrical qualities, which are excellent, its use is not being materially extended. The insulation obtained is never less than 200 megohms per kilometer, while the capacity does not exceed ·05 microfarad per kilometer. The cable now chiefly employed is insulated with paper, and made in the workshops of M. Georges Aboilard, Avenue de Breteuil. While possessing (with No. 20 wire) a capacity of ·055 mf. per kilometer, fifty-two twisted pairs occupy a diameter of only forty-three millimeters, including the leaden protection. An insulation resistance of 6,000 megohms per kilometer is easily attained. The paper employed is of French manufacture, and before being used is severely tested for strength, a strip fifteen millimeters wide and one meter long being required to support a weight of seven kilogrammes and to resist twisting round eight times. The paper strip is very rapidly wound spirally on the conductor by special machinery in such a way that an air space is left between the wire and its covering. A second spiral in the reverse direction is then added, the process resulting in the formation of an almost perfect paper tube, round which a light cotton thread is wound to keep it in position. The conductors are then twisted in pairs, and made up into cores containing two, seven, twenty-eight, or fifty-six pairs, which are kept together by a spiralling of cotton threads. The core is wound on iron drums and dried in an oven at a temperature of 110° Centigrade for twenty-four hours before receiving its coating of pure lead. This it does by means of an hydraulic press, through which the cable passes, the molten lead which is fed to the press being somewhat cooled by water. The finished cable emerges cased in a leaden tube three millimeters thick, which lies directly on the core, the intervening layers of jute, &c., employed by British and German makers being dispensed with, as is also the usual steel armour. But then it must be remembered that this French cable is laid in troughs or trenches, and not drawn into conduits. The different-sized cores are employed according to the distance from the switch-room : thus a 56-pair cable leaves a station and drops, say, twenty-one pairs at the first junction box ; thence it is continued by a 28-pair and a seven-pair, and finally by a seven-pair alone, until all the pairs have been disposed of. Cable of this kind is used for the interior wiring of the central station at the Rue Gutenberg, as well as for the outside work. Capacious as the sewers are, their resources are not inexhaustible, while the necessity of protecting the cables laid therein in costly iron troughs renders sewer work somewhat expensive. These considerations have led to cables containing junction lines between some of the switch-rooms being laid in trenches beneath the street pavements. In one such trench, one meter deep, twenty 52-pair lead-covered cables are laid without any protection other than a galvanised iron netting placed some inches above them, designed to give warning of their existence to strange workmen who may open the ground. An admirable feature of these paper cables is the fact that they cannot be spoiled by access of moisture. The ends are not sealed in any way, and should water get in through a fault, even to the extent of short-circuiting all the wires, it may be driven out and the insulation raised again to its normal figure of 6,000 megohms per kilometer by forcing dry air, not necessarily warm, into one end of the cable, under a pressure of two kilogrammes per square millimeter. This air gradually makes its way through the cable, whatever its length may be (from seven to eight kilometers have actually been operated on), carrying with it to the further end all the moisture within it. For some hours after the application of the pressure no improvement is noticeable ; then the insulation begins to go up slowly, but at an ever-increasing ratio, until at the end of some twenty-four hours the mending proceeds with great rapidity, so that thirty hours of pressure usually suffices to restore what had appeared to be a hopelessly bad cable to full working efficiency. If it is not convenient to look for and remove the fault, the application of pressure continuously, or for a few hours every day, will keep the cable going without disturbing the subscribers. When the fault is looked for, its position is first determined as nearly as possible by electrical test, and the pressure then turned on. Usually the workmen find the fault by the sound of air issuing from it, or by simple inspection, and it may then be effectually cured by wrapping a piece of sheet lead round, and soldering it to, the tube. The air is dried by being forced through tubes containing sodium chloride before entering the cable; if made to pass through similar tubes at the further end, the amount of moisture removed may be ascertained by weighing the salt. It is said that a pint of water was on one occasion poured into a cable and all removed in a few hours. This process, which was invented by M. Aboilard, is so commonly employed in Paris that nozzles have been fitted to the cable-heads at the different switch-rooms, so that air pressure may be applied to any cable at any time. Of course, it is not necessary to disconnect any wires or stop any communications, and therein lies the great utility and beauty of the plan.

At Lyons, where sewers similar to the Parisian ones exist, the work is mostly underground, and generally on the same plan as in the capital; in Bordeaux there is a certain amount of underground wiring, but in all other towns the construction is either entirely aerial or nearly so.

In Versailles, St.-Ouen, St.-Denis, and other suburbs of Paris, the overhead wires are of 1·1 mm. bronze, supported on small double-shed insulators which, like most of those used in France, are provided with projections or ears (fig. 47) for the purpose of retaining the wire should it break from its fastenings. An extensive use of bracket standards attached to the fronts of the houses is made; indeed, it would seem that it was at one time thought that such contrivances would prove permanently sufficient, as even the original exchange fixtures at Versailles and St.-Denis were of this type; but standards attached to gable ends, chimneys, and roofs are now being erected in Versailles. The attachments to fronts of houses are naturally of restricted capacity, thirty to thirty-six insulators being carried at the most, while the wires are subjected to interference from the windows, and the low elevation at

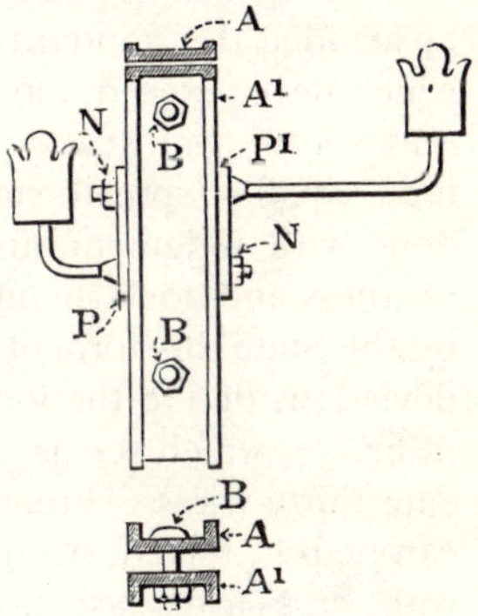

FIG. 47

which they cross the side streets must impede the transit of fire-escapes. Some of these bracket standards are of wood, round or square, but the more recent ones are built up of two lengths of channel iron, placed back to back and bolted together at intervals with a space of two or three centimeters between them, through which the insulator stalks are passed. Fig. 47 shows the details of this arrangement, A A[1] being the two pieces of iron kept together by the bolts B. The insulators are fixed in pairs on reverse sides, the stalks passing through iron plates, P P[1], which have generally a leaden sheet sandwiched for the purpose of moderating vibration, and being screwed up by the nuts N N. Figs. 48 to 51 show different forms of bracket standards in use; when fixed to houses the short-stalked insulators are always on the inside, as in figs. 48 and 49; but when they project above the roofs the insulators usually alternate, as in figs. 50 and 51. In some cases a small platform or stand for the workmen is attached to the lower bracket. On crowded routes, double standards of the form shown in fig. 52 are beginning to appear; they are simply two uprights like that in fig. 51 tied together by two horizontal rods. Standards are never fixed to a roof if a gable, wall, or chimney is available, as vibration is still a serious bugbear in France; when a roof fixture cannot be avoided the standard is bolted to the rafters. These standards, which never exceed ten or twelve feet in height, are, as a rule, only stayed against the pull on angles, but occasionally one with four equally spread stays is observed. The staying is always done with judgment, and the work generally is commendable for neatness and good maintenance. In the country towns exploited by the State this form of construction also obtains, with occasional deviations due to the local engineers. Such a deviation is shown in fig. 53, which seems a needlessly roundabout way to accommodate thirty wires. Frenchmen never resort to cross-arms if they can help it, but M. André has erected at Rheims double standards with cross-arms as shown in fig. 54. At Lille and Amiens the author also observed iron double standards with arms sandwiched between the uprights, but not quite like those of M. André. At Lille, a town exploited in the first instance by the Société Générale des Téléphones, most of the standards are of wood, generally with two or three uprights with cross-arms of planks, to

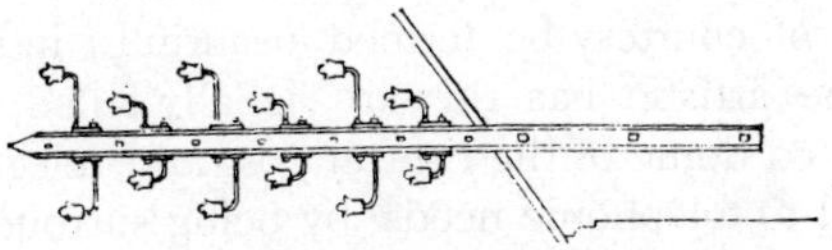

FIG. 51

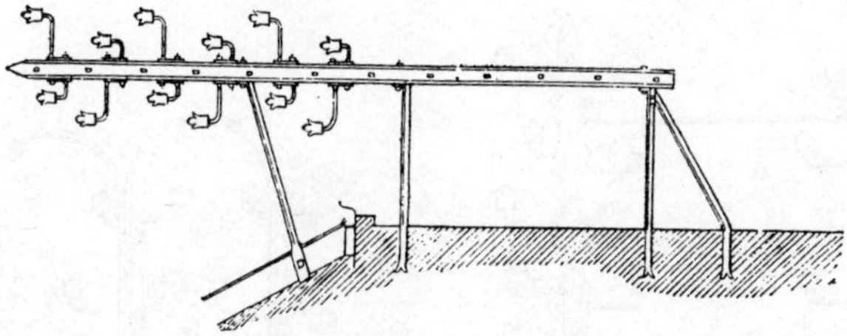

FIG. 50

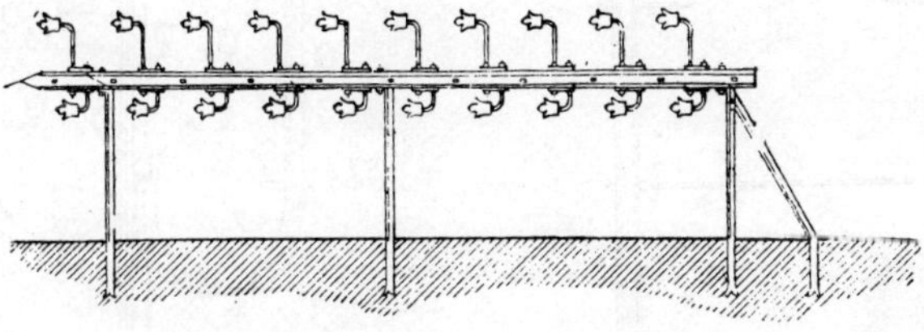

FIG. 49

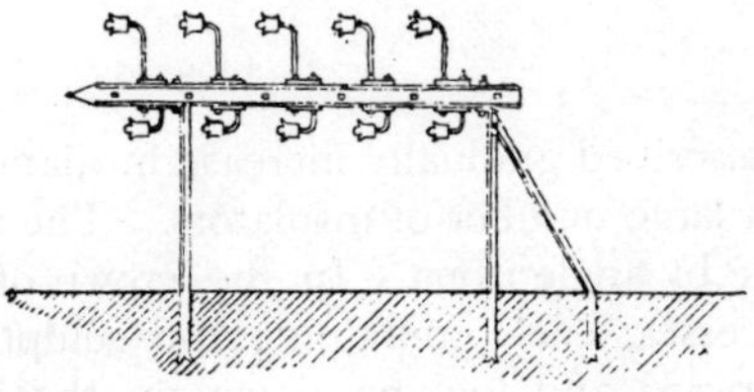

FIG. 48

which the insulators are attached as in fig. 55. The planks are sometimes sandwiched between double uprights. These fixtures cannot by any stretch of courtesy be termed beautiful; indeed, the French sense of the artistic has therein signally failed. At Lille the handsome slated dome of the central post and telegraph office has been adapted to telephonic needs by being surrounded by ten circles of wooden arms, bent to the contour of the dome, and supported on brackets attached to its framework. Commencing

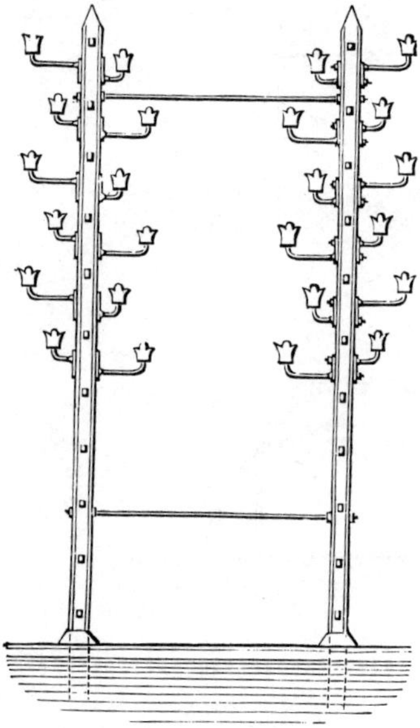

FIG. 52

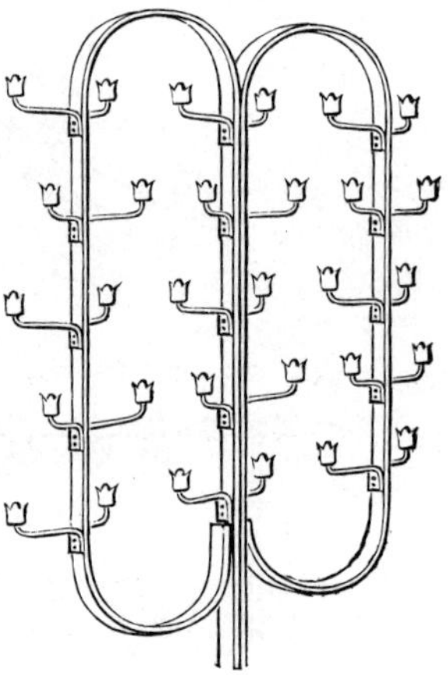

FIG. 53

near the top, the circles described gradually increase in diameter, and space is afforded for a large number of insulators. The arms are stiffened on the outside by angle irons. In the crown of the dome there are eight recesses, each containing the sculptured head of a satyr leaning forward and looking down on the insulators and wires beneath, as though engaged in a perpetual watch for contacts. Surmounted by a flagstaff, and of graceful proportions, the dome looks well from a distance; near at hand it is seen that the arms have warped, and are no longer symmetrical. At Amiens the central station fixture is one of the towers designed by M. Belz, a specimen of which was shown at the Paris Exhibition of 1889, erected on a red brick turret. Taken as a whole, the French overhouse construction must be adjudged deficient in capacity, although strong and well executed. When the French subscribers begin to come on more freely than they have (outside Paris) hitherto done, present methods will not suffice, and a new departure will have to be taken. The ground pole work in

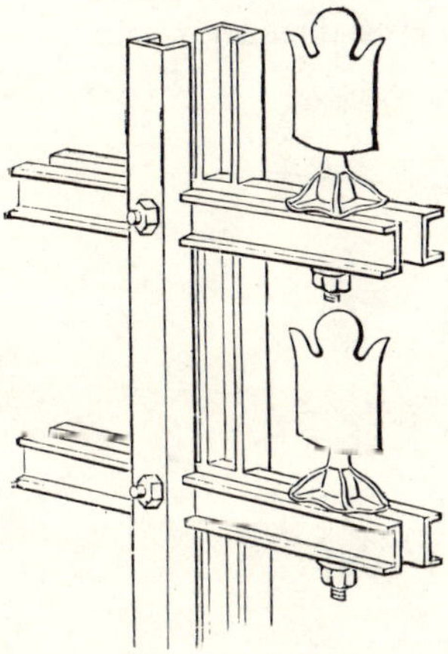

FIG. 54

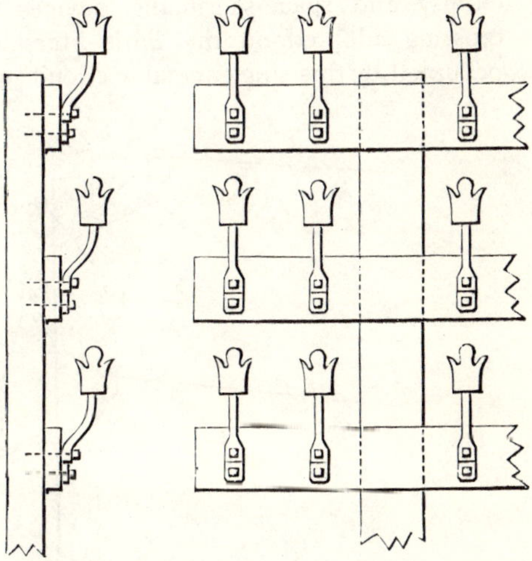

FIG. 55

France, so far as the author's observation went, has attained no abnormal development whatever. The poles are simply the familiar erections of the French Telegraph Department and the French railway companies. A common form of suburban telephone route is composed of two 18 or 20 feet wooden poles tied together as in fig. 52, with insulators arranged in precisely the same fashion. When the poles are straight, well dressed, and well matched—which is not always the case, however—with the insulators properly spaced, such a route is not wanting in picturesqueness, but it is wofully deficient in carrying capacity.

OUTSIDE WORK (TRUNK)

The wire used for the trunk line in France was at first galvanised iron of 4 to 5 mm., but recently nothing but high conductivity bronze or hard copper has been erected. This varies from 3 mm. diameter on the shorter lines (Paris–Brussels, 320 kilometers) to 5 mm. on the longer (Paris–Marseilles, 1,000 kilometers; and Paris–London, 501 kilometers). Trunk lines are crossed, not twisted, but the non-use of cross-arms leads to the adoption of clumsy and space-sacrificing devices. Fig. 56 represents the crossing adopted on the Paris–Marseilles trunk. In the space occupied by this single metallic circuit two or even three arms, each

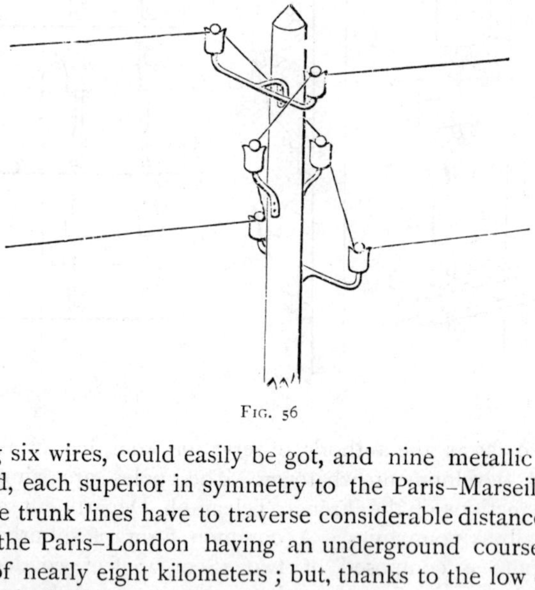

Fig. 56

carrying six wires, could easily be got, and nine metallic circuits obtained, each superior in symmetry to the Paris–Marseilles. In Paris the trunk lines have to traverse considerable distances in the sewers, the Paris–London having an underground course of this nature of nearly eight kilometers; but, thanks to the low capacity of the Aboilard and Fortin-Hermann cables, no inconvenience results. Many of the French trunks are worked simultaneously as telegraph lines on the Van Rysselberghe, Picard, and Cailho systems, but notwithstanding this, the speaking attains a high figure of merit.

PAYMENT OF WORKMEN

The workmen are divided into 'commissioned' and 'non-commissioned,' the former class being retained in the service under all circumstances, the latter only while sufficient work exists. The two classes do not differ materially in skill and experience. In Paris (where the rates of pay are higher than in the country) foremen receive from 92*l.* to 112*l.* per annum, with 46*l.* extra for expenses. Commissioned wiremen get from 56*l.* to 88*l.* per annum, with 32*l.* extra for expenses. Non-commissioned men are paid by the week at the rate of from 4*s.* 9½*d.* to 6*s.* per day, according to skill. In the provinces these rates are reduced by 10 per cent. to 20 per cent., according to locality.

PAYMENT OF OPERATORS

After successfully passing a probationary period, during which nothing is paid, girls, who must not be younger than seventeen, receive 1*s.* 11¾*d.* per day, with 9·6*d.* for luncheon. The next step is to 50*l.* per annum in Paris and 40*l.* in the country, also with a luncheon allowance of 9·6*d.* Subsequently they rise by increments of 8*l.* every three years to a maximum (in Paris) of 74*l.* per annum. Lady superintendents are selected for ability, not by seniority. The working hours are eight per day, out of which one is allowed for luncheon and recreation.

STATISTICS

The latest detailed return of the number of centres and subscribers in France is dated as far back as the end of 1891, but a return of the collective numbers up to the end of 1892 has been issued. The only figures obtainable for 1893 and 1894 are the budget estimates for those years. This is a pity, since the development prior to 1893 was insignificant compared with the progress made since, especially in the provinces. At the end of 1892 the number of exchanges in operation was 207, with a total of 220 switch-rooms, 201 public stations, and 22,918 subscribers' instruments. The length of the local routes was, underground 7,585 kilometers, and aerial 4,415 kilometers; and of the local wires, underground 43,239 kilometers, aerial 16,389 kilometers. The excess of underground mileage is due to the preponderance of Paris, which at this date had nearly three-fourths of the total subscribers. Of internal trunks there were 201, of international trunks 8; with a total length, in routes of 11,428, and in wire of 22,856 kilometers. It will be noticed that the length of wire is that of the routes doubled, which throws doubt on the accuracy of the return, there being certainly more than one metallic circuit in the Paris–Brussels and Paris–London routes if nowhere else. The number of local conversations between subscribers is returned at 19,000,000; between public stations and subscribers at half a million: over trunk lines, 542,910. The number of telegrams telephoned was, outward 385,785, homeward 200,993; and of messages telephoned for local delivery, 1,354. The receipts from all sources amounted to 10,307,823 francs, and the expenses to 9,869,108 francs, leaving a profit of 438,715 francs, or 17,548*l.*

The number of subscribers in the principal towns was stated by a high official to be roughly as follows, in January 1895:

Paris (town)	12,500	Marseilles	1,000
Paris (suburbs)	1,500	Le Havre	1,000
Lyons	1,200	Rouen	600
Bordeaux	1,200		

With the exception of the capital, therefore, it is evident that the French cities are far behind even the English in development.

IX. GERMAN EMPIRE

(EXCLUSIVE OF BAVARIA AND WÜRTEMBERG)

HISTORY AND PRESENT POSITION

BAVARIA and Würtemberg are the only members of the German Empire which have preserved their posts, telegraphs, and telephones in any way independent of the Imperial Post Office; Saxony, Baden, Hesse, and the rest being, in this respect, as essentially Prussian as is any suburb of Berlin. As securing uniformity of practice over a vast area this arrangement commends itself to the practical man, but it of course depends upon the quality of the uniformity obtained as to whether the results to the public are beneficial or otherwise. On this point it must be said that in many respects the arrangements, especially in regard to tariffs in the larger cities and to services rendered, are distinctly good and liberal; on the other hand, it is impossible to pretend that the technical and engineering plans (with a few exceptions) are otherwise than rudimentary and disappointing.

The history of telephony in Germany bears a certain resemblance to our own. At first the Imperial Post Office doubted both the utility and practicability of telephone exchanges. The next stage was the refusal of licences to the International Bell Telephone Company. Time went on, and public opinion calling for exchanges, the Government itself undertook the work. The official appreciation of the nature of the problem and of what was required for a smart telephonic service may be gauged by the fact that the first exchange operators were recruited from the ranks of the superannuated postmen. For many years after starting, the

Government engineers declined to have anything to do with microphonic transmitters, and until 1888 insisted upon supplying their subscribers with nothing but a push-button and battery, a trembling bell, and two receivers, one to speak to, the other to listen by, in spite of the fact that all the lines were single and subject to ininductive disturbances. These receivers were both attached by long cords, so that a subscriber had to hold one to his ear and the other before his face, somewhat in the attitude of mermaid and looking-glass. With both hands so engaged, the taking of notes or holding of papers was of course impracticable. When at last, in 1888, they were compelled by public clamour to provide microphones, the type chosen was a kind of Crossley mounted vertically, and known as the Mix & Genest transmitter. Magneto ringers they would not have at any price until last year, when Berlin and Hamburg were provided with them, all the rest of the Imperial towns being still worked with batteries and pushes. In Berlin and Hamburg the old battery instruments have to a large extent been converted to magnetos at an expense—said to amount to 65 marks (shillings) per instrument—exceeding the cost at which new magneto instruments of really efficient design could have been purchased. The Imperial Post Office still adheres to single wires with earth return, and has not expressed, or given evidence of—the latest multiple boards being made for single wires—any intention of an ultimate conversion to double, although the speaking over the trunk lines, as between subscriber and subscriber, at least, is already far from satisfactory. The enormous expense of such a change is assigned as a reason, but it is an inadequate and ludicrous one in face of the facts that the General Telephone Company of Stockholm has actually converted its system within the last two years, and that its example is being followed by other companies and by several Governments. At least, new exchanges might be run with metallic circuits, and the area over which the inevitable change will have to be made thereby limited. As it is, subscribers are crowding on in all parts of Germany, and the public money is being spent in connecting them in a manner which is already recognised nearly everywhere else—even in Servia, Bulgaria, and Roumania—as obsolete. In a few years more the machine will have become so huge and clumsy, and the trunk-line speaking so immeasurably inferior to that which will prevail in neighbouring States, that an entire reconstruction will have to be undertaken at enormous cost.

The author visited several of the principal cities both in the north and south of the Imperial postal district, including the chief towns of Baden, Hesse, Alsace-Lorraine, Saxony, and Hanover, with the view of obtaining a just idea of the whole and of avoiding the danger of generalising from only local experiences. There were but few differences to note. The outside construction is practically the same everywhere, better done in some of the towns than in others, but always on the same plan; the subscribers' instruments (excepting in Berlin and Hamburg) are identical. Only the switch-boards and exchange fixtures differ. In all the towns the author took great pains and disbursed divers marks with the object of testing the service, especially that over the trunk lines, from a subscriber's point of view. All the hotels of any note are connected in the various towns, the instruments being usually under the care of the hall porters, invariably men of intelligence and practised in the manipulation of their telephones. Under these circumstances it was found a good plan to get through to hotels in other towns and inquire after supposititious letters. This was not an expensive amusement, inasmuch as a three-minute talk between any two connected parts of the Imperial postal district costs only one shilling (this is one of the points on which the Administration is deserving of earnest commendation); but it required a good fund of perseverance and patience, since the lines, when first asked for, were invariably engaged, and the precincts of the instrument had to be haunted until—perhaps after some twenty or thirty minutes—the notification of connection came. The result of this experience (October 1894) was decidedly disappointing, for on no single occasion did the author succeed in obtaining a trunk communication that was even tolerably good. The best (and yet indifferent) were between Frankfort-on-Main and Mannheim, and between Leipzig and Berlin. The worst between Berlin (Central Hotel) and Hamburg (Hamburger Hof), excepting that between Berlin and Cologne, which had to be abandoned as hopeless. To compare any Imperial

German speaking (as between subscribers) with that between Brussels and Paris, Paris and Marseilles, or Stockholm and Gothenburg, would be absurd : there is no similitude. The local service in Berlin is slow, but faster than that of Paris. The Central Hotel, Berlin, has a telephone-room, in charge of an attendant, containing three instruments in connection with the exchange, which, during the busy hours, especially the forenoon, are in incessant request by commercial travellers and others staying in the house, would-be users waiting their turns sometimes several deep. It is under such circumstances as these that a good system shines and a bad one breaks down. In that Berlin telephone-room the only thing that shone was the patience, under long suffering, of the attendant and customers. At the same time it must not be overlooked that the Berlin exchange is the largest in Europe, if not in the world, counting, as it does, some 25,000 connected instruments in the city itself and nearly 3,000 more in the suburban area. The problem that presents itself for solution in the Prussian capital is consequently unique, and it would be unfair and ungenerous to underrate its difficulties. But it is reasonable to argue that methods which give bad results with 500 subscribers cannot possibly prove satisfactory with 25,000, and it is on the score of persistence in rudimentary forms when an advanced stage of development has been reached that fault may most justly be found with the Imperial Post Office. The overhearing on some of the single wires is very pronounced. At Frankfort-on-Main, the hotel porter, in describing his telephone and the uses he put it to, remarked that before ringing for a connection to his fishmonger he always lifted the telephone off its hook and listened, because if the fishmonger was talking to anybody else he could always distinguish his voice and so knew that it was useless to ring just then. If, on the other hand, the familiar tones were absent, he knew that the connection could be got.

There is some official predilection in Germany towards an eventual abolition of inclusive annual subscriptions in favour of the Swiss plan of a small annual payment and a fee for each connection asked for over a certain number. It is considered that an automatic register of the communications had, to be placed in the subscriber's office, is necessary to the success of such a plan, and some experiments are being conducted with meters invented by Messrs. Mix & Genest and by an official of the Imperial Administration. Such registers, however, unless very complicated (in which case the expense of their introduction and maintenance would outweigh all advantages), could not supersede the operators' notes, since they would not differentiate between the numerous classes of connections, local, suburban, short- and long-distance trunk, telegrams, matter to be mailed, &c., that may be asked for. A simple record of the number of connections would help but little, and if the operators' notes must be preserved at all, they had better accomplish the whole task as in Switzerland and Stockholm. In the latter city these reasons have led to counters, efficient as such, being abandoned after extensive use. The Imperial Administration deserves praise for the manner in which it has consistently supported home manufacturers. It has taken as little of its apparatus from abroad as possible, even multiple switch-boards, the most complicated of all telephonic mechanism, having been, whenever possible, procured in Germany. The gratifying result is that, although the native instruments may be somewhat lacking in design, a school has been founded which is rapidly becoming equal to all demands. At present it is traversing ground which has been already exploited elsewhere, making the same mistakes and acquiring the same experience. As regards workmanship, the productions of the three chief firms—Siemens & Halske, Mix & Genest, and R. Stock & Co.—leave nothing to be desired.

SERVICES RENDERED TO THE PUBLIC

1. **Intercourse between the subscribers and public stations of the same town.**—The rate is universally 7*l.* 10*s.* per annum, irrespective of the size of the town, and includes connections of any length up to five kilometers. This rate is too high, notwithstanding the long length given without extra charge, for small towns. In such places the vast majority of the lines are much less than half a mile in length, and 90 per cent. less than one mile. A more equitable figure would be 4*l.* or, at most, 5*l.*, for connections not

exceeding one and a half kilometers, with an ascending scale for the exceptionally longer lines. On the other hand, for cities like Berlin and Hamburg 7*l.* 10*s.* may be admitted as reasonable; but the fact only accentuates the injustice done to the inhabitants of small towns and villages, whose telephones must necessarily be much less valuable than those of the Berliners and Hamburgers. When a 5*l.* rate is found sufficient in Stuttgart, the capital of a German State, there is certainly ground for complaint under the Prussian rule. The efforts made in Würtemberg to restrict the user of telephones to their actual hirers are not made by the Imperial authorities, whose official instructions to the subscribers are silent on the point, perhaps wisely, for when such restrictions are imposed they soon become dead letters. The subscribers get annoyed at what they regard as an unjust and unreasonable regulation, while the officials become tired of trying to enforce rules which produce nothing but ill-temper and friction. Imperial subscribers are simply prohibited from accepting payments from outsiders for the use of their telephones. Subscribers are bound to insure their instruments, together with all leading wires and fixtures connected with them, against fire. Would-be subscribers must produce a written way-leave from their landlord authorising the fixing of all necessary wires and apparatus; in the absence of such a way-leave no person is accepted as a subscriber. Subscribers whose communication has been interrupted for more than four weeks are allowed a proportionate rebate. Subscriptions will also be refunded should the Administration, in the exercise of the powers conferred by Parliament, close any exchange or line permanently or temporarily. Subscribers removing are liberally dealt with, no charge being made unless the new premises come under a more expensive section of the tariff. Peremptory powers to remove instruments are possessed in the event of non-payment of subscriptions when due, damage to apparatus, and improper language addressed to the operators. The proprietor of a building let off as dwellings or workshops to different tenants may pay for a wire to the exchange under the usual tariff, and by providing an attendant at his own expense to operate a switch-board supplied by the Administration is allowed to have instruments fixed in any or all of his tenants' places and to give them exchange communication through this switch-board. There is a special tariff (see *Tariffs*) for such extensions. The proprietor renders himself responsible for all payments, and collects subscriptions from his tenants. If any of them neglects to pay he is the loser.

2. **Intercommunication between a town and its suburbs** and, in some cases, other small towns not very far removed. For example, the Berlin suburban intercourse includes Spandau (8 miles), Kopenick (9 miles), and Potsdam (15 miles); the Leipzig includes Markranstadt (8 miles); the Frankfort-on-Main includes Homburg (10 miles), Hanau (13 miles), and Mayence (20 miles). For this suburban intercourse an additional yearly subscription or a fee per communication has to be paid. The connecting lines between these district centres are metallic circuits.

3. **Long-distance internal trunk communication.**—Herein the policy of the Imperial Administration must be acknowledged to be most liberal and praiseworthy. The charge for three minutes is 50 pfennige (5*d.*) up to about thirty kilometers—the exact distance varying in different districts—and 1 mark (1*s.*) for any distance beyond. This means that between any two connected points of the German Empire (excepting Bavaria and Würtemberg) a three minute conversation may be had for one shilling. The trunk system is already very extensive, and is growing every month. It has penetrated to every corner of Germany, from the Baltic to the Neckar, and from Saxony to the North Sea and the frontiers of France. Already the distances which may be spoken over exceed 450 miles.

The Imperial Administration admits urgent or express talks over the trunk lines at triple the unit charge. No talk may be prolonged beyond three minutes if the line is wanted by others. When orders given for trunk communications cannot be executed for reasons beyond the control of the Administration the caller must pay a whole unit fee. Such reasons include the failure of the called subscriber to answer, or the absence of the caller at the moment when the connection is ready. When a communication cannot be given at once, the caller may cancel it at any time before the operator has asked the distant station for it; if that stage has been reached, the caller must pay whether he speaks or not.

Subscribers have, as a rule, to accept the operators' records as to duration of talks, &c., as correct ; but complaints of error or overcharge are investigated, and if discovered to be reasonably well founded, admitted.

4. **International trunk communication.**—The telephone has crossed the frontiers at several points. Reichenberg-Zittau and Warnsdorf-Grossschönau, both in Saxony, have communication with a few of the nearest Austrian towns. Würtemberg and Bavaria (see those sections), which, although members of the German Empire, possess independent postal and telegraph administrations, have both effected junctions—the former, *viâ* Pforzheim and Heidelberg, with Baden and the south-west of Germany ; the latter with Frankfort-on-Main and the south-west of Germany *viâ* Aschaffenburg, and with Berlin *viâ* Hof. The isolated Bavarian Palatinate of the Rhine, which possesses exchanges at Ludwigshafen, Kaiserslautern, Neustadt, and Speÿer, is also connected to the Imperial Post Office territory, *viâ* Mannheim. The tariff from Berlin to Bavaria is two marks, or shillings, per three minutes, double that which obtains within the limits of the Imperial Administration. The distance from Berlin to Munich, in the direct air line, is 310 miles, for which the charge under the proposed British Post Office scale would be 4*s*. 6*d*. The speaking on the loop is loud. Berlin is also connected with Vienna, distant 616 kilometers. At present the communication is limited to the Bourses and to such lines as are metallic circuits. Communication existed for a time between Mulhouse and the south of Alsace and Switzerland, but was discontinued by orders from Berlin. An agreement has twice been all but concluded with Belgium, but broken off at the instance of the German Political Bureau. A trunk line from Berlin and Hamburg to Copenhagen is now spoken of. Urgent talks at triple fee are admitted to Munich and Vienna.

5. **Public telephone stations.**—These are fairly numerous. There are twenty-nine in Berlin itself, and thirty-one in its suburbs, all at post or telegraph offices. Other towns are not so well provided, but still one can always be found at the central, and mostly also at the chief branch, post offices. Automatic boxes for checking payments are not used, attendants being always provided, to whom

fees are payable. Complaints have been made of delay in obtaining communication from these stations, due to the amount of preliminary ceremony that has to be gone through. A would-be talker has to fill up a form with the name, list number, and switch-room number of the person he wants. To this form, which he must also sign, he has to affix postage-stamps to the value of the communication demanded. The attendant then checks the form, enters the particulars in a book, and finally permits access to the instrument. In some towns local subscribers may use the public stations free for local talks in the absence of any paying customer ; a demand for the line from such a person leads to the free talk being interrupted without ceremony. The attendants are instructed to receive complaints of interruption, &c., from subscribers, and to telephone them on to the proper office. The services from the public stations are limited to speaking over the local, suburban, and trunk lines, telephoning of telegrams and mail matter being inadmissible. They are consequently of less public utility than those, for instance, of Denmark and Switzer land ; but yet they are recognised public institutions which the people know where to find and how to use. Germany is consequently far in advance of Great Britain, where the Post Office has ever made it a rule to forbid the establishment of public telephone stations at the post and telegraph offices, or anywhere within the bounds of the postal authority.

6. **Telephoning of telegrams.**—Subscribers may telephone their telegrams to the local telegraph office to be forwarded, and also receive those arriving for them through their own instruments.

7. **Telephoning of mail matter.**—Subscribers may telephone messages to the central office to be written down and put in the post as letters or post-cards. This is a very handy and useful arrangement, as it virtually extends the time of closing the mail, which may frequently be caught by a telephoned message when an ordinary letter posted by hand would certainly miss. More especially is this the case with suburban subscribers, who may neglect the hour of closing of, say, the English mail at their local post office, and get a telephoned message through to the head office in Berlin two or three hours later in time to be included.

The fees charged being in addition to the ordinary letter postage, it pays the Administration as well as benefits the subscribers.

8. **Telephoning messages for local delivery.**—As in most continental countries, subscribers may dictate messages for non-subscribers resident in the same town to the central office, where they are written down and delivered immediately by messenger.

TARIFFS

1. **Rates for local exchange communication.**—Uniformity in this respect prevails throughout the Imperial Administration. For a distance not exceeding five kilometers (2 miles 1,480 yards) measured direct, the charge, per annum, is 7*l.* 10*s.* When the distance exceeds five kilometers the annual charge is increased by 3*s.* per 100 meters. When the distance exceeds ten kilometers a further additional charge, payable only once, not annually, of 10*s.* per 100 meters is exigible. Should it be necessary to employ cables or other works of a specially expensive character, power is reserved to make such further charges as may be deemed equitable.

A second instrument attached to the same line is also charged 7*l.* 10*s.* per annum, provided the deviation necessary to include it does not exceed 500 meters; if more wire is necessary, the excess rate of 3*s.* per 100 meters comes into play.

For extra instruments let out to tenants of one proprietor and communicating with the exchange through that proprietor's line, 5*l.* per annum per instrument, with a minimum of 10*l.*

Extra instruments for the use of one subscriber:

If within the same building as the exchange instrument, per instrument per annum	2*l.*
If in another building, but on the same property . . .	5*l.*

Extra bells are charged 5*s.* per annum. Any special works or deviations from ordinary practice desired by a subscriber have to be paid for, and become the property of the subscriber.

Charges are usually payable annually in advance, but the Administration may collect quarterly if it judges expedient.

Agreements are for one year only, and continue from year to year, subject to three months' notice.

2. **Rates for suburban connections.**—Suburban subscribers pay the local rate for connection to their local exchange, and communication within their own suburb or group of suburbs (each large town has two or more groups in its vicinity, particulars of which are given in the local lists); but in order to communicate with the town, or with other suburbs not scheduled as being within their own group, they must pay an additional annual subscription of 5*l*., or 3*d*. or 5*d*. per three minutes' talk. These charges are equally due by town subscribers who wish suburban communication. Any person paying the extra 5*l*. annual charge is not only entitled to call any subscriber on the list, but also to be rung up freely by everybody, whether they also pay the extra rate or not. The three-minute rate depends on the distance of the suburban group from the town. For instance, the charge is 3*d*. between Berlin and Group I., which comprises Charlottenburg, Rixdorf, Friedenau, Pankow, Rummelsburg, Schöneberg, Weissensee, and Westend, none of them very far away; and 5*d*. between Berlin and Group II., or between Groups I. and II. The latter includes Potsdam, Spandau, Kopenick, and some twenty other places comprised within a radius of fifteen or sixteen miles. In the case of Leipzig there are three so-called suburban groups, the most distant comprising Chemnitz, 48 miles away. 5*d*. per three minutes, or 5*l*. per annum, is the uniform rate. Under such a rule at home, Brighton would be considered a suburb of London and brought within the scope of an extra 5*l*. annual payment. The arrangements at Frankfort-on Main are equally liberal, the 5*d*. per three minutes, or 5*l*. per annum, covering Mayence (20 miles), Rudesheim (33 miles), Hanau, Homburg, and many other towns.

3. **Rates for long-distance internal trunk communication.**—These are simplicity itself. For distances up to about thirty kilometers (the practice varies somewhat in different districts, and is sometimes modified by the inclusion of towns nearly fifty miles distant in suburban groups) the charge per three minutes is 5*d*.; for all other distances, 1*s*. Express talks are allowed at triple rate. No talk may exceed three minutes if others are waiting to use the line. If a communication asked for cannot be got through from some cause beyond the control of the Administration, the caller is charged a unit fee.

4. **Rates for international trunk communication**:

	s.	d.
Between Berlin and Vienna, per three minutes . . .	2	6
,, ,, Munich, ,, ,, ...	2	0
,, Mannheim or Heidelberg and Würtemberg, per three minutes	1	0
,, other places in Baden and Würtemberg, per five minutes	1	0

Between Mannheim and Ludwigshafen (Bavaria), 5*l.* per annum, or 3*d.* per three minutes.

Urgent or express talks are allowed on payment of triple unit charge.

5. **Rates affecting public telephone stations**:

	s.	d.
Three minutes' local talk	0	$2\frac{1}{2}$
,, suburban talk	0	5
,, short trunk talk (up to about 30 kilometers) .	0	5
,, long trunk talk (any distance exceeding 30 kilometers)	1	0

In Frankfort-on-Main and some other towns subscribers may use the public stations locally free of charge in the absence of any paying customer.

6. **Rates for telephoning of telegrams.**—For each telegram forwarded or delivered by telephone, a foundation charge is made of 1*d.*, with $\frac{1}{10}$*d.* per word added. Telegram accounts must be covered by deposit and settled monthly, or, if desired by the Administration, as soon as they amount to 10*s.*

7 and 8. **Rates for telephoning of mail matter and of messages for local delivery.**—In addition to the postage or cost of special messenger, the telegram charge of 1*d.* for each message, with $\frac{1}{10}$*d.* for each word, applies also to these services.

WAY-LEAVES

The Imperial German Administration has been specially credited in Great Britain with being possessed of quite Gargantuan powers in the direction of autocratic way-leaves. On examination, however, the fairy vision vanishes. The plain fact is that, apart from the clause, which, like the National Telephone Company, it inserts in its subscribers' agreements, the German

Government has no control over private or municipal property whatever. No subscriber is connected to the exchange unless he undertakes to give (or, if the property is not his own, obtain) permission to erect on his building fixtures and wires for the common use of the exchange as well as his own. That is an inflexible rule, which is acted upon, and naturally produces good results. The National Telephone Company compels its subscribers to sign a similar agreement, but does not press for its observance if any reluctance to comply with it is shown ; the results obtained are consequently inferior to the German. A new Telegraph Act was passed as recently as April 6, 1892, by which the Government was given various additional powers in connection with telegraphs and telephones. The last clause of this Act declares, 'The Imperial Government does not acquire through this law any powers in excess of those presently existing with regard to private lands or public roads and streets.' The Administration has to take property owners and public authorities along with it in everything it does. The author has been informed by German subscribers that once telephonic communication has been established a subscriber cannot be deprived of it, even if he gives notice to take away any standard or wires that have been erected on his property other than for his own accommodation, unless the Government can show to the satisfaction of the proper tribunal that no other means exist of getting his wire in. Subscribers have been known, it is said, to consent to the Government way-leave clause, get in their telephones, and as soon as practicable thereafter to give the stipulated notice to take away all fixtures but their own, and to have, nevertheless, succeeded in retaining their connections. The German Government is stated (in Great Britain) to make a practice of coercing property owners who refuse the use of their roofs by planting enormous poles opposite their doors, or by suddenly discovering that their drains are faulty and must be renewed ! The author could not succeed in hearing of such a case in Germany. Apart from the unlikelihood of such undignified proceedings being permitted by the Government, such poles could not be erected under the Act without the co-operation of the local authorities, who would scarcely connive at an outrage on a townsman. In the matter of way-leaves Imperial

Germany is less autocratic than Republican France. Certainly the possession of most of the railways gives the State a great pull in way-leave facilities over an English telephone company, but that is a matter apart from streets and private houses.

SWITCHING ARRANGEMENTS

The multiple switch-boards in use are of three types, manufactured respectively by the Western Electric Company, Mix & Genest, and R. Stock & Co. The former company has supplied single-cord boards of a total capacity of 24,200 lines to six of the Berlin switch-rooms, and a single-cord board for 5,400 lines to Hamburg. Double-cord boards have been supplied to Frankfort-on-Main (2,800 lines), Cologne (2,200 lines), Breslau (2,000 lines), and Mannheim (1,000 lines).

Messrs. Mix & Genest, of Berlin, have supplied their type of board to Hamburg (2,800 lines), Stettin (2,000 lines), Düsseldorf (1,600 lines), Crefeld (1,200 lines), Barmen (1,200 lines), Cassel (1,000 lines), Dortmund (600 lines), and Bochum (600 lines).

Messrs. R. Stock & Co., of Berlin, have supplied boards to Berlin Moabit (6,000 lines), Dresden (5,000 lines), Leipzig (3,200 lines), Altona (2,000 lines), and Hanover (2,000 lines). Messrs. Stock have also supplied two single-cord boards, each of 2,000 lines, to Hamburg, and have extended the Western Electric board at Frankfort-on-Main to 6,000 lines. Experimentally, a flat board has been fitted up at Berlin Moabit by the same firm.

The Western Electric boards are of that company's well-known type, and call for no special mention.

The original form of Messrs. Mix & Genest's multiple, which was designed by Mr. D. Oesterreich, has also been often described and illustrated. Its principal feature was the saving of the usual test wires by causing a voltaic current, too weak to actuate the call bells, to flow from a central battery at the exchange continuously over all the subscribers' lines to earth. The jacks being in series, it was discovered whether a wire asked for was engaged or not by inserting a double-contact plug in one of the jacks. A sensitive galvanometer was looped in the test cord, and, if the wire was free, revealed the test current circulating ; if, on the other hand, the line was engaged, no current passed the galvanometer, since, if the connection had been made in front of the jack tested, one side of the galvanometer was insulated, although the other was joined to the battery ; while if the connection was on behind the test point, the galvanometer was cut off from the battery altogether. As now used, in addition to the test battery at the exchange, there is a Daniell cell in each subscriber's office, which sends a current to line as long as the

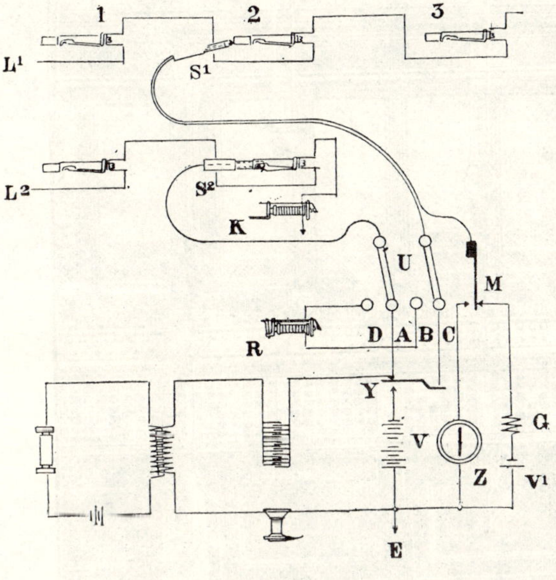

FIG. 57

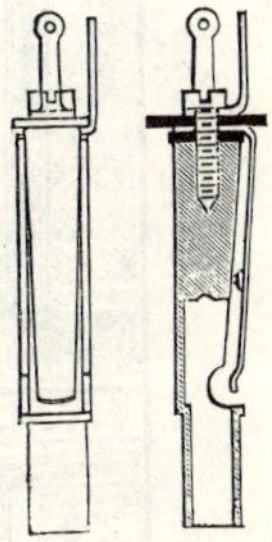

FIG. 58

receiver is off the hook, but not at other times. The connections are arranged as in fig. 57, in which L[1] L[2] are two subscribers' lines joined through the series jacks 1, 2, 3, and to earth through the calling indicators K. In the circuit of each pair of plugs and cords there is a switch U (in practice combined in a single lever), making contact with A and C or with B and D, according to position. When on A, C, the speaking set is brought into play, together with a ringing key Y and battery V ; when on B, D, the ring-off drop R is looped into the cord. There is also a Morse key switch M, having its lever connected through the cord to the tip (which is insulated from the body) of the plug S^1; its back stop to a test battery of one Daniell cell V^1 through an adjustable resistance G,

FIG. 59

and its bottom stop to earth through a 150-ohm galvanoscope Z. Subscriber L^2, in taking his receiver off its hook when making a call, puts his test cell in connection with the line and immediately blocks it against intrusion, since an operator testing by applying the tip of plug S^1 to any of his jacks and pressing the key M would get a current on the galvanoscope Z. L^1, the line asked for, being found free by pressing the tip of the plug S^1 against a jack and depressing the key M, is connected by pushing the plug home; when this has been done, the portion of L^1 to the right of the connection is guarded by the test battery V^1 acting through its separate conductor in the cord and the insulated tip, but until the subscriber takes off his phone, his line to the left of the connection is not guarded, and another connection may consequently be unwittingly popped on in the interval between the call and the reply. When L^1 has answered, the switch U is put over to B, D, and the subscribers left talking through the ring-off drop R. The insulated tip of the plug S^2 has no connecting wire in the cord, for, as this plug is always used with the answering jack, the function of the tip is simply to cut off the calling indicator K and earth. After connection, the whole of L^2 and the portion of L^1 to the left of the jack used is guarded by the subscribers' test cells, and the portion of L^1 to the right of the jack used by the exchange test cell V^1. A section and top plan of the Mix & Genest spring-jack are shown in fig. 58, and a view and end section of their table in figs. 59 and 60.

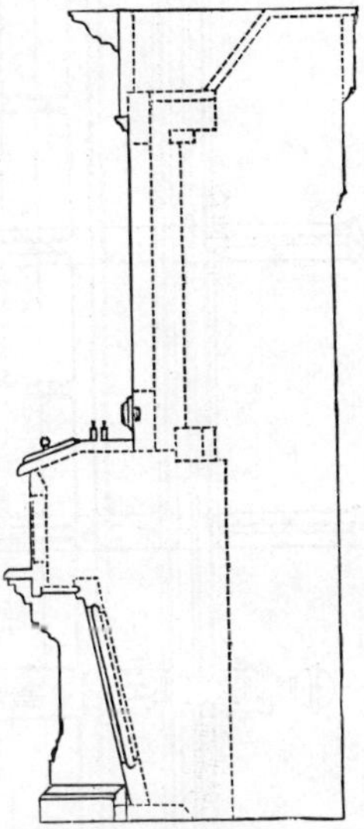

FIG. 60

A front view and cross section of Messrs. Stock & Co.'s latest Berlin Moabit, single-wire, double-cord board are shown in figs. 61 and 62, which explain themselves. It will be seen that it differs in plan from a Western Electric board only in matters of detail. Each switching section accommodates 200 subscribers, and can be served by three operators. The return cables go to the intermediate field, thence to the answering jacks, and finally to springs against which suitable contact pieces in connection with the indicator coils press when the drops are in place. This absence of soldering greatly facilitates the withdrawal of drops for inspection and repair. There is nothing

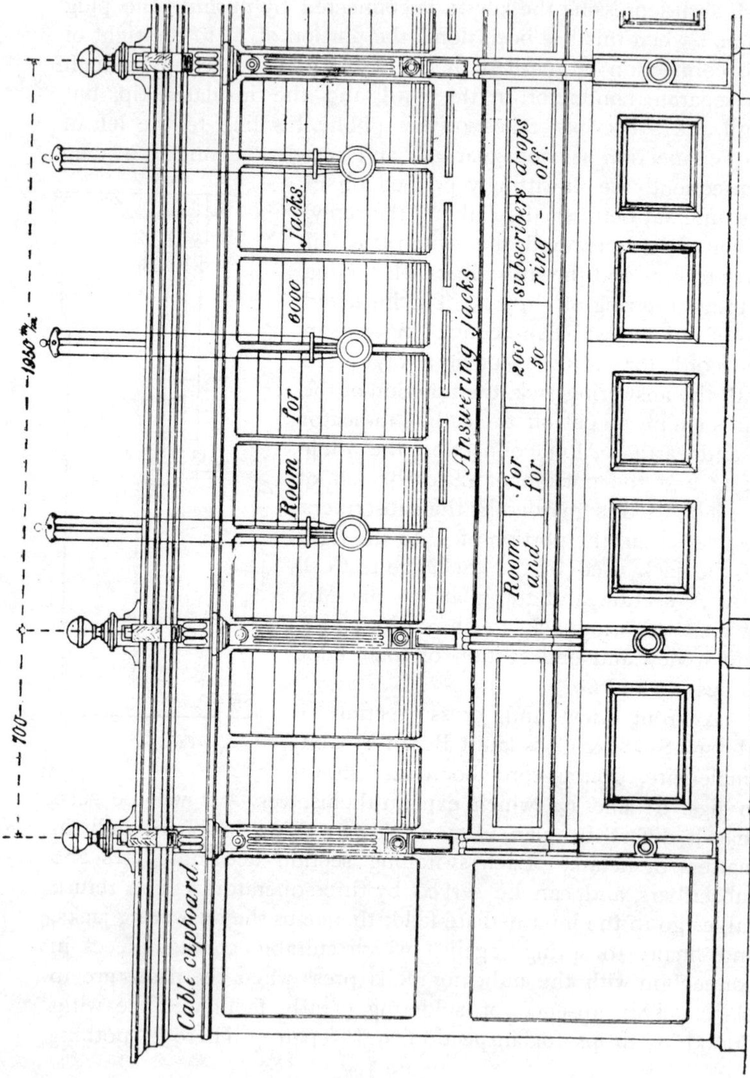

1950 m/m
700
Cable cupboard.
Room for 6000 jacks.
Answering jacks.
Room for 200 subscribers' drops
and for 50 ring - off.

FIG. 61

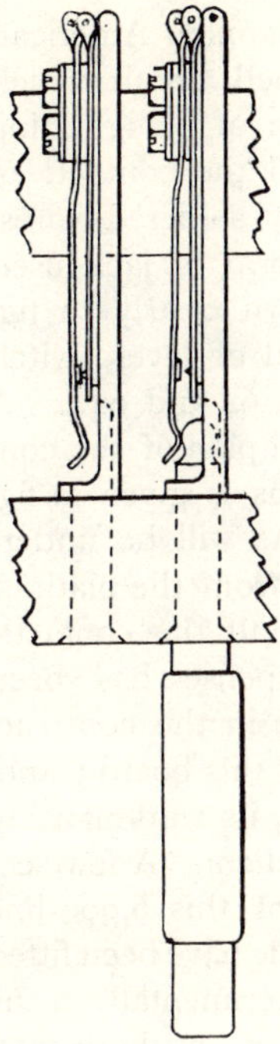

Fig. 63

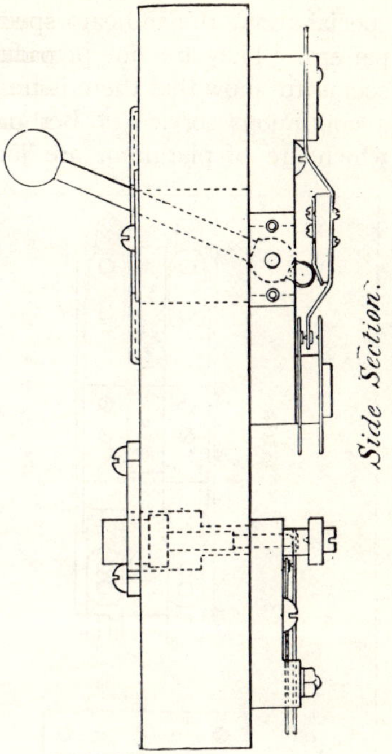

Fig. 64

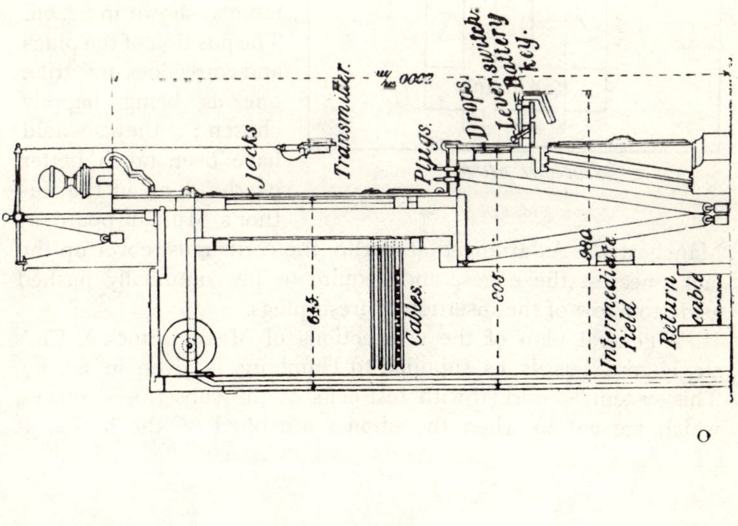

Fig. 62

special about the indicators, which are of the familiar American pattern. They are not provided with a night-bell circuit, which seems to show that there is no present intention of inaugurating a continuous service in Berlin. The jacks, all the contacts of which are of platinum, are joined in series by soldered wires. The form of jack used is shown clearly in fig. 63, and of lever switch in figs. 64 and 64A. A general plan of the connections is given in fig. 65. As will be understood from the platinising of all the contacts, no expense has been spared in the construction of this board; and, in fact, its workmanship is excellent. A few sections of this 6,000-line multiple have been fitted up experimentally in the form of a horizontal table as shown in fig. 66. The position of the plugs and cords does not strike one as being happily chosen; they would have been much better overhead, as in the author's Mutual board at Manchester. As arranged at Berlin, the cords must cover up the jacks nearest the edges, and require to be continually pushed aside to allow of the insertion of fresh plugs.

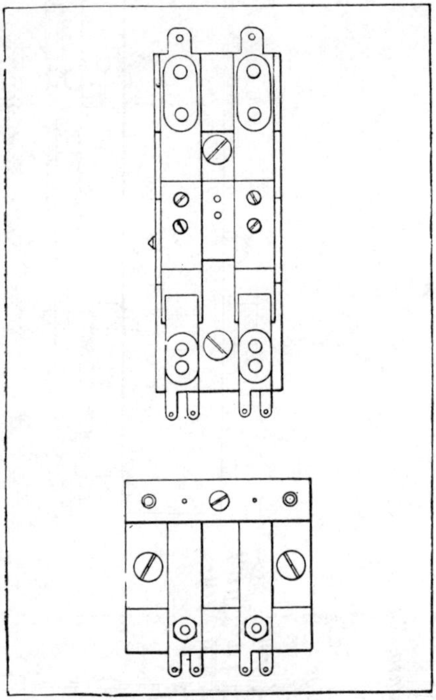

Bottom plan.

FIG. 64A

A general plan of the connections of Messrs. Stock & Co.'s single-cord boards, as supplied to Hamburg, is given in fig. 67. This system is worked with test cells at the subscribers' offices, which are cut in when the phones are lifted off the hooks, as

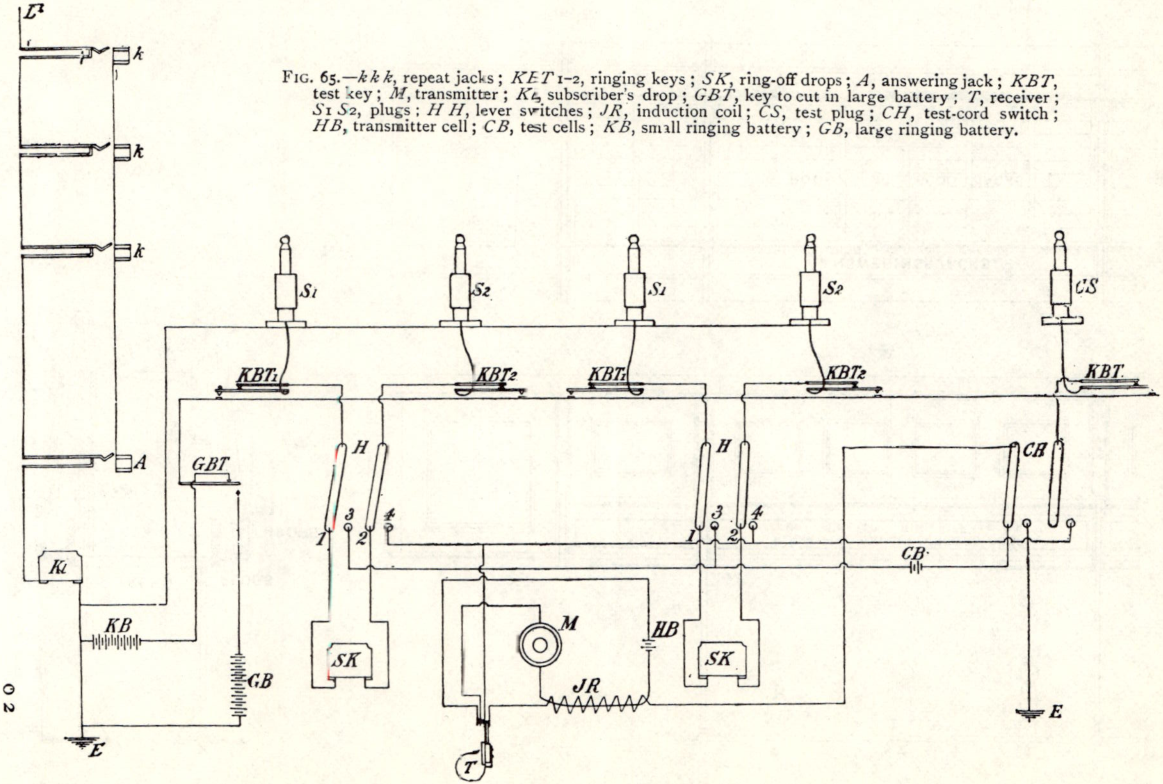

FIG. 65.—*k k k*, repeat jacks; *KET* 1–2, ringing keys; *SK*, ring-off drops; *A*, answering jack; *KBT*, test key; *M*, transmitter; *Kl*, subscriber's drop; *GBT*, key to cut in large battery; *T*, receiver; *S*1 *S*2, plugs; *H H*, lever switches; *JR*, induction coil; *CS*, test plug; *CH*, test-cord switch; *HB*, transmitter cell; *CB*, test cells; *KB*, small ringing battery; *GB*, large ringing battery.

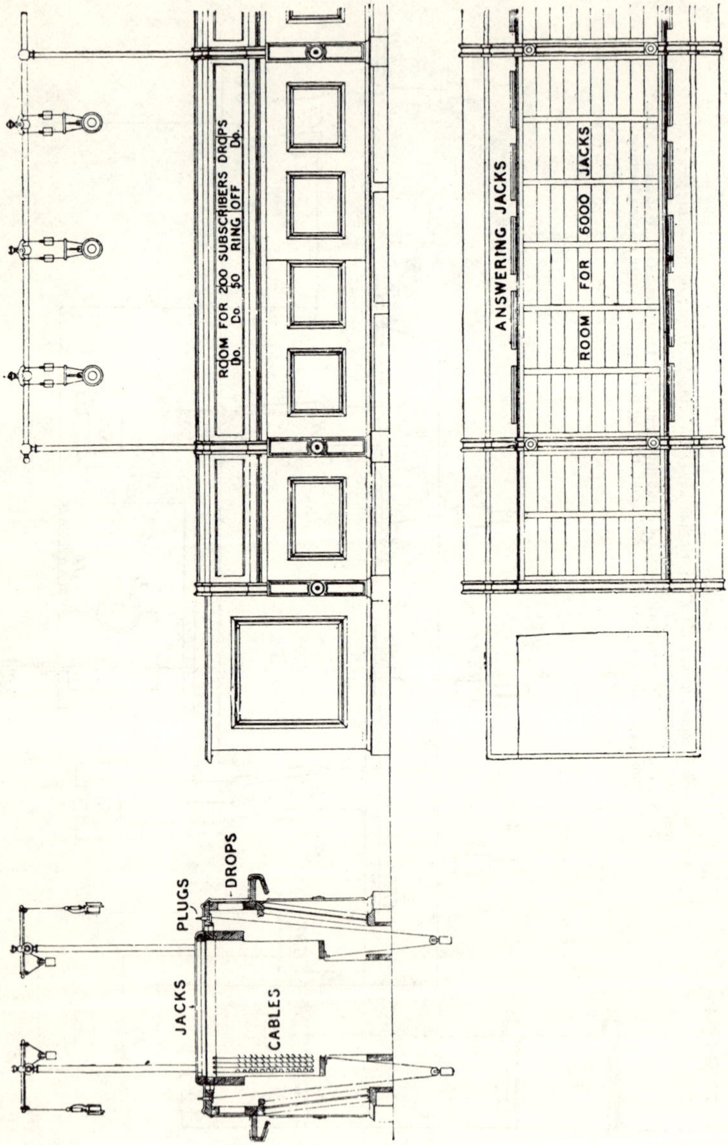
ROOM FOR 200 SUBSCRIBERS DROPS
Do. Do. 50 RING OFF Do.
ANSWERING JACKS
ROOM FOR 6000 JACKS
DROPS
PLUGS
JACKS
CABLES

Fig. 66

described in connection with Mix & Genest's board, in addition to a test battery at the exchange. A calling current (fig. 67) passes by *a b*, K*l.*, plug L.S. to earth. When the plug is lifted the phone K*t.* is cut in viâ *c d e k i*, J.R., key T*k.*, test key CT, test battery, and earth. Test is made by applying L.S. to the socket *h*; line being free, connection is established by pushing L.S. home in the desired jack. The calling battery W.B. is divided into two parts for short and long line ringing. To ring on a short line the key U. is depressed, bringing *l* in contact with *g* and the cord of the plug L.S. For a long line the key G.B.T. is depressed simultaneously, and the whole battery brought in. After connection is ascertained to be satisfactorily through by the presence on the line of a current from one or both of the subscribers' test cells, the phone is cut out by pushing down U., and so separating the contacts *i* and *k*. Key CT is used to cut out the exchange test cell momentarily when currents from the subscribers' cells are being tested for. In addition to the single cords, there are a few double cords with ring-off drops and keys kept in reserve. These are shown at SK*l.*, U., T.', T''. Each pair of double cords has a jack *m* to receive connections from the next table when necessary. *n* is in connection with the calling battery, and the key K.B.T. is used for ringing through the plug C.S

The multiple boards in the remaining six Berlin switch-rooms are of Western Electric Company's manufacture. One of them is, for want of room for lateral extension, arranged in two tiers or stories, the operators of the upper tier sitting some six feet above the level of the heads of those below. This is ingenious, and saves space, but is not conducive to health. The lady superintendents, familiar in other countries, are dispensed with; the girl operators, who, as German State officials, are of course in uniform, being kept up to the mark by mature gentlemen of severe and martial aspect. Should the British Post Office take over the telephone exchanges in 1897, a new field for employment would be open to the army reserve men were Parliament to sanction the adoption of the Prussian corporal plan. When located in old buildings, the German switch-rooms sometimes lack cubic content and ventilation; but when opportunity offers,

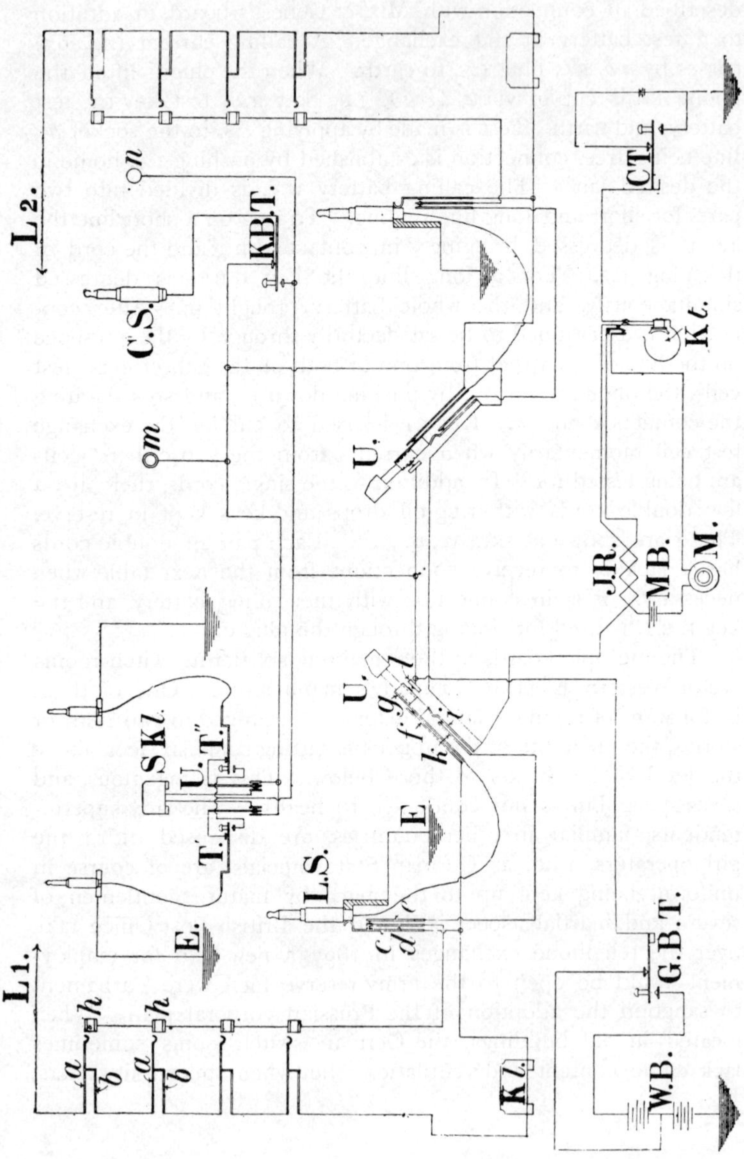

FIG. 67

as at Moabit, Breslau, Frankfort-on-Main, &c., the architecture, decorations, and accommodation are worthy of all praise.

There are seven switch-rooms in Berlin, arranged in an irregular circle round the centre of the city. Each has direct junction lines to every other, there being some 700 wires so employed, without counting those going to the suburban rooms. All these junctions are single and erected overhead. The trunks all come into one switch-room, and are multipled over small tables, divided from each other by partitions, each of which accommodates two trunk lines and is attended to by one operator. These trunk tables are a speciality of Messrs. Mix & Genest, who have supplied nearly 300 to the Imperial Administration for use in different towns. Fig. 68 shows their general appearance. Each section is fitted with answering jacks for the trunk and intermediate board wires, together with forty repeat jacks and the necessary indicators; also a metallic circuit on which branch switch-rooms may be put through to the trunks without the intervention of a translator. The local operators notify trunk

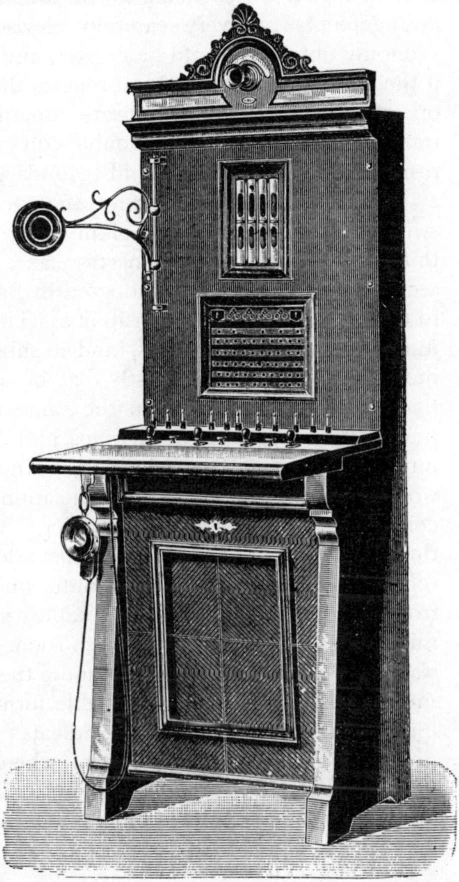

Fig. 68

calls to the small boards, and the connections are completed through an intermediate section on which all the local lines are multipled. The trunk tables are provided with sand-glasses on the Swiss plan for checking the duration of conversations. The arrangements are very carefully devised, but the speed and economy obtained would be greater, and the chance of error less, if the trunk girls had the local repeats directly at command. One operator to two trunks appears superfluously luxurious. The translators used are of the double-coil type with yoked cores, the resistance of both primary and secondary being 170 ohms.

To those who understand the possibilities of telephonic switching in the direction of rapidity, and are accustomed to think of demand and connection as a matter of three or four seconds only, the methods adopted in Berlin appear strange, even to the verge of incomprehensibility. The seven switch-rooms are numbered from 1 upwards, and a subscriber is represented in the list by two numbers, firstly that of his switch-room, secondly that of his line, so that in the same town the same series of numerals is repeated seven times and distinguished by an index number, like so many logarithms. Indices, consisting of short words differing widely in pronunciation—such as the names of colours, of jewels, of rivers, anything—would be much more distinctive and less liable to be misunderstood than a constant repetition of numerals. That confusion is apt to arise is obvious from the rule which enjoins the calling subscriber to mention the number *and* name of the switch-room to which the person he wants is connected. Thus, to quote the rule, No. 3 switch-room must be asked for in a ten-syllable formula, 'Amt drei, Oranienburgerstrasse.' The following indicates the steps of a Berlin connection through *one* switch-room when the fates are propitious and the course of telephony runs smooth. A wants B.

Operation 1.—A takes one of his two telephones off its hook and applies it to an ear.

[He is instructed to do this, but is not told *which*. If he happens to take the left-hand one—and a stranger would be as likely as not to do so—he cannot ring the exchange, and naturally does not get any answer. It is true that in another part of the instructions he is advised to leave both telephones in their places when not corresponding, and in any case to leave the one on the movable hook, as otherwise the bell cannot be rung ; but this is not in the specific directions for obtaining a connection.]

Operation 2.—A turns the crank of his magneto 'slowly and at most once.'

[The instructions are emphatic as to the necessity of ringing slowly and only once, 'in order not to hurt any officers or subscribers.' It seems that some of the instruments are arranged so that people handling them are apt to get their bodies into circuit, and that when the magnetos were first put in, divers subscribers were unwittingly almost electrocuted by their friends. One is said to have gone to answer a call from a debtor whom he was pressing for payment and received a shock, which for a time he persisted in regarding as intentional and designed to close the account even more summarily than he was proposing to do.]

Operation 3.—A takes off the second telephone and applies it to his other ear.

[The Berlin telephones weigh nearly two pounds each.]

Operation 4.—Fräulein (answering ring): 'Here office.'

[The operators are habitually addressed as 'Fräulein.']

Operation 5.—A (who has all the time kept both phones to his ear): '9014, Verwaltung des Ritterguts.'

[The subscribers are directed to state the number *and* name of the person they want.]

Operation 6.—Fräulein : 'Please call.'

Operation 7.—A hangs up one phone, keeping the other to his ear.

Operation 8.—He turns his crank 'slowly and at most once.'

[As it is the left phone he must hang up, and is instructed to keep the other to his ear, he is necessarily compelled to turn the crank with his left hand.]

Operation 9.—B: 'Here Verwaltung des Ritterguts; who there? Please answer.'

[Subscribers are recommended to close every remark with the words 'please answer' until they reach the final one, which should be followed by 'finished.']

Operation 10.—A takes off his second phone and commences talk.

Operation 11.—(After conversation.) A and B hang up both their phones.

Operation 12.—A and B now each turn their cranks 'three times, by jerks, very quickly.'

[This they do regardless of consequences to officers and to each other, and yet with fear and trembling, for have they not been already told in plain German black and white that 'in order not to hurt officers and subscribers' they must ring 'slowly and only once'? Is it possible that, like Genesis, the Berlin book of instructions has been written by two authorities, the one oblivious of what the other has said?]

When, as in about five cases out of seven, the connection has to pass through two switch-rooms, A has to ask his operator for the room to which his client is joined in these terms: 'Office three, Oranienburgerstrasse,' or 'Office seven, Blankenfelden-strasse.' The first operator thereupon rings the second upon one of the junction wires between the two rooms, and A, upon finding himself through, prefers his request for the person he wants to the second girl.

If on the completion of a conversation another connection is wanted, half a minute must (according to the regulations) elapse after the ring-off is given before the operator can be rung up again. Such a regulation is a practical admission of the unsuitability of the system employed for a busy telephone exchange. With the Mann system, as used by the Mutual Telephone Company at Manchester, two separate connections could be obtained and got rid of within the half-minute so lightly wasted at Berlin, a short conversation being held on each occasion. But in practice, according to the author's observation, this regulation is neglected. In the telephone-room at the Central Hotel, already alluded to, a fresh customer seizes the crank, and oblivious of consequences to officers and subscribers alike, begins to twirl it vigorously as soon as the place is vacated by his predecessor, although much more than half a minute frequently elapses before any tangible result is obtained.

In the other towns the method of procedure is much the same, but (except in Hamburg) the battery press-button takes the place of the magneto.

In the suburban intercourse the calling subscriber is put

through to the town in which his client is located, and asks the connection from the operator there. In the trunk service he gives all the particulars to his own operator, and is rung up by her as soon as the connection is ready.

During thunderstorms traffic is suspended. The subscribers are recommended not to touch their instruments, and the operators are forbidden to answer any calls while a storm continues.

HOURS OF SERVICE

In this matter Germany is very far behind Great Britain and the age generally, Berlin being open only from 7 A.M. till 10 P.M. The principal suburban switch-rooms have the same service; others are open from 7 A.M. till 9 P.M., and others again from 7 or 8 A.M. (according to the season) till 9 P.M. In the provinces, the hours in the larger towns are from 7 A.M. (summer) or 8 A.M. (winter) till 9 P.M. These arrangements mean that for nine or ten hours out of every twenty-four the vast capital sunk in the German exchanges and trunk lines is lying idle and unproductive, while the subscribers are deprived of some of the most important of all the applications of the telephone. Other countries can find traffic for their lines during the night, and so, no doubt, could Germany, if the effort were made, or even if the opportunity were afforded and the effort left to the public.

SUBSCRIBERS' INSTRUMENTS

These generally consist of a battery-push and trembling bell microphonic transmitter, two spoon-shaped double-pole receivers (weighing from 23 ozs. to 2 lbs. each), and a separate battery-box or cupboard; but in Berlin and Hamburg magneto ringers have replaced the battery-pushes to a large extent, although the trembling bells are for the most part still retained. The general appearance and internal arrangements may be gathered from figs. 69, 69A, and 70, which represent wall- and table-instruments respectively. The battery instruments are similar in appearance, a push-button occupying the place of the magneto spindle. The instruments represented are by Messrs. R. Stock & Co., but the design is that of the Imperial Post Office, and Messrs. Siemens & Halske, Mix & Genest, C. F. Lewert, and others supply instruments of exactly the same type. The workmanship in every case is superior; let the design make such impression as it may on telephone engineers. Many of the instruments have been converted from the battery form at a cost, the author was told, of 65 marks (3*l.* 5*s.*) apiece. The German Government could have been supplied with new and complete instruments, comprising magneto, battery-box, backboard, good carbon transmitter, double-pole receiver and cord, of better design and equal workmanship from England, America, Belgium, or Sweden, delivered free in Berlin, for 3*l.* 3*s.*, or even less. Sometimes the conversion has been effected by placing a magneto in a separate box on the top of the battery instrument; in these cases the crank-handle is at the right-hand side of the instrument, as it should be, but too high up, while the appearance is ungainly. The introduction of magnetos was strongly objected to by the subscribers, who found that they often got unpleasant shocks from them. That there was something more than imagination in this appears evident from the instruction in the Berlin list, already quoted, to 'ring slowly and only once to avoid injuring officers and subscribers'! It is not often that the comic element intrudes into telephone subscribers' lists, and we are here

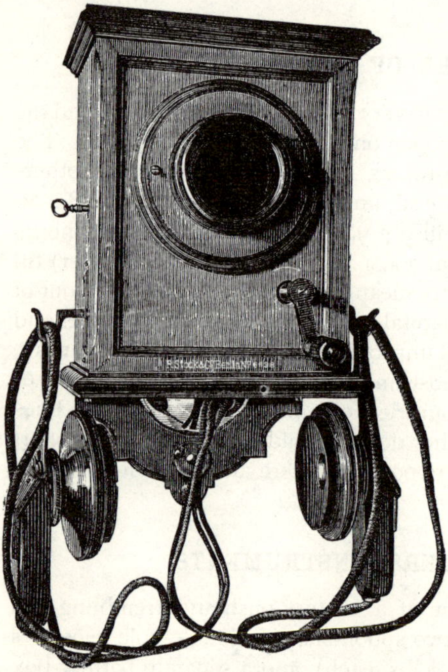

Fig. 69

Fig. 70

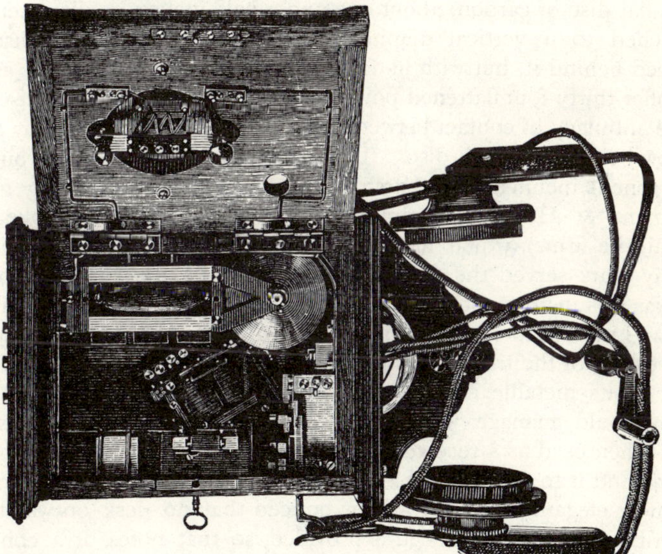

Fig. 69A

under a distinct obligation to concede a 'record' to our Berlin friends. As the difficulty is not one that causes trouble elsewhere, it is presumably due to faulty arrangement of parts in the German instruments. When ringing batteries are used, the cells, eight to twelve in number, are contained in a small cupboard placed on the floor immediately below the instrument. The cupboard is about two feet high, and has a veneered front of decorative wood, with ornamental mouldings, so as to look somewhat like a piece of ordinary furniture. The automatic switch is of Morse-key pattern, with top and bottom anvil contacts, a form which, it will be remembered, was adopted in the American instruments of 1879 and 1880, and which was speedily abandoned in favour of rubbing surfaces owing to the facility with which the original ones choked with dust. The transmitters are most often of the familiar Mix & Genest type, two carbon blocks, mounted on a vertical wooden diaphragm, carrying three horizontal pencils backed by silk or felt packing and an adjustable spring ; but there is also a transmitter by Siemens & Halske, which consists of a flat disc of carbon, about one and a half inches in diameter, attached to a vertical diaphragm and touching a similar disc placed behind it, but with its face cut into lozenge pattern so as to offer thirty-four flattened points to the pressure of the front disc. The intimacy of contact between the two plates is adjustable by a screw behind the back disc. This transmitter speaks loudly, but the tone is inclined to be harsh. The receivers are universally of Siemens & Halske's admirable double-pole type (but supplied by all the firms), which has been often described, and which for many years served the German Post Office as transmitters also. It was this instrument which enabled the Mutual Telephone Company, Limited, to open its Manchester exchange in 1891, before the expiry of the transmitter patents, and to obtain better speaking on its metallic circuits than the National Telephone Company could manage with Blake microphones and single wires. But when used as a receiver its weight (23 ozs.) and shape do not commend it to those who have been accustomed to light receivers of more elegant form. It will be noticed that no desk on which a writing pad can be placed is provided, so that notes of a conversation cannot be taken, and the use of reference books or papers is rendered difficult. The instruments have the following resistances: Induction coil, 1 and 200 ohms; receiver, 200 ohms; generator armature, 200 ohms; trembling bell, 170 ohms.

OUTSIDE WORK (LOCAL)

A feature of the German outside work is the manner in which the central stations are often adorned, or at least made striking in appearance, by special and costly domes and towers in iron or steel, the number of which is constantly increasing as new stations are opened or old ones rebuilt. Fig. 71 represents a wire fixture of this nature, and gives a good idea, although there is some differences in detail, of the telephone dome at the Oranienburgerstrasse switch-room, Berlin. It is erected on a tasteful brick turret at the corner of the Artilleriestrasse. Painted green picked out with gold and studded with white insulators, the whole produces an effect which is decidedly pleasing. At the Moabit switch-room there is a somewhat similar fixture, but the ironwork is in the form of a square steeple and not domed. The remaining five central station fixtures in Berlin are ordinary affairs enough; the only one worthy of any remark, and that only on account of its size, is at Blankenfeldenstrasse, which is an immense oblong, containing forty-two wooden uprights connected by iron bars.

FIG. 71

Fig. 72 shows the telephone tower at the new postal buildings at Frankfort-on-Main, which few, perhaps, will consider beautiful. It is not quite certain whether the attachment of hundreds

Fig. 72

of aerial wires to the summits of brick or masonry towers is not open to objection. The vibration is not only great, but incessant, while it is difficult, in most cases impossible, to secure an equal stress all round. When this cannot be done there is a permanent strain on the tower. The great structure at Stockholm is built on steel pillars carried down to the ground in order to avoid trusting to brickwork or masonry.

Having given the Imperial Administration every credit for the enterprise and ability which stand revealed in its exchange fixtures, the author is constrained to lament that the same class of work has not been considered necessary for the ordinary overhouse standards. These are decidedly wanting in the most important of all qualities—strength. Figs. 73, 73A, 74, and 74A show the single and double standards respectively with their fittings and details to scale. There are also standards with three and even four uprights, but these are simply extensions of the double. The standards consist of iron or steel tubes, three inches in external diameter, which are bolted or clamped to the rafters or other suitable portions of the roof. The arms are formed of two flat iron bars riveted together, the rivets passing through spacing rings, and having a stiffening piece cut out to fit the circumference of the tube, fastened at the middle by two rivets which pass through the stiffening

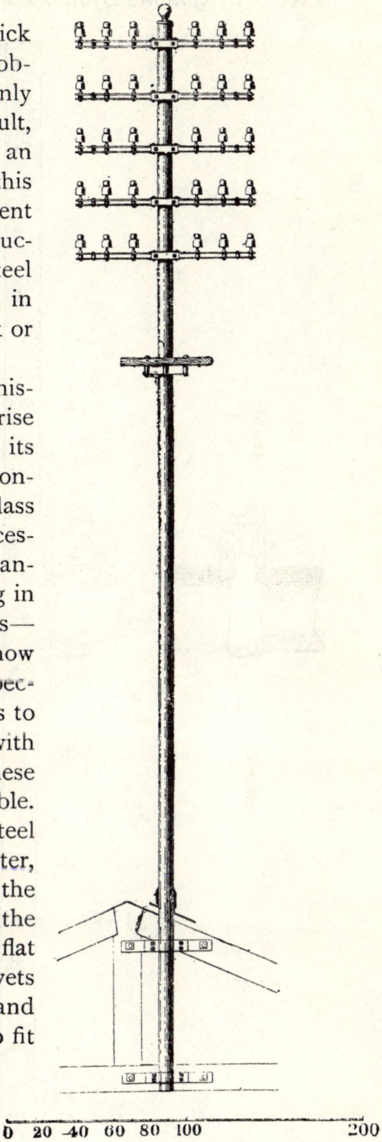

FIG. 73.—Scale of 200 centimeters.

P

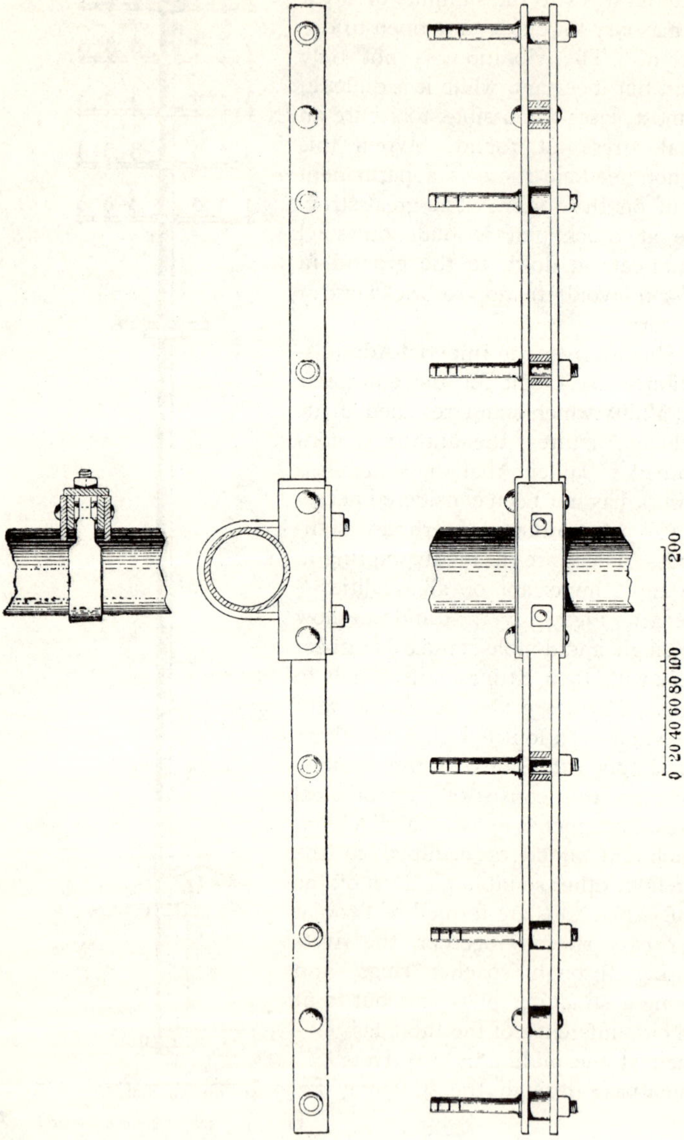

Fig. 73A.—Scale of 200 millimeters.

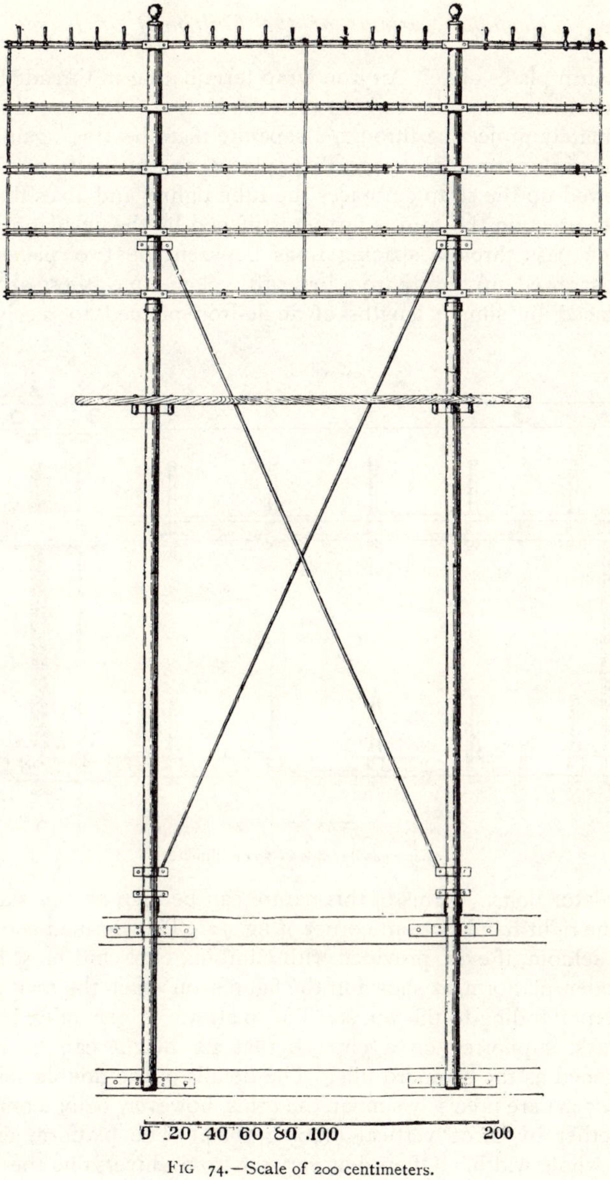

FIG 74.—Scale of 200 centimeters.

P 2

and arm plates alike. An iron strap terminating in threaded bolts passes round the tube and between the plates of the arm, the bolts ultimately projecting through a separate plate bearing against the front of the arm. Nuts are then placed on the bolts, and being screwed up the strap embraces the tube tightly and fixes the arm. After erection the arm is further stiffened by the insulator bolts, which pass through spacing rings between the two plates, and are screwed up tightly from beneath. Sometimes these arms are replaced by simple lengths of angle-iron pierced to receive the

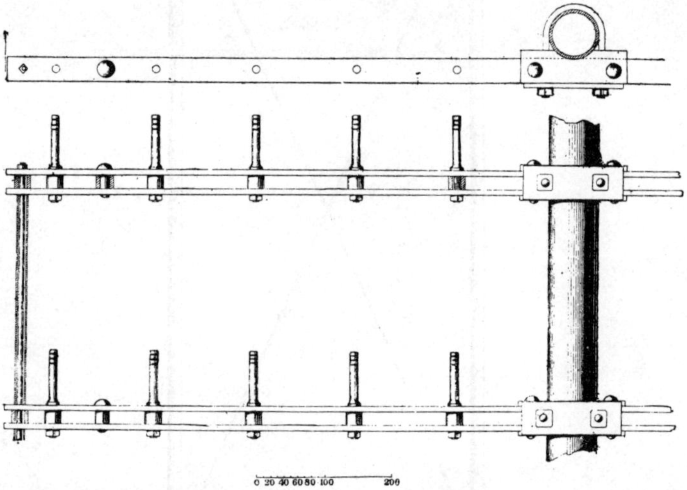

FIG. 74A.—Scale of 200 millimeters.

insulator bolts. Arms of this nature can be seen on the standard in the right-hand bottom corner of fig. 72. The German standards are seldom, if ever, provided with climbing clips, but most have a wooden platform, as shown in the figures, on which the man stands when attending to the wires. The platform is generally, but not always, supported on a clip, so that its height can be readily lessened as the standard fills. The details of the double standard (fig. 74A) are precisely similar, the arms, however, being connected together by three vertical bracing rods. The platform extends the whole width. Often, but very far from universally, the cross-braces shown are added, but with variations, as they frequently do not extend up nearly so far as shown. In Mannheim and Frankfort-on-Main cross-braces are generally present ; in Berlin and Leipzig they are mostly wanting. The standards, which seldom exceed eighteen or twenty feet in height, have a neat appearance when newly erected, but they are always most inadequately stayed and frequently have no stays at all, although loaded sometimes with over 200 wires. They are kept straight at first by adjusting the tension of the wires on either side, but often, as might be expected, fall out of shape. There is (October 16, 1894) a single three-inch tube on 23 Kaiserstrasse, Frankfort-on-Main, carrying seven arms and thirty wires, without a stay of any kind ; it contains more curves and angles than a box of drawing instruments. A few roofs off, on No. 27, there is another, carrying two arms and seven wires, almost as bad. On 34 Französischestrasse, Berlin, there is a triple standard, carrying eleven long and three short arms and 336 wires, provided with only two inadequate and wrongly-placed stays. The tubes are badly bent and leaning in various directions, and the arms are all awry. As seen from 49 Markgrafenstrasse, this standard reminds one of the human figure—there is not a straight line in it. Such instances might be multiplied. He would be a bold man who would insure the Prussian overhouse system against a winter's storm accompanied by damp snow, or even an hour's downfall of damp snow unaccompanied by wind. A visitation of damp snow followed by a gale would certainly lay the whole in ruin. There is no provision against the destruction of a span by fire or tempest. The wires are made to balance, one span against another, and there is nothing to save the standards in the event of the stress on one side becoming suddenly much greater than that on the other. In such a contingency the three-inch tubes would collapse like paper and crumple up. The Dutch, who use the same type of fixture, are much wiser in this respect (fig. 83, Dutch section). The work is, nevertheless, very pretty to see. On 31 Kl. Fleischergasse, Leipzig, there is a double standard carrying ten long and one short arm and 202 wires, practically without stays, which is perfectly straight and regular. In Berlin, where the overhead wires are necessarily extremely numerous, junction standards consisting of eight or twelve three-inch tubes arranged in a square and connected by long arms are sometimes placed at the meeting of two or more routes, the wires being joined through between the different sides of the square by insulated leads going down boxing on one side and up on the other. These structures are necessarily much stronger than simple double or triple standards, but the tubes are unstayed and not braced together except by the arms, so that the sudden destruction of several hundred wires on one side would probably cause a collapse or at least a severe distortion. Trunk lines are frequently carried on short arms attached to one or both tubes of a double standard above the long arms. The standards are sometimes connected to earth as a precaution against lightning. Noise and vibration seem to be experienced in the houses carrying standards, as the wires are frequently provided with dampers in the form of pieces of lead clamped on the wires two or three feet from the support. This may be due to the bolting of the tubes rigidly to the rafters, instead of allowing them to sit in a socket without any rigid fastening as is practised in Great Britain. The appearance of a standard carrying, perhaps, 200 dampers on either side of the insulators is more peculiar than pleasing. In ground pole work the author saw nothing striking in Germany. The poles appear to be uniformly of wood; frequently, when additional height is wanted, a tube is fastened to the top of a wooden pole, and in some instances double fixtures are treated in the same way, as shown in fig. 75. The top of the pole is grooved out for some two feet, the tube is laid in the groove, and iron clamps placed round both and tightly screwed up with bolts and nuts. Arms are either of the double-bar type (fig. 73A), or simple lengths of angle-iron. Ground poles are not earth-wired. Bronze wire of 1·25 mm. to 1·5 mm. gauge, supported on small double-shed white porcelain insulators, is now used for town work. Wires are led into subscribers' premises at the back whenever possible, joint cups being sometimes used. Underground work is being undertaken in Berlin, Hamburg, Cologne, Frankfort-on-Main, and other towns. The conduits are simply iron pipes, into which the cables are drawn, connecting draw-boxes and manholes placed from 100 to 150 meters apart. Numerous types of cable have been tried, mostly insulated with india-rubber or gutta-percha served with metal foil for earthing. In some cables the wires have been placed parallel, but in later types twisting in pairs or in fours has been introduced, together with, in some cases, paper insulation. The underground work, so far, is understood not to have been an unalloyed success, which is not surprising when the plan usually followed has been to suppress one evil—overhearing—by exaggerating another—capacity. The growing importance of the trunk system will eventually force a resort to metallic circuits, and then the want of foresight which has prevailed will be deplored. The cables have been supplied chiefly by Siemens & Halske, Felten & Guilleaume, Western Electric Company, and Franz Clouth; the workmanship in every case may be pronounced excellent. One of the cables employed has a conductor composed of three tinned copper strands of 5 mm. diameter, insulated with one layer of white Para rubber and one layer of vulcanised, then wrapped in prepared tape, and all vulcanised together. Afterwards, each wire is taped with tin-foil. The cable consists of seven bunches of four wires, coloured blue, green, red, and white, each bunch arranged round a bare copper wire of 1 mm. diameter, with which the foil comes in contact. The whole is wrapped in impregnated tape and drawn into a leaden tube of 1·5 mm. thickness. The copper resistance per kilometer is 31 ohms; capacity, ·25 mf.; and insula-

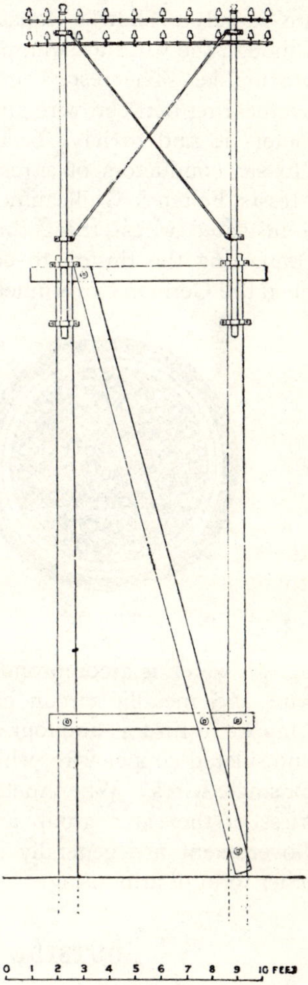

FIG. 75

tion, 250 megohms. Mr. Clouth has recently supplied several cables of this nature for use under the streets of Cologne, which are stated to show a capacity of only ·075 mf. per kilometer, although the wires are wrapped in tin-foil. The foil is the thinnest procurable. The results are said to be excellent. There is no overhearing between wire and wire, and a distance of sixty kilometers is said to have been spoken over. This cable contains fifty-six conductors of a resistance of 21·5 ohms per kilometer. Messrs. Felten & Guilleaume's cable of this type has generally four uninsulated wires strung through it (fig. 76) for the purpose of connecting the tin-foil to earth. This latter firm has also supplied the German Government with cable of the kind illustrated in

FIG. 76

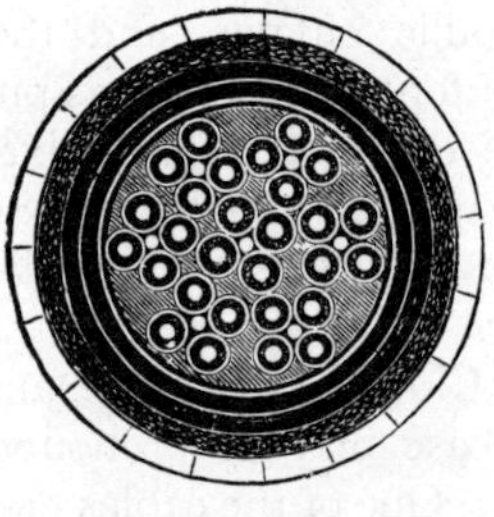

FIG. 77

fig. 77, which is a compromise between the anti-induction single wire and metallic circuit classes. Each insulated conductor is wrapped in tin-foil, and four such conductors are twisted round an uninsulated copper wire, which is earthed when the cable is used for single wires. When metallic circuits are required the opposite wires of the same group are looped. Cables for the German Government are generally sheathed in flat iron wires or some other form of armouring.

OUTSIDE WORK (TRUNK)

There is little calling for remark about the trunk line work. The wire used is generally 3 mm. copper, but for the long lines, like the Berlin-Cologne, Berlin-Munich, Berlin-Vienna, and

Berlin–Memel, the gauge is 4 and 4·5 mm. The insulators are large double-shed of white porcelain of German manufacture. The trunks generally follow the railways and are supported on ordinary wooden poles, the wires being crossed at intervals.

STATISTICS

The union of the telephone with the telegraph is so intimate in Germany that no separate account is kept, or at least published, of the exclusively telephonic receipts and working expenses. It is consequently impossible to know whether the system is remunerative or the reverse.

December 31, 1893, is the date of the following—the latest—figures relating to lines, instruments, and volume of traffic.

Exchange areas	366
Switch-rooms	384
Exchange subscribers	75,121
,, subscribers' instruments	80,782
Official and service instruments	12,349
Exchange instruments of all kinds in connection	93,131
Public telephone stations	164
Instruments in stock exchanges	106
Trunk lines	432
Length of local or town routes, kilometers	13,162
,, wire of all descriptions, kilometers	142,269

Number of talks for year

Local	313,628,062	372,710,240
Trunk	59,082,178	

At the end of 1894 the exchange instruments working in the chief towns numbered approximately :—

Berlin	25,000	Frankfort-on-Main	2,700
Breslau	2,300	Hamburg	9,200
Cologne	2,800	Leipzig	3,320
Dresden	3,300		

X. GREECE

To date of writing (March 1895) no telephone exchange has been opened for public use in Greece, but a small one exists for police purposes only between Athens and the Piræus. A law was, however, passed in 1893 reserving the establishment of a public exchange in Athens and the Piræus to the State, but authorising the granting of concessions for the other towns to individuals or private companies.

XI. HOLLAND

HISTORY AND PRESENT POSITION

TELEPHONICALLY, as in other respects, Holland is one of the most interesting countries on the Continent. The industry and the proverbial ability of the Dutch to adapt means to ends have resulted in the telephone being brought, and that without State intervention, within the reach of all, for surely that point has been nearly approached when annual subscriptions have been reduced as low as *2l. 9s. 7d.*, including the supply and maintenance of wires, apparatus, and all expenses. For a parallel it is necessary to go to Scandinavia, and it is worthy of remark that the lowest rates are everywhere associated with companies, not with Government administrations. The sole exception is the case of Switzerland, but in that instance the rates are low only for those who use their telephones but little : for the busy firms the $\frac{1}{2}$*d.* per call mounts up during the year to a total that exceeds anything known in Holland or Scandinavia. That is, of course, as it should be ; the important firms paying, as they can well afford to do, in proportion to their actual needs. When an all-round rate exists the poorer folk are really taxed for the benefit of their richer brethren, and such a rate possesses no other merit than convenience.

The Dutch Government, until the advent of the era of trunk lines, did not attempt to participate at all in the telephonic game. It granted concessions to companies and, in some instances, to private firms and even individuals, for definite towns and districts, within which they were secured from competition. The International Bell Telephone Company obtained Amsterdam, which it subsequently handed over to a local association, the Netherlands Bell Telephone Company, to which fifteen of the other chief towns have since been conceded. Messrs. Ribbink, van Bork & Co., manufacturing electricians of Breda and Amsterdam, hold and work concessions for eleven of the smaller towns, the exchanges in which, under the fostering influence of a 2*l.* 17*s.* 10*d.* rate, have obtained respectable proportions. The historic town of Zutphen, population 17,004, has a model exchange of 141 instruments on the same subscription. Maastricht is worked by the Maastricht Telephone Company, also on 2*l.* 17*s.* 10*d.* Nijmegen, which, with a population of 34,128 and a 2*l.* 17*s.* 10*d.* rate, has 450 subscribers, belongs to Mr. J. W. Kaijser. Alkmaar and Helder are in the hands of Mynheer Jan Sōt, who carries off the palm for low subscriptions with 2*l.* 9*s.* 7*d.* per annum, everything included. It is perhaps superfluous to remark that Mynheer Jan Sōt possesses none of those autocratic powers in respect to way-leaves which apologists in this country have so liberally, if gratuitously, endowed foreign telephonists generally by way of accounting for the low rates on which they are able to live and thrive.

The concessionaries have to obtain licences both from the State and the local authorities, power being reserved to the State to revoke its grant at any time. The municipal licences are for from fifteen to twenty-five years. The concessionaries' tenure is therefore somewhat uncertain, but so far the State has not intervened anywhere. No royalty is payable to the Government unless a subscriber's line exceeds five kilometers in length. It is then deemed to partake of the nature of a trunk line, and the State makes an annual charge of 1*l.* 13*s.* for the sixth and 16*s.* 6*d.* for each additional kilometer. The municipalities generally stipulate for a few free connections in return for their licence (which, however, usually carries with it valuable way-leave privileges); the Town Council of Amsterdam alone exacts a money payment, and this is no less than 2*l.* 1*s.* 9*d.* per annum on every primary subscription of 9*l.* 14*s.* 2½*d.* obtained by the company in Amsterdam. If a subscriber for any reason pays more than the unit rate, the company keeps the whole of the excess. In addition, the company has to give the Amsterdam Corporation no less than thirty-one free connections and a reduction of 50 per cent. on any above that number. In return, way-leave is granted for the streets and public buildings.

The history of the Dutch trunk lines is rather involved. The Government had conceived the idea at an early date that trunks meant ruin to telegraphic traffic, and fell into the usual fallacy that because the telegraph system belonged to the public it was necessary and essential to protect it against the public. That is to say, that which was no longer the best and fittest for certain purposes must, in the interests of the nation, be fostered and protected by artificial means to the damage of the new and worthier method of communication, because, forsooth, the public had originally paid for the obsolete system.

As a consequence, the action of the Dutch Government was not encouraging. Owing to financial or other reasons it was not at that time deemed politic for the State to undertake the construction of the trunks; but it was not till 1887, when the commercial community had long been clamouring for communication, that it was resolved to allow the Netherlands Bell Company to connect Amsterdam with Haarlem. The conditions imposed were sufficiently onerous. The company was to erect and maintain the line, pay over half the profits to the State, and, moreover, undertake to make good the full value of any diminution of telegraphic traffic that might occur between the points connected. The telegraphic traffic was further protected by the imposition of high rates. Messages were not to be paid for singly, but all users of the trunks were to pay an annual rate equal to the local subscription in the towns to which they spoke. Did not the trunk make a Haarlem man virtually a member of the Amsterdam exchange, and an Amsterdam man a participator in that of Haarlem? Then let the Haarlem subscriber pay the Amsterdam rate and the Amsterdam subscriber the Haarlem rate in addition to his own, and ends would meet. Notwithstanding these conditions traffic flourished and, strange to say, without producing any marked effect on the telegraphic revenue. At the end of the first year the company paid a small sum to the State to put the telegraphic receipts on a level with those of the previous year; but during the second year the telegraph recovered itself, and no further payment was demanded. Then the Government acquired a little courage and consented to Amsterdam being connected with the Hague and Rotterdam, a work which the company successfully achieved in the face of considerable difficulties. The local authorities along the route raised many objections to the planting of the poles, and no less than seven submarine cables had to be laid across the intervening rivers and canals. Experience again demonstrated that, although the telephonic traffic was considerable, the effect on the telegraphic revenue was both slight and transitory, and the Government at last determined to yield to public opinion and bring about the linking up of the other principal towns. But, although the company had proved at its own expense and risk the existence of a telephonic demand and the practicability of satisfying it, the Government determined to keep the trunks so far as possible in its own hands. Apparently there were obstacles to such a policy being given effect to openly and without reserve ; so it was decided to allow the Netherlands Bell Company to continue constructing and working, on the understanding that the State should supply the material and the company the labour, the company receiving 4 per cent. per annum on the cost of their share of the work by way of interest, and agreeing to make over the lines to the State at any time on reimbursement of their outlay, the amount of which was to be determined and certified on the completion of each trunk. This is a good bargain for the company, since it gets back the full value of its work, whatever the state of the lines may be when eventually taken over. At the same time (November 1889) the annual trunk subscriptions were abolished and the present payment per time unit substituted. The trunk lines go straight into the company's exchanges and are worked by its employees without interference of any kind. The lines, however, are maintained by the State. The receipts are divided, 75 per cent. going to the State and 25 per cent. to the company. This policy has resulted in the linking up of all the sixteen towns conceded to the Netherlands Bell Company and one other.

The trunk traffic is large, but the State officials are not now disposed to say that it has any bad effect on the telegraph revenue. The impression rather prevails that the efficacy of the telephone as a general feeder and stimulant over the whole system compensates for any diversion of traffic between particular points. Exact comparisons are not possible, as, since the telephone trunks came into operation, the telegram tariff has been reduced and receipts have fallen, although messages have multiplied. The Dutch internal telegram tariff is 4·95*d.* for ten words, with ·59*d.* for each additional word ; but for telegrams passing between parts of the same town the charge is only 2·97*d.* for ten words, with ·198*d.* for each extra word.

The subscribers' lines in all the large towns are single, but the Netherlands Bell Company recognises the superiority of the metallic circuits, and some of its recently constructed exchanges have been fitted with it, as all future ones will also be. The Zutphen Company has adopted the metallic circuit; but the other concessionaries continue to run single wires. In Amsterdam there is a considerable amount of underground work, the extent of which is growing rapidly. To date of writing, no international trunk lines actually exist, but an agreement has been signed with Belgium by which the Dutch and Belgian centres will be brought into communication at as early a date as possible. The rate agreed upon, as between Amsterdam and Rotterdam on the one hand, and Antwerp and Brussels on the other, is 2*s.* 5*d.* per three minutes. Last autumn experiments were tried with the view of establishing telephonic communication with England by means of a direct cable, the Dutch being averse to adopting a route *viâ* Belgium. It was found possible to telephone fairly well, using ordinary instruments, through the old telegraph cables between Lowestoft and Zandvoort and Benacre and Zandvoort, so that, given a special telephonic cable, the practicability of the scheme is beyond doubt. The Dutch Government has given the promise of a concession to Dr. Hubrecht, managing director of the Netherlands Bell Telephone Company, for the works on the Dutch side, and that gentleman proposes that an Anglo-Dutch company shall be formed to lay a cable between Aldborough in Suffolk and the Hook of Holland, and establish the necessary connecting lines on both sides. But nothing can be done without the consent of the British Post Office, which now has several memorials on the subject before it.

SERVICES RENDERED BY THE NETHERLANDS BELL TELEPHONE COMPANY

1. **Local intercourse between the subscribers and public telephone stations of the same town.**

2. **Internal trunk line communication.**—Seventeen towns, with a total of 4,700 subscribers, had been put into communication at the end of 1894, these, with the exception of Nijmegen, being all those conceded to the Netherlands Bell Company. The number of trunk messages exchanged during 1892 was 71,833; during 1893, 79,424; and during 1894, 85,142. The trunk regulations are in some respects peculiar to Holland. For instance, subscribers who use the trunks pay an annual subscription of 16*s.* 6¼*d.* in addition to the charge per connection, which, for the distances spoken over, is high—9·9*d.* per three minutes—compared with that which obtains in some other countries. When a called subscriber does not answer within one minute, the caller is debited with half a fee, 4·95*d.* Express talks are allowed, a subscriber being given precedence over any others who may be waiting their turn in return for a double fee; but no connection must exceed six minutes in duration if others are waiting. Deposits to cover conversations must be made in advance, the minimum deposit accepted being 4*l.* 2*s.* 3½*d.*

3. **Public telephone stations.**—Of these there are eight in Amsterdam, six in Rotterdam, six in the Hague, four each in Groningen and Utrecht, and from one to two in each of the smaller towns. These stations are frequently situated in the booking halls of the railway stations and at the post and telegraph offices, and are available both for local and trunk talks. Automatic boxes for checking payments are not used, the charges being payable to an attendant in cash or in tickets. At the Amsterdam and Rotterdam Bourses messengers are in attendance to fetch to the telephone station members who may be asked for. Persons so called, if they come, have to pay the tariff charges. To facilitate this fetching system a plan of the Bourse, on which each member's place is indicated by a number, is printed in the subscribers' lists, and the number of the member wanted must be mentioned when asking for him. The messenger hands the member called a dated

and timed ticket bearing the name and telephone number of the person who wants him. Trunk talks are subject to the same charges and regulations as those made from subscribers' offices.

4. **Telephoning of telegrams.**—This is an important service, but, owing perhaps to the higher charges and less elastic regulations, the traffic does not attain the proportions reached in the neighbouring kingdom of Belgium. In 1894 the total number of telegrams handled by the Netherlands Bell Company was 104,367, of which Amsterdam was responsible for 66,348. Senders of telegrams have to deposit the estimated value of their traffic in advance, and are not allowed to outrun their deposits. The company's operators attend at the telegraph office to receive and transmit telegrams by telephone ; the State charges nothing for the space occupied, nor for lighting or warming. In connection with the State telegraphs there is a little facility granted to the public which appears peculiar to Holland. Senders of telegrams from any of the Dutch towns, when addressing a telephone subscriber in any of those towns in which the telegraph office is connected to the telephone exchange, may order their message to be telephoned on its arrival to its addressee even when the latter does not subscribe to the ordinary telegram service. To take advantage of this regulation it is only necessary to write the letters T. B. in brackets before the address and pay for them as two words. Should it not be possible to get the addressee to answer his bell, the message is delivered by messenger in the ordinary way.

5. **Time service.**—All the Netherlands Company's exchanges receive the correct time from Amsterdam Observatory once a day. Subscribers wishing to regulate their clocks are told the time on demand. Nothing is charged for this service. It is nevertheless not without an importance to those subscribers who use the trunks a good deal and like their monthly accounts, made up from the operators' registers, to tally with their own notes.

TARIFFS [1]

1. **Rates for local exchange communication.**—The rates levied by the Netherlands Bell Company were approved by Royal

[1] One florin = 1*s.* $7\frac{3}{4}$*d.*

Resolutions in 1881 and subsequent years. Instead of increasing in proportion to the mileage beyond a defined radius, as in most countries, a system of division into districts has been effected, and a definite rate allotted to each district. Thus in Amsterdam there are three grades of subscription:—

	Per annum £	s.	d.
Subscribers located within Amsterdam proper	9	14	2½
,, ,, in Nieuwer-Amstel	12	6	10½
,, ,, Ouder-Amstel	20	11	5½

In Rotterdam:—

	£	s.	d.
Within the city	9	14	2½
In Kralingen	12	6	10½

In Dordrecht:—

	£	s.	d.
Within the town	4	3	11
In Zwijndrecht	8	4	7

The remaining towns of the Netherlands Bell Company have a single tariff. They are:—

	Per annum £	s.	d.
The Hague	9	1	1
Arnhem, Baarn, Bussum, Groningen, Haarlem, Maassluis	4	19	1½
Schiedam, Utrecht, Zaandam	4	19	1½
Amersfoort, Hilversum, Vlaardingen	3	6	8

The towns worked by Messrs. Ribbink, van Bork & Co. are:—

	Per annum £	s.	d.
Breda, Deventer, Enschede, s'Hertogenbosch, Leeuwarden	2	17	10
Leyden, Middelburg, Tilburg, Flushing, Zwolle	2	17	10

Mr. Kaijser has one exchange:—

	Per annum £	s.	d.
Nijmegen	2	17	10

The Zutphen Telephone Company has one exchange :—

	Per annum £ s. d.
Zutphen	2 17 10

The Maastricht Telephone Company has one exchange :—

	£ s. d.
Maastricht	2 17 10

Mr. Jan Sōt has two exchanges :—

Alkmaar	2 9 7
Helder	

The Dutch rates cover all expenses of installation and maintenance. They do not, at least to an unprejudiced or disinterested outsider, appear remarkable for extravagance or oppressiveness ; but such is the frailty of human nature, which for ever yearns for something not yet within its grasp, the subscribers are not satisfied, and hope to obtain better terms when the present concessions expire. On the other hand, the concessionaries appear quite satisfied. The Zutphen Company is making money, and Messrs. Ribbink, van Bork & Co. deplore the fact that Holland, telephonically speaking, is, at least pending the reclamation of the Zuyder Zee, nearly used up, and but few towns worth mentioning remain to be telephoned. This firm assured the author that their rate of thirty-five florins (2*l.* 17*s.* 10*d.*) pays them satisfactorily, and that they are willing to take as many new towns as they can get on the same terms, and would even agree, if the Government wished, to put in metallic circuits. It is well to state, however, that the firm are manufacturing electricians and, there being no patent laws in Holland, make all the switch-boards and instruments they require in their own shops. Something, the manufacturer's profit, is saved in this way on the first cost of their exchanges. But the Zutphen Company without this advantage, and with first-class construction and instruments, contrives to make a profit on the same rate.

2. **Rates for internal trunk communication.**—To acquire the right to use the trunks a subscriber must agree to pay 16*s.* 6¼*d.* per annum in advance in addition to his local subscription.

Besides, each trunk talk must be paid for at the rate of 9·9*d*. per three minutes, irrespective of distance.

Express or urgent communications, by which the caller is given precedence of any others who may be waiting, are charged double rates. A half-fee is exigible for a connection asked for, but which cannot be had through no fault of the company. The right to use the trunks for a stated daily period may be acquired by annual subscription. Fifteen minutes' daily use costs 41*l*. 13*s*. 4*d*. per annum ; some newspapers subscribe as much as 500*l*. in this way.

3. **Rates at public telephone stations.**—No distinction is made between subscribers and strangers.

Local talk, 5 minutes	4·95*d*.
,, 10 ,,	9·90*d*.
,, 13 ,,	14·85*d*.
Each additional 3 minutes	4·95*d*.

Talks must not last longer than ten minutes if others are waiting their turn.

Trunk talk, per 3 minutes	9·9*d*.

The charge is irrespective of distance. Talks must not exceed six minutes in duration if the line is otherwise wanted. If the called subscriber does not answer within one minute, or if the connection cannot be had through no fault of the company, the caller must pay half-fee. Payments may be made in cash, or by tickets which are sold by agents appointed by the Company at a reduction of 20 per cent. Express talks are admitted on payment of double rate.

4. **Rates for telephoning telegrams.**—To enjoy this service, subscribers must pay 8*s*. 3*d*. annually in addition to their ordinary subscriptions ; this charge is, however, remitted to those who subscribe to the trunk service.

Each telephone despatched or received by telephone is charged ·99*d*., irrespective of the number of words. Copies of telegrams telephoned to subscribers are not delivered unless specially desired. In that case a copy is posted and the addressee debited with ·495*d*. If he does not care to wait for the post, he can have a copy immediately by sending to the telegraph office and paying ·495*d*.

WAY-LEAVES

Neither the Netherlands Bell Telephone Company nor any other of the concessionaries possesses compulsory powers: they have to beg and pay their way in the same fashion as the English companies. The Netherlands Bell Company inserts a clause in its agreements by which subscribers bind themselves to grant way-leave facilities on their premises, but it has not been found politic to enforce it strictly. The same company pays the Amsterdam Municipality no less that *2l. 1s. 9d.* per subscriber per annum, and provides no less than thirty-one free connections for the right to erect poles and lay cables in the streets and public places and to fix wires on public buildings. This does not obviate the necessity of going on private property, a privilege which has to be bought occasionally with a free exchange connection or payment of one florin (*1s. 7¾d.*) per wire per annum. The provincial towns deal more liberally with the company than Amsterdam does, and corresponding rights-of-way are usually granted in return for a few free connections to the municipal offices.

ROYALTIES

None are payable to the State unless a subscriber's line exceeds five kilometers in length. In such a case *1l. 13s.* is charged for the sixth, and *16s. 6d.* for each succeeding kilometer per annum.

SWITCHING ARRANGEMENTS

The switch-boards in Holland are not of the latest type; Amsterdam with nearly 1,700, and Rotterdam with nearly 1,000 subscribers being still worked with Gilliland boards. The reason is the company's undefined position in respect to the State. A new post and telegraph office is to be built in Amsterdam, to which it is proposed that the telephone exchange shall be removed. With such a shift in prospect, it is not to be expected that the company would go to the great expense involved in fitting a modern multiple board on its old premises. Much the same state of matters exists at Rotterdam. At Amsterdam, where the trunks chiefly concentrate, there is a special trunk table fitted for fifty lines. The key-board is shown in fig. 78. It is mounted with ten pairs of double-conductor plugs and cords, I to X, each pair being connected to a key—1 to 10. Six of these keys—1, 2, 3, and 8, 9, 10—are joined to the keys T1 T2 T3 and T8 T9 T10, which bring the translators (of the Landrath pattern) into circuit. By turning down the switches A and C the key-board is divided into two sections and may be attended to by two operators; when A and C are up and B down one operator can control the whole. L1 to L4 are listening keys, and cut off one side of a connection when the plugs are in; R1 to R4 are ringing keys. Fig. 79 (with

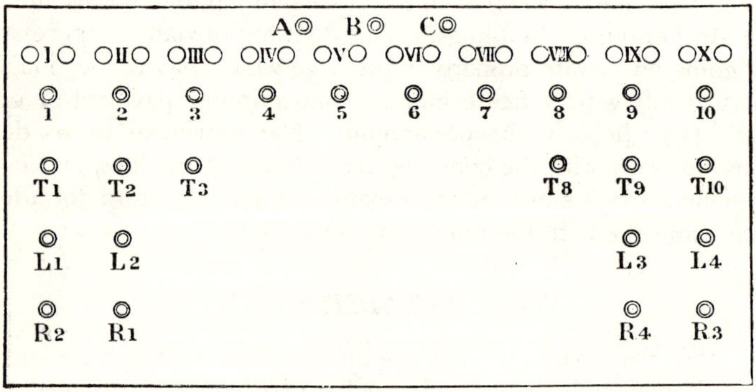

Fig. 7

the same reference letters) is a diagram of the general connections. The spring-jacks may have attached to them either metallic circuits or single earthed wires; thus W Y Z are metallic trunks and X an earthed wire going to the main switch-board for joining to the subscribers' single lines. It will be seen that the table allows of all necessary combinations—i.e. direct connection of two metallic circuits, of two single wires, and of a metallic with a single either through a translator or direct.

Subscribers are asked for by numbers; after receiving the operator's intimation that the connection has been made, the caller hangs up his phone, and himself rings his client's bell. After the talk is over he rings off in the ordinary way. When a subscriber is called he takes down his phone and speaks without ringing back. It is hard to accept this system as satisfactory. If a called person does not answer immediately, the caller continues

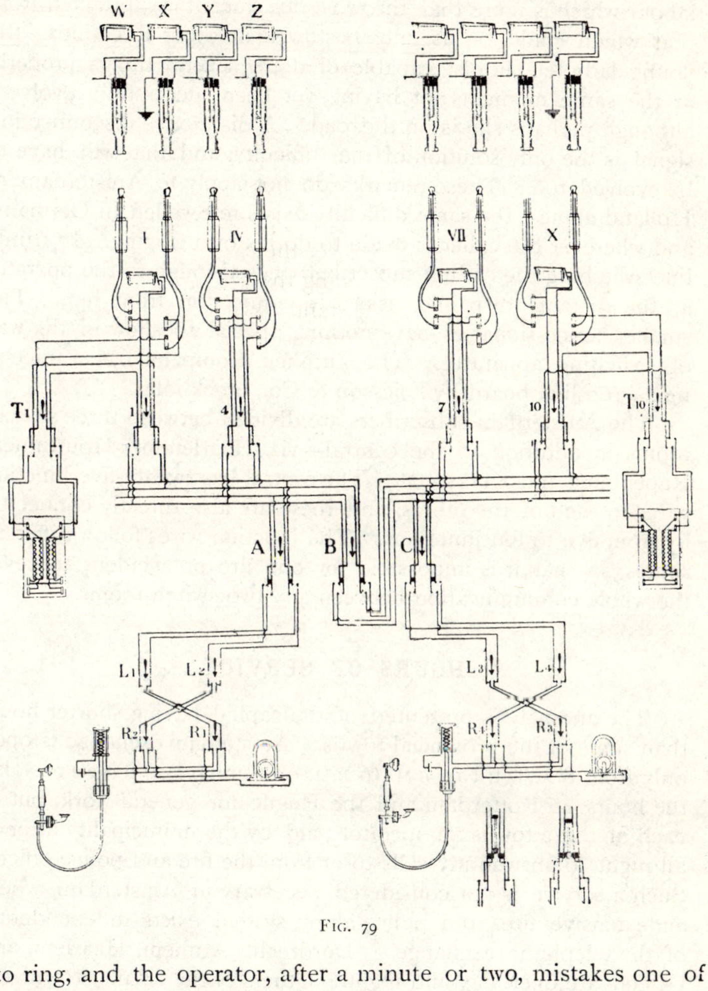

FIG. 79

to ring, and the operator, after a minute or two, mistakes one of these rings for a ring-off and disconnects, with the usual effect on the tempers of all concerned. If by force of tapping she learns the true position of affairs and does not disconnect, she has the useless labour of restoring the ring-off shutter every time it falls, labour which is more than thrown away, since it is subtracted from that which could be usefully bestowed in other directions—the young lady telephonist capable of doing several things properly at the same moment not having yet been successfully evolved, although perhaps she is on the road. A distinctive disconnection signal is the only solution of the difficulty, and that will have to be evolved too. These remarks do not apply to Amsterdam or Holland alone; the same difficulty exists in Sweden, in Germany, and wherever the caller is made to do his own ringing. In trunk-line switching the calling subscriber rings through to the operator at the distant town and asks his connection from her. The smaller concessionaries have nothing special to show in the way of switching apparatus. The Zutphen Company has a nicely-made 160-line board by Ericsson & Co., Stockholm.

The Amsterdam subscribers are divided between three switch-rooms in addition to the central—viz. Haarlemmer-Houttuinen, Rapenburg, and Kerkstraat. The central has twenty-five junction wires to each of the others, and these are also directly connected by from five to ten junctions. The junction wires follow different routes, so that it is impossible for one fire or accident to sever the whole communication between any two switch-rooms.

HOURS OF SERVICE

The anomaly is presented of the capital having shorter hours than some of the provincial towns. Amsterdam exchange is open only from 8 A.M. till 10 P.M. (6 P.M. on Sundays). These are also the hours at Rotterdam and the Hague for general work, but in each of these towns an operator paid by the municipality attends all night to answer any calls to or from the fire and police offices. Such a service is not considered necessary in Amsterdam, where an extensive fire- and police-alarm system exists independently of the telephone exchange. Dordrecht, Arnhem, Haarlem, and Utrecht are open day and night. Others of the smaller towns are closed during the day to allow the operator away for meals; thus at Zaandam the hours are 8.30 A.M. till noon, 1 P.M. till 5 P.M., and 7 P.M. till 9 P.M. ; and at Hilversum, 8 A.M. till 5 P.M., and 7 P.M. till 8.30 P.M. Messrs. Ribbink, van Bork & Co.'s exchanges are open from 7 or 8 A.M. till 9 or 10 P.M., according to local requirements. They all have, however, a night service for the fire and police offices and doctors. The Zutphen exchange is kept open continuously for all kinds of traffic.

SUBSCRIBERS' INSTRUMENTS

The Netherlands Bell Company now employ magneto ringers of substantial, but not uncommon, design, together with double-pole receivers and the Groof form of Hunnings transmitter, all of Antwerp manufacture. There is, however, a goodly number of Blake transmitters and single-pole receivers still in use. Messrs. Ribbink, van Bork & Co., at their centres, use magneto ringers, double-pole receivers, and a modified form of Berliner transmitter, all manufactured by themselves. The Zutphen Telephone Company use magneto ringers, double-pole receivers, and transmitters manufactured by Messrs. Ericsson & Co., Stockholm.

OUTSIDE WORK (LOCAL)

1·5 mm. bronze wire, supported on small double-shed porcelain insulators, is now used in the towns instead of the original galvanised steel. With few exceptions, the subscribers' lines in Amsterdam and the other principal towns are single with earth return ; but Vlaardingen and Amersfoort, the two latest centres of the Netherlands Bell Company, are metallic circuit, and it has been determined that all future ones shall be so likewise. The pole work of the Netherlands Bell Company is exceedingly well executed. In Belgium and Switzerland much attention is given to the design of poles of the largest size—from 50 to 80 feet—which are often both handsome and substantial, while their smaller work partakes of the commonplace ; in Holland the design of the small poles receives as much attention as that of the large, with the result that the citizens do not complain of being affronted by ugly and evil-smelling creosoted posts, such as are mostly affected in the United Kingdom. Along the canals in Amsterdam and in the suburbs (as well as in most of the other towns) one sees far-stretching routes of supports of the design shown in figs. 80 and 81. Essentially the poles are but the familiar iron lattice signal-post of the

Fig. 80

British railways; their attractiveness lies in the tasteful arrangement of the cross-arms, insulators, and finials. When nicely painted, with clean insulators and well-regulated wires, they look extremely well, and give one the impression that the company in

FIG. 81

erecting them has done its duty, both to the citizens and to its shareholders—for they are strong and durable withal, and go far to disarm grumblers. To give a firm hold on the ground the inside of the pole is filled to just above the ground level with concrete. The taller poles are of quite a different type, although they too are handsomely got up. They are of wood, painted with preservative compound; the pole is encased in a square wooden box from the butt to some three feet above the ground level, the space between the pole and the box being tightly rammed with clean dry sand. The box is closed with a moulded lid, and lends a finish to the appearance of the pole; but it is intended primarily by its deviser, Dr. Hubrecht, the general manager of the Netherlands Bell Telephone Company, to prevent the decay which invariably attacks wooden poles at or near the ground line. When so fitted it is the box, which can be readily renewed, which decays; while the pole, embedded in dry sand, lasts an indefinite period. The square box furthermore affords the pole a better hold in the ground than the rounded butt could give. Fig. 82 shows such a pole, 75 feet high and carrying 150 bronze wires. The climbing steps on these poles are riveted to long strips of iron, which are spiked or screwed to the poles on either side; this form of construction was adopted owing to steps working loose when fastened individually direct to the wood. In the suburbs light telescopic iron tubular poles are employed for branch routes of six or eight wires; they occupy little room and look well. The Dutch do not earth-wire their wooden poles. The more recent standards are of German type, consisting of one or more tubes fastened rigidly at their lower extremities to some part of the roof, and fitted with cross-arms consisting of strips of iron connected by rivets and by the insulator bolts. Such arms are cheap—one to carry six insulators costing 80 cents (1*s.* 3½*d.*), and one to carry twenty insulators only 1·80 florins (2*s.* 9*d.*) They are not, however, nearly so strong as the channel-iron arms designed by the author for the National Telephone Company, and now exclusively employed in the United Kingdom. The details of these standards are given in figs. 73 and 74 (German section). But, although identical in design, there is an important difference between the methods of erection in Germany and Holland. In the former country stays are rarely employed, and scarcely ever in an efficient manner, even when there are 200 wires attached; but in Holland there is no false economy in the matter of stays, and the standards are treated as though the destruction of the span of

FIG. 82

wires on one side by fire or storm is not altogether an impossible contingency. In a word, the Dutch work is far superior to the

German. At the same time, it must be said that the Dutch standards are not nearly so well calculated to withstand the vicissitudes of accident and tempest as are the Belgian ; but that is a fault of the design, not of erection. They are not earth-wired. Fig. 83 gives a good idea of an Amsterdam double standard.

The numerous rivers and canals in Amsterdam and elsewhere compel the frequent use of lengths of submarine cable. Originally, indiarubber-covered wires encased in lead were put down, but did not stand. Now a regular type of armoured submarine cable containing guttapercha-covered wires is employed.

Underground work has not been neglected, there being 11·6 miles of cable already down in Amsterdam. It is chiefly designed to get past the crowding of overhouse wires around the exchange, and the cables usually lead to a terminal pole in some secluded corner whence the wires are distributed overhead. No attempt has been made to serve the subscribers directly underground ; the Dutch towns do not lend themselves to such a method, the cost of which would be prohibitive. The cables usually contain fourteen pairs of twisted wires insulated with paper, each pair being spiralled with metal foil for earthing. One wire of each pair is tinned, and the identification of the pairs is assisted by two adjacent pairs in each layer being coloured respectively blue and pink. The cable is first covered with plain linen and then by a leaden tube, which in its turn is covered by a layer of impregnated jute and another of impregnated linen, the whole being protected by flat steel wires laid on spirally. These cables are supplied by Messrs. Felten & Guilleaume. It will be seen that the ultimate adoption of the metallic circuit is borne well in mind. The cable is laid in sand contained in a closed trough of creosoted wood, access boxes being placed every fifty meters or so, to facilitate distribution should it be found desirable at any future time to erect additional poles and terminate thereat some of the spare cable wires. The engineers appear to have confidence in this method of laying, no accidents from the picks of strange workmen having been experienced, and the cables maintaining their electrical conditions well. The plan involves the reopening of the ground whenever the spares on a route become exhausted, so a good deal of capital has to be buried in the shape of wires that may not be required for a considerable time. Great pains are taken at the junction of the underground with the overhead wires. Whenever space permits, a small hut (fig. 84) is built at or near the base of the distributing pole and fitted most efficiently

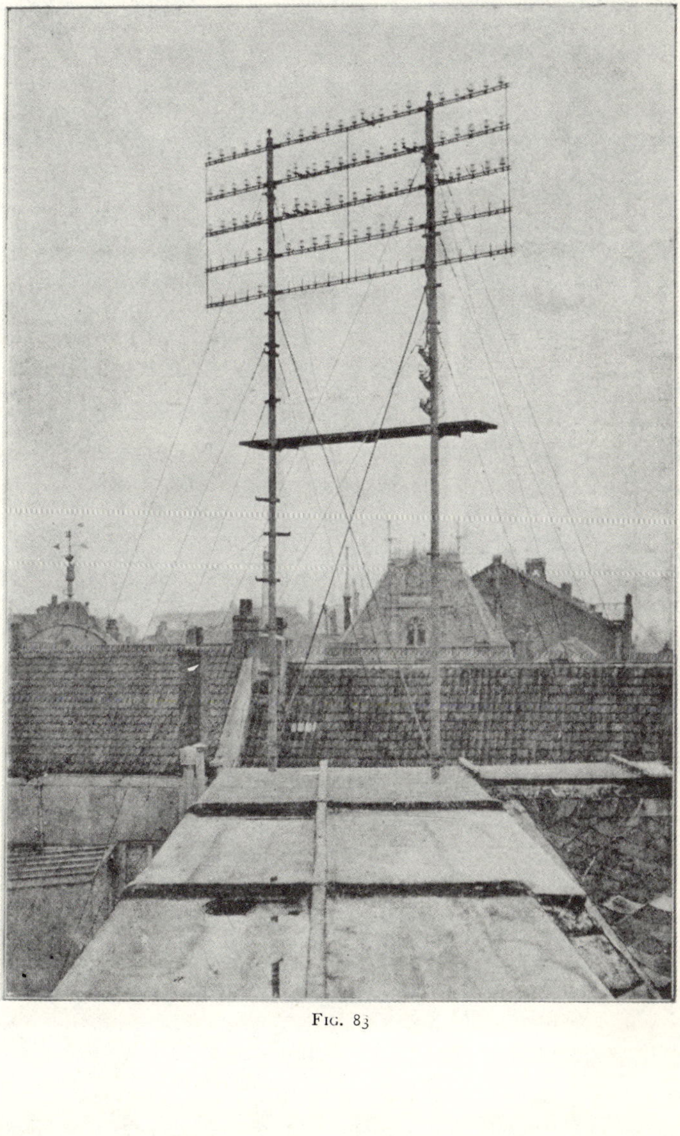

Fig. 83

Fig. 84

with cross-connecting terminals and lightning-guards. The cable ends are of course sealed in insulating material, the

Fig. 85

junction between underground and overhead being effected by an intermediate cable insulated with india-rubber or gutta-percha.

Fig. 85 shows a public telephone station of the Netherlands Company at Baarn.

Messrs. Ribbink, van Bork & Co.'s methods of construction are in no wise noteworthy. Although some bronze wire has been erected, their subscribers' lines are run chiefly with steel of 1·8 mm. gauge, of a breaking strain of 300 kilogrammes per square millimeter.

The Zutphen Telephone Company is remarkable in many ways. Its rate, *2l.* 17*s.* 10*d.* per annum, is not the lowest in Holland—Mynheer Jan Sōt takes care of that—but no attempt has been made elsewhere to give metallic circuits, the best of modern instruments, and a perpetual service for such a moderate sum. But they do it at Zutphen, and, what is stranger still, find it pays. The originator of the company and its present manager, Mr. C. J. van Bueren, a retired (Dutch) East Indian merchant, resident at Zutphen, applied, in conjunction with Mr. Carel Henny, for the concession (as much with the idea of passing the time as anything else), and having obtained it for the town and five kilometers around, succeeded in forming a company to work it. Mr. van Bueren knew nothing about telephone work at the time, but determined that he would have the best system and best workmanship procurable for his exchange, and, after due inquiry, placed a contract with the Netherlands Bell Telephone Company for its construction. All materials were to be of the best, and, with a view to ultimate connection with the Dutch trunk wire system, all lines were to be double and of 1·5 mm. bronze, having a breaking strain of 120 kilogrammes and a conductivity of 60 per cent. of pure copper. The exchange was opened on July 1, 1893, with 107 subscribers (the population of Zutphen is 17,004), and as these all had Ericsson transmitters, double-pole receivers, and metallic circuits, the speaking was as near perfection as well could be. By December 31, 1894, the instruments connected had increased in number to 141, with many more in prospect. The company enjoys free premises at the town hall, with the use

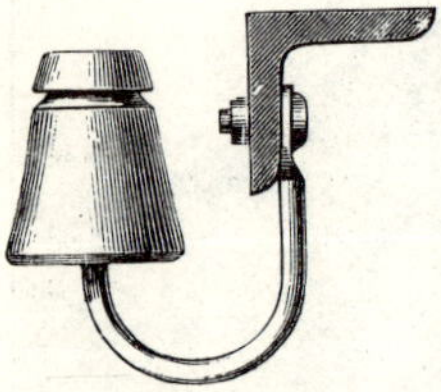

FIG. 86

of the roof, in return for four free connections given respectively to the burgomaster, town hall, and to the fire and police offices. As all these are contained within the walls of the town hall, the company may be adjudged to have made a very good bargain. Owing to the configuration of the town hall roof and the existence of a steeple, two separate fixtures had to be erected. These are substantially built of angle-iron, the larger consisting of eight uprights arranged in a square of 3·3 meters and connected by nineteen cross-arms. The uprights are fastened solidly to the roof, and the whole stands without the aid of stays. The fixtures are joined to the lightning conductor of the neighbouring steeple, and, in addition, have a special conductor and earth of their own. All the other standards in the town are carefully earthed and each metallic circuit has a lightning-guard, not only at the exchange, but at the premises of the subscriber served by it. Fig. 86 shows the method of attaching the insulators to the exchange cross-arms. Fig. 87 shows one of the standards used through the town, with dimensions. The tubes are continued through the roofs, and are bolted or strapped to the woodwork. The finials are provided with holes through which, when subscribers exist in a building on which

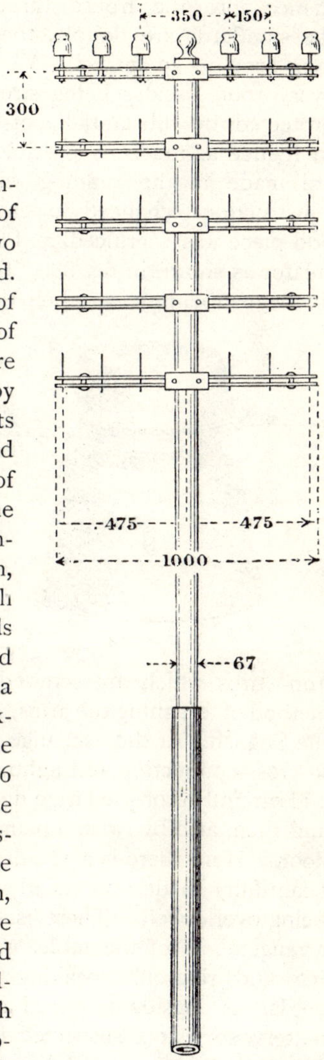

FIG. 87.—Dimensions in millimeters.

a standard is erected, the leading-in wires may be taken, passing thence into the house through the tube and roof. Some of the standards are double, the tubes then being connected by long arms as in fig. 83. All fixtures are carefully painted, and every roof, besides being strengthened under the standard, is protected by substantial foot-boards. The spans are short, and to reduce as much as possible the chance of contact, no joints are made in the running wire; when a coil of wire, during construction, chanced to end in the middle of a span the odd piece was sacrificed, and the joint made at the preceding insulator as shown in fig. 88. There are no joints, therefore, in the line wires themselves to help them to hang together during a gale.

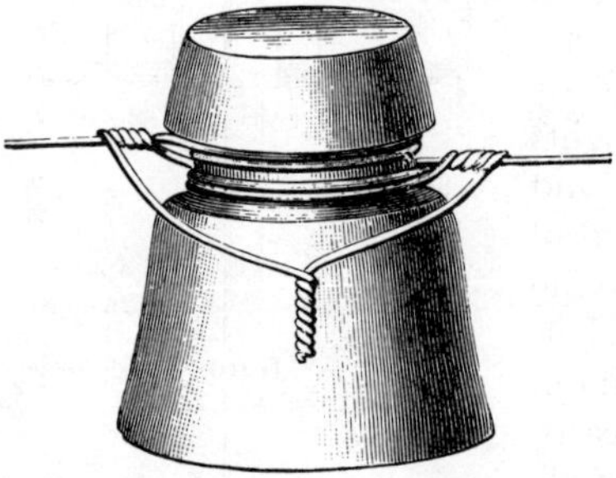

FIG. 88

All joints throughout the system are soldered with resin. Standards are used only when it is impossible to manage with poles. Of these last there are a good many, ranging from fifty-eight to seventy-five feet in height. They are of fir, pickled, and in every case well erected and carefully fitted. The climbing steps, as in Amsterdam, are riveted to iron strips which are screwed to the poles. Fig. 89 shows the method of attaching the arms, which differs in several respects from the English. At the exchange the wires are first led by twisted pairs to cross-connecting and lightning-guard boards placed in an attic and carefully protected from dust by wooden casing with glass doors, and then, also by twisted pairs, to the switch-room on the ground floor. Here there is a 160-drop table by Ericsson, of Stockholm, beautifully made and neatly fitted, no detail, however trifling, being overlooked. There is, however, nothing special about the arrangements of the table, which has the usual indicators, calling and ring-off, speaking and ringing keys, and weighted cords. Alongside it is fixed a testing galvanometer with keys and battery, so that a suspected line may be tested for earth or disconnection at once. Adjacent to the switch-room is a public telephone station containing an American 'long-distance' desk set of the kind designed by Mr. Thomas D. Lockwood, of Boston, in 1888. It consists of an elegant table, on which are conveniently mounted the transmitter, receiver, and ringer, with every convenience for writing. There is not a bit of scamped work from beginning to end, and Zutphen constitutes a really model exchange, to which it would pay certain telephone administrations and companies, English not excepted, to send their engineers as to a school. To the date of opening, the installation, including the preliminary expenses, had cost 20,000 florins (1,650*l.*).

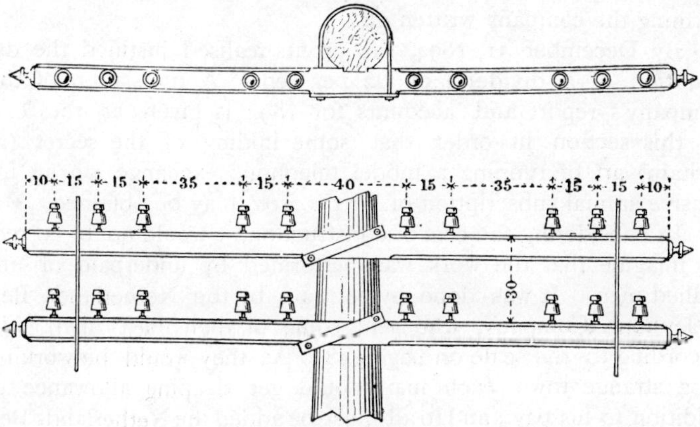

FIG. 89.—Dimensions in centimeters.

As there were 107 subscribers to start with, this amounts to 15*l.* 8*s.* 5*d.* per line, but plenty of spare room for future expansion was provided at the exchange fixture and on the poles and standards. Inspection and repairs have cost since the opening from 2*l.* 1*s.* 3*d.* to 2*l.* 9*s.* 6*d.* per month. Way-leaves cost about 24*l.* per annum. Day operating costs 12*s.* 6*d.* per week (one girl relieved for meals by a younger under-study who is competent to take her place on holidays or in case of sickness); and night operating, 16*l.* 9*s.* 0*d.* per annum This is performed by a young man, otherwise engaged during the day, who sleeps in the switch-room with an alarm bell worked by the indicator shutters over his head. The manager receives only 24*l.* 13*s.* 9*d.* per annum by way of salary, but he is entitled to a preferential percentage of the profits, and is, besides, a shareholder. At Deventer and Enschede, neighbouring small towns in which similar—but single-wire—exchanges exist, the managers are a master plumber and an insurance agent respectively. New construction and repairs are contracted for with the Netherlands Bell Telephone Company at fixed rates. Inspection and testing is performed by the manager. As a result of the first year's working, to June 30, 1894, all expenses to date were paid, and the costs of obtaining the concession and forming the company written off.

By December 31, 1894, the profits realised justified the declaration of a dividend of 4·2 per cent. A translation of the company's report and accounts for 1894 is given at the end of this section in order that some inkling of the secret (in Britain) art of running a model telephone exchange on an inclusive annual subscription of 2*l.* 17*s.* 10*d.* may be obtained.

In considering the cost of construction, it would not be correct to imagine that the work was performed by underpaid or unskilled men. It was done by contract by the Netherlands Bell Telephone Company, who sent some of their best men, paid according to the scale on page 248. As they would be working in a strange town, each man would get sleeping allowance in addition to his pay; and to all must be added the Netherlands Bell Company's profit on the contract. The author does not profess to regard the manager's salary as sufficient, nor the provision for reserve and deterioration adequate, but an advance of the subscription to 4*l.* 5*s.* per annum would afford ample margin for these items. With this reservation there is no reason why, under similar conditions, the Zutphen results should not be obtained in English towns of the same size; and the author does not doubt its practicability in many cases, especially if undertaken by the municipal authorities.

OUTSIDE WORK (TRUNK)

There is not much to remark about the Dutch trunk line work except that it is generally very well done. The first lines were erected along the roads, the railways being avoided, as it was feared that the strong currents in the telegraph wires would inter-

FIG. 90.—A Dutch trunk line.

fere in a degree even with metallic circuits. That theory is, of course, now disproved, and was known to be groundless in Great

Britain at least as early as 1881, six years before any trunks were erected in Holland. Advantage has been taken of the railways for the later extensions. The Netherlands Bell Company likes tall poles for its trunks, and on some routes there are long stretches of 50-feet poles, which lift the wires well above the trees. Fig. 90 shows a Dutch trunk route with 54-feet poles. The wire used is 3 mm. hard copper, and the insulators are large double-shed white porcelain. The wires are crossed, not twisted, but the Dutch Government is understood to contemplate the twisting of the projected international trunk line to Belgium as far as the frontier. The Netherlands Bell Company, which is to construct the line, well aware of the uselessness and drawbacks of such a proceeding, has protested and may succeed in getting the intention altered. The speaking over the trunks is very good, but the distances are not, of course, great.

PAYMENT OF WORKMEN

Foremen receive from 3 to 4 florins (4*s.* 11*d.* to 6*s.* 7*d.*) per day; skilled wiremen, 4*s.* 2*d.*; and labourers, 3·96*d.* per hour. When working away from home the men's actual expenses are paid. Working hours are from 7 A.M. till 6 P.M., with one and a half hours off for meals.

PAYMENT OF OPERATORS

Girls, when taken on at the age of seventeen years, receive 6*s.* 7*d.* per week, and rise by degrees to 9*s.* 10½*d.* as ordinary operators. The average pay of this class at Amsterdam is at present 8*s.* 3*d.* per week. The trunk operators and those who attend at the telegraph office for the telephoning of telegrams are required to understand English, German, and French in addition to their own language, and are paid from 16*s.* 6*d.* to 19*s.* 9*d.* per week, according to length of service. These amounts include a small premium payable on each telegram handled without error. Applicants for vacancies must produce high-school certificates of intelligence and industry.

STATISTICS

At the end of 1894 there were in Holland 7,263 subscribers distributed as follows :—

Owner	Town	Number of subscribers	Population
Netherlands Bell Telephone Company	Amsterdam	1,752	426,914
	Arnhem	284	51,105
	Amersfoort	36	14,182
	Baarn	10	—
	Bussum	5	—
	Dordrecht	252	34,125
	Groningen	173	57,967
	Haarlem	165	55,311
	Hague	381	169,828
	Hilversum	47	12,199
	Maassluis	2	—
	Rotterdam	961	222,233
	Schiedam	54	25,280
	Utrecht	214	89,436
	Vlaardingen	24	12,059
	Zaandam	13	14,545
		4,373	
Ribbink, van Bork & Co.	Breda	202	22,987
	Deventer	198	23,351
	Enschede	196	5,664
	s'Hertogenbosch	200	27,594
	Leeuwarden	197	30,712
	Leyden	300	44,198
	Middelburg	150	16,455
	Tilburg	195	35,068
	Flushing	100	12,565
	Zwolle	200	27,706
		1,938	
J. W. Kaijser	Nijmegen	450	34,128
Maastricht Telephone Company	Maastricht	225	32,757
Zutphen Telephone Company	Zutphen	141	17,004
Jan Sŏt	Alkmaar	73	14,048
	Helder	63	23,145

The number of local connections is unfortunately not given, but the chief exchanges are undoubtedly very busy. In Amsterdam as many as 254 connections have been given to one instrument in one day. On January 29, 1895, seven Amsterdam subscribers asked for over 200 connections each, an eighth for 184, and a ninth for 167; and this traffic is not exceptional. During 1893 100,311 telegrams were forwarded from, or received at, subscribers' offices by telephone. For the year 1894 the number of trunk connections was 85,142. The Netherlands Bell Telephone Company has a capital of 600,000 florins (49,375*l.*), the whole of which, together with its reserve fund, has been expended in constructing its system. A special reserve fund is provided, out of which the cost of improvements and renewals is defrayed. Last year a dividend of 9 per cent. was paid. Seeing that the company's effective rate in its chief centre, Amsterdam, is only 7*l.* 12*s.* 9½*d.*, this result must be admitted as very satisfactory.

SECOND ANNUAL REPORT OF THE ZUTPHEN TELEPHONE COMPANY

It was with much pleasure that I acquainted the shareholders last year that the company's undertaking had been successfully launched. On the present occasion I also feel satisfaction in being able to report that the favourable expectations held out last year have been realised; that the number of subscribers has gradually increased, whilst the establishment and its working have been satisfactorily maintained.

The number of faults has been small and less than last year, viz.:—

Disturbances of wires	68
,, ,, instruments	46
Total	114

The company's system now comprises:—

Free connections given in terms of concession	4
Service connections	3
Free connections in part payment of way-leaves	2
,, ,, complete ,, ,,	6
Paying subscribers	126
Total	141

In the course of the year nineteen new subscribers joined and four gave notice, two on account of leaving the town. At the commencement of the current year two more also gave notice. The construction of a connection to the Waterworks has been commenced, and one to the Netherlands Industrial School will also be put in hand shortly: these are certain to lead to further developments.

For these and other new lines some additional capital will be necessary, in connection with which proposals will be laid before the shareholders.

Although the number of calls is liable to be influenced by several circumstances, a steady increase is observed, viz.:—

Total calls for second half of 1893	30,653
,, whole of 1894	73,270

The telegraph station has not yet been connected; this, however, may be expected shortly.

The costs of repairs and maintenance of lines and instruments amounted to Fl. 315.

It will be seen from the small number of faults that the company's system is very efficient; also that the repairs have been done very cheaply. The shareholders will remember that repairs are done for us by the Netherlands Bell Telephone Company from its Arnhem centre.

I am also pleased to report that the employees have done their duty with diligence and exactitude. Efficient substitutes are provided against sickness or holidays.

For the financial position of the company I have the honour to refer to the annexed accounts.

[Signed] C. J. VAN BUEREN, Managing Director.

ZUTPHEN: *February* 18, 1895.

VALUE OF THE COMPANY'S PROPERTY AT DECEMBER 31, 1894

1 florin = 1*s.* $7\frac{3}{4}$*d.*

	Fl.
Exchange system	19,994·38[5]
Central office	621·44[5]
Office furniture	279·76[5]
Materials on hand	473·25[5]
Tools ,,	81·99
	Fl. 21,450·84

Dr. PROFIT AND LOSS ACCOUNT **Cr.**

	Fl.		Fl.
Debit from 1893	300·24	Subscriptions and various receipts	4,447·82
Capital Account	977·26		
Head office	29·71		
Office furniture	13·87⁵		
Materials	13·38		
Tools	2·21		
Interest Account	22·27		
Stationery	81·61		
General expenses	401·09		
Salaries	1,070·50		
Repairs to system	315·34⁵		
Advertisements	67·45		
Way-leaves	237·50		
Dividend	900·00		
Balance, Profit and Loss, 1893	15·38		
	Fl. 4,447·82		Fl. 4,447·82

BALANCE SHEET, DECEMBER 31, 1894

Assets	Fl.	*Liabilities*	Fl.
Cash in hand	57·53	Capital	20,000·00
Value of exchange system	19,994·38⁵	Netherlands Bell Telephone Company	655·92⁵
Head office	621·44⁵	Sundry creditors	93·69⁵
Office furniture	279·76⁵	Dividend	900·00
Materials	473·25⁵	Profit and Loss	15·38
Tools	81·99		
At Banker's	156·63		
	Fl. 21,665·00		Fl. 21,665·00

The above dividend of Fl. 900 to be divided according to Article 21 of the Rules, and will be payable at the company's office at the rate of Fl. 4·20.

C. J. VAN BUEREN, Managing Director.
C. SCHILLEMANS } Directors.
CAREL HENNY }

ZUTPHEN: *February* 18, 1895.

XII. HUNGARY

THE establishment and working of telephone exchanges has been declared a privilege of the State in Hungary; but before the Government had determined to enter the field actively, some concessions for thirty years had been granted to private persons, and the telephone system of the country is now divided between the Government and several companies. Trunk lines between Buda-Pesth and the chief towns have recently been commenced, and in some instances completed, and these belong exclusively to the State, and are, indeed, intended primarily for State use, public traffic being only a secondary consideration. There is also a trunk route consisting of seven metallic circuits from Buda-Pesth to Vienna, over which Szegedin, Temesvar, Arad, Raab, Pressburg, and a few other towns can communicate with Austria. An international line to Odessa has been proposed.

SERVICES RENDERED AND TARIFFS

1. **Local exchange intercommunication.**—Subscriptions are of two classes: (1) for instruments located actually within a town, and (2) for instruments located in the suburbs.

In Buda-Pesth the annual rates are:—

	Per annum £	s.	d.
CLASS 1	12	10	0
CLASS 2.—If not more than ½ kilometer beyond the town boundary	12	13	4
If between ½ kilometer and 5 kilometers .	12	18	4
For each additional kilometer . . .	0	1	8

In other towns :—

	Per annum £	*s.*	*d.*
CLASS 1	5	0	0
CLASS 2.—If not more than ½ kilometer beyond the town boundary	5	3	4
If between ½ kilometer and 5 kilometers .	5	8	4
For each additional kilometer	0	1	8

These rates cover all expenses of installation and maintenance. Hotels, restaurants, clubs, and other places where the public have access to the instruments are charged 50 per cent. extra ; on the other hand, State, municipal, church, and charitable institutions enjoy a reduction of one-half.

2. **Rural exchange communication.**—This corresponds to the German 'vicinity' intercourse and the French 'annexes,' and is intended for extra-suburban villages in the neighbourhood of towns which possess an exchange. The subscription depends on the facilities required. A subscriber desiring only power to call the other subscribers in his own village pays 5*l.* per annum ; if he wishes to ring up the town subscribers also, he is charged 10*l.* This, however, only applies when the State owns both the town and the village exchange. When a company owns the town exchange the village subscriber who wants the town subscribers must pay 10*l.* + 1*l.* 5*s.* = 11*l.* 5*s.* if the town is a country one, and 10*l.* + 5*l.* = 15*l.* in the case of Buda-Pesth.

3. **County or departmental exchanges.**—These serve the purely country districts, and are intended to connect one or more villages with the chief village of a parish or ward. Such an exchange may be connected with a similar one situated in another parish or ward, whether of the same or of an adjoining county or department, and also with a town exchange within its own county. It is likewise permissible to join it to a town exchange in a neighbouring county, provided this town is situated near the boundary between the two counties. Subscriptions vary with the service required. A subscriber calling only those connected to his own village switch-room pays 2*l.* 10*s.* per annum ; if he would be free to call through all the village exchanges in his group he pays double—5*l.* ; if he would wander telephonically at will over villages of the adjoining county also, his rate is 6*l.*, which also

entitles him to originate conversations with one town exchange, situated either in his own county or near its boundary. For all other connections (except trunk ones, which are denied him under any circumstances) he must pay per five minutes' talk according to the public telephone station scale, but speaking from his own instrument. Town subscribers who would call through the county village exchanges must pay 1*l.* per annum in addition to the town subscription.

4. **Internal trunk line communication.**—A uniform trunk rate of 1*s.* 8*d.* per three minutes has been fixed for the whole country. Express or urgent talks are admitted at double rates.

5. **International trunk line communication.**—These actually exist only with Austria, the rates being the same as for the interior of Hungary.

6. **Telephoning of telegrams.**—The Buda-Pesth subscribers may forward and receive their telegrams by telephone at a charge of 2*d.* per message, irrespective of length. A similar facility is accorded to some of the provincial towns, and even to some of the villages, at 1*d.* per message.

7. **Public telephone stations.**—These exist in the towns and departmental districts, but not in the rural. The time unit is five minutes. A town subscriber or non-subscriber pays 2*d.* for a local talk. In the departments the charge is 2*d.* for speaking within the same ward; 6*d.* for a call to other wards of the same or adjoining department; and 10*d.* for communicating through a town exchange of the same department, or of a neighbouring department if situated near the boundary.

WORK

No information of importance can be given on this head, promised details not having arrived in time for inclusion. The Buda-Pesth exchange is worked with two double-cord, single-wire, series multiples supplied by the Western Electric Company. The Hungarian system is, with the exception of the trunks, single wire throughout. It is, for the most part, aerial; but some underground work, with cable supplied by Messrs. Felten & Guilleaume, exists in the capital.

STATISTICS

No figures dealing with a later period than 1892 are available. At the end of that year the State owned 14 out of a total of 23 exchanges; 16 out of 25 switch-rooms; 59 out of 71 public stations; and 2,988 out of 3,952 subscribers. There were then no trunk lines in operation.

XIII. ITALY

HISTORY AND PRESENT POSITION

THE Italian telephone system is worked entirely by concessionary firms or companies under the regulations imposed by the law of April 7, 1892. This law reserves absolute power to the State to forbid the erection of even private wires, unless confined entirely to the property of the constructors, without its formal sanction, and empowers it to exact an annual payment of 16*s.* for each private wire, and 4*s.* for each instrument in excess of two used in connection with it, besides an extra charge if such a private line should exceed three kilometers in length.

With regard to exchange communication, the State reserves right to work exchanges itself, and to grant more than one concession for the same town or district should it deem such a course desirable. The maximum term for any concession is twenty-five years, but the State may purchase the system after twelve years on giving one year's notice. In such a contingency the price, failing mutual agreement, is to be fixed, without right of appeal, by three arbitrators, named respectively by the Government, the concessionary, and the president of the court by which such a dispute would ordinarily fall to be tried. But in any case the price is not to exceed the mean of the last three years' profits multiplied by the unexpired years of the concession. Profits are defined as meaning the gross receipts less the ordinary working expenses and Government taxes. Should the Government not purchase at the end of twelve years, the concessionary will retain possession for the whole term of twenty-five years; but on the expiry of that period the system becomes the property of the State without any payment whatever.

Concessionaries must therefore arrange matters, if they would avoid loss, so as not only to make a living out of the business during their term of occupancy, but to get back the whole of the capital invested before the time for relinquishing comes. This is unquestionably a bad system. It simply means that the subscribers pay both principal and interest, and that during the concluding years of the concession improvements will be tabooed and the service starved.

On local exchange communication an annual tax of 10 per cent. on the tariff charges is imposed, plus an annual charge of 2*l.* for each public telephone station opened. On trunk communication the tax is 5 per cent. of the gross receipts. These taxes are payable by the concessionary. The Italian Government appears to have taken the British Post Office as its model in this matter, although the Italian tax is not quite so onerous as the British, which is 10 per cent. on the trunk as well as on the local gross receipts. The law further provides that should the Government itself undertake the construction and working of trunk lines the whole of the receipts will belong to it, giving the companies nothing for the use and operating of the terminal wires. When trunk lines are erected and worked by concessionaries, the receipts less 5 per cent. will belong to them, but they must guarantee the Government the average of the previous three years' receipts for telegrams between the two points connected. Parishes which erect telephone lines to Government telegraph offices at their own expense, with the object of participating in the telegraph service, are exempt from all these payments.

The maximum tariffs which concessionaries may charge to their subscribers are fixed by the law, but these have proved too high for the pockets of the people, and except in the largest towns—Venice, Turin, Genoa, and Milan—are not applied. In Rome there is competition between a company and a co-operative society, and the rates are consequently lower than in the towns just mentioned. The legal maximum tariff is as follows :—

For each subscriber's line within a radius of three kilometers of the central station, 8*l.* per annum if aerial, and 12*l.* if underground. Excess distance to be charged at the rate of 4*s.* 9½*d.* and 6*s.* 5*d.* respectively for each additional 200 meters or fraction thereof.

For each conversation from a public telephone station, 2·88*d.* over a line not exceeding three kilometers in length, the charge to be increased at the rate of 48*d.* for each additional kilometer. The time unit to be five minutes.

For trunk communication the charge fixed is 2*s.* 5*d.* for distances not exceeding 500 kilometers, with increments of 5·76*d.* for each additional 100 kilometers or fraction thereof, the time unit being five minutes.

The only reduction authorised to ordinary subscribers is one not exceeding 20 per cent. on each instrument taken in excess of the first. Concessionaries are authorised to require from each subscriber a first-and-last payment, not exceeding one-fifth of his annual rental, as a contribution to the cost of his line. This regulation is permissive, not obligatory. Concessionaries are bound to connect Government, municipal, and parochial offices at half-rates, but such connections are freed from the usual taxes. They are also bound to permit Government, at its own expense, to join its post and telegraph offices to their exchanges free of charge.

The chief fault of this tariff is that it possesses no elasticity. The rates are made the same for the capital and the villages, and there is no distinction between trunks fifty kilometers long and five hundred.

The lot of the telephone concessionary in Italy is not, on the whole, a happy one. In addition to the legal obligations already enumerated, he has to deposit, as security for due payment of the Government taxes, a sum equal to 10 per cent. of the maximum legal tariff multiplied by two for each thousand inhabitants of the locality to which his concession applies. Should he contemplate dabbling in trunk lines he must deposit a further sum equal to 50 per cent. of the annual telegraphic receipts between the two points connected, based on the average of the last three years. He must pay his taxes monthly at the nearest telegraph office. If the concession is worked by a company, copies of its articles of association, proceedings at its general and special meetings, of its balance-sheets, and of its directors' and auditor's reports, must be regularly furnished to the Minister of Posts and Telegraphs. Then the concessionary is bound to reimburse to his subscribers charges collected for conversations that could not be held. If a line is interrupted for more than three days *from any cause whatsoever*, a proportionate part of the annual subscription must be returned to the subscriber; if the interruption is one which might have been avoided by care and attention, the subscription for its whole duration must be refunded. If such an interruption continues more than ten consecutive days, the subscriber may claim damages to the tune of double his subscription for the period of the interruption; and if it lasts fifteen days he may, if he chooses, terminate his agreement as well. These regulations are certainly calculated to engender a sense of responsibility and to conduce to careful construction and good maintenance, but at the same time their enforcement in the case of interruptions due to fire, floods, snow, or extraordinary tempests is unjust to the concessionaries, and cannot be productive of good.

A Swiss company, with headquarters at Zürich, is the owner of thirteen concessions, while a good many have been taken up by French companies, and a few by co-operative societies. The capabilities of the telephone, as measured by the services rendered to the public, have not yet been exhausted in Italy. The internal trunks are yet on paper; the international ones have scarcely reached even that stage; there is no telephoning of messages for local delivery or for mailing; the public telephone stations are few, and there appears to be no messenger organisation. With the exception of Brescia, all the Italian exchanges are run on the single-wire plan, and, again with the exception of the Brescia, are exclusively overhead.

SERVICES RENDERED AND TARIFFS

1. **Local exchange communication.**—The rates charged by the different concessionaries vary greatly. Some of them have made a uniform price for connection within the legal three-kilometer radius; others have divided that radius into two, and others again into three zones, taking care that the maximum charge does not exceed that fixed by law.

Town	Population	Annual subscription	Remarks
Rome . . . ,, . . .	407,936	6*l.* 14*s.* 5*d.* 5*l.* 10*s.* 5*d.*	Two competing systems: Società Romana di Telefoni and Società Anonima Co-operativa dei Telefoni
Naples . . .	536,000	8*l.*	
Milan . . .	426,500	8*l.*	
Palermo . . .	273,000	8*l.*	
Turin . . .	230,183	8*l.*	Reduced to 6*l.* 8*s.* for private houses, doctors, and druggists
Genoa . . .	212,500	8*l.*	
Florence . . .	197,000	6*l.* 8*s.*; 7*l.* 4*s.*; 8*l.*	Three zones
Venice . . .	149,500	8*l.*	
Bologna . . .	147,000	6*l.* 8*s.*; 6*l.* 16*s.*; 8*l.*	Three zones
Messina . . .	142,000	7*l.* 4*s.*	
Leghorn . . .	106,000	6*l.* 8*s.*; 7*l.* 4*s.*; 8*l.*	Three zones
Padua . . .	79,500	6*l.*; 7*l.* 4*s.*	Two zones
Verona . . .	69,500	4*l.* 16*s.*; 6*l.*	Two zones
Bari . . .	58,266	6*l.*	
Parma . . .	44,492	6*l.*	
Brescia . . .	43,354	5*l.* 12*s.*; 6*l.*; 7*l.* 4*s.*	Three zones
Pisa . . .	37,704	4*l.* 16*s.*	
Pavia . . .	29,836	4*l.* 16*s.*	
Vicenza . . .	27,694	6*l.*	
Mantua . . .	28,000	4*l.*	
Perugia . . .	17,395	4*l.*	
Piacenza . . .	35,000	4*l.*	
Casale Monferrat .	17,096	3*l.* 12*s.*	
Biella . . .	(?)	2*l.* 16*s.*	

It will be seen that competition has given the capital lower rates than prevail in the chief provincial towns; also that the endeavours of the concessionaries to adapt themselves to local circumstances have brought about a nearly regularly descending scale of subscriptions in sympathy with the population, until, in the small towns, the point reached is almost Norwegian or Dutch-like in its moderation.

2. **Public telephone stations.**—These are not numerous in Italy, the Government tax of 2*l.* per annum for each station deterring the concessionaries from opening any that are not quite certain to pay. In Rome there are eight; in Milan two; in

Turin four ; in Verona four ; in Venice five ; in Genoa three. In Naples, Bologna, Palermo, Messina, and many other towns there are none at all. The legal maximum tariff is 2·88*d.* for five minutes, but this is imposed in two towns only, Leghorn and Venice. In other towns possessing public stations five minutes' local talk costs as follows :—

Rome :		Turin	2·4*d.*
Società Romana	1·44*d.*	Genoa	·96*d.*
Società Co-operativa	·96*d.*	Padua	·96*d.*
Milan	1·92*d.*	Verona	·96*d.*

In some towns subscribers use the public stations free of charge, but the more usual plan is to make everybody pay.

3. **Internal trunk lines.**—These have, so far, attained but little development. Milan is connected with Monza, and a line from Milan to Legnano is in course of erection. At the date of writing (February 1895) none of the chief towns are in regular telephonic correspondence, but the Italian Government has prepared a very large scheme which, when given effect to, will place all the business centres in communication. The trunk rates have been fixed in anticipation by law, as already stated (p. 259).

4. **International trunk lines.**—The Italian Government has approached the French, Austrian, and Swiss Governments with proposals for international lines, but the schemes have yet to be matured.

5. **Telephoning of telegrams.**—This traffic is not large. The direct connection of telephone exchanges with telegraph offices for the transaction of the subscribers' business appears not to be practised. Thus at Milan, the second largest telephone centre in Italy, the subscribers' telegrams are taken down at the central office and sent across to the telegraph station by messenger ; conversely, telegrams arriving for subscribers are delivered at the telephone office and thence dictated to the addressees. For this service the company charges 1·92*d.* per message, irrespective of the number of words.

WAY-LEAVES

The law of 1892, which hits the concessionary very hard in most directions, comes to his aid a little in the matter of way-leaves, for it decrees that telephone wires may be passed *without fixing* over both public and private lands and properties, or in front of buildings provided the view from windows or other openings is not interfered with. But no wires may be fixed to a building without the consent of the proprietor interested, while the local authority is given power to rate such fixtures for the benefit of its funds. Concessionaries are warned that when it is necessary to fix telephone wires to public monuments which have an artistic or historical value, it will be necessary to take steps to protect the said monuments from damage, and to preserve their artistic effect. Evidently all faith in human nature has not departed from the Italian Parliament when it is willing to trust its public monuments to the artistic taste of telephone men, even though they be countrymen of Michael Angelo, in want of a way-leave.

SWITCHING ARRANGEMENTS AND SUBSCRIBERS' INSTRUMENTS

With a separate company in almost every town, the practice as regards switch-boards and instruments is naturally very mixed. French apparatus is used to a considerable extent, many of the concessionary companies being of French origin; but there is also much of English, American, Swiss, Belgian, and German manufacture. The Società Telefonica Lombarda (Telephone Company of Lombardy), one of the largest and most progressive of the companies, has a multiple board for 1,600 single lines, supplied by the Western Electric Company, at its Milan exchange. The board, which possesses no special features, is now (February 1895) nearly full, there being 1,450 subscribers connected to it. The same company at its Como and Monza exchanges has non-multiple boards made by the Officina Elettrica de Milano after English models. The subscribers' instruments in these towns comprise magneto, back-board, battery-box, Blake transmitter and Bell receiver, all of the type and arrangement familiar in Great Britain. Called subscribers are rung by the operator. At Brescia, where there are metallic circuits, the Hipp form of Hunnings transmitter, without induction coil, is used. The operators are

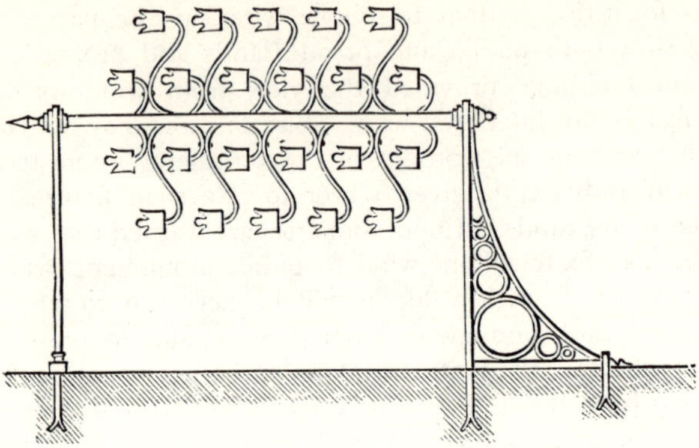

FIG. 94

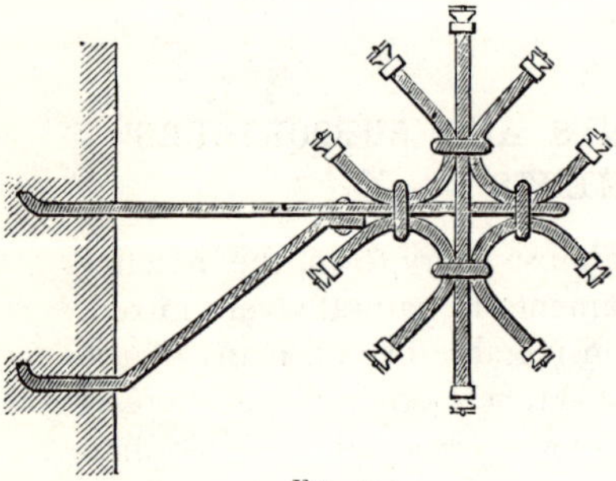

FIG. 92

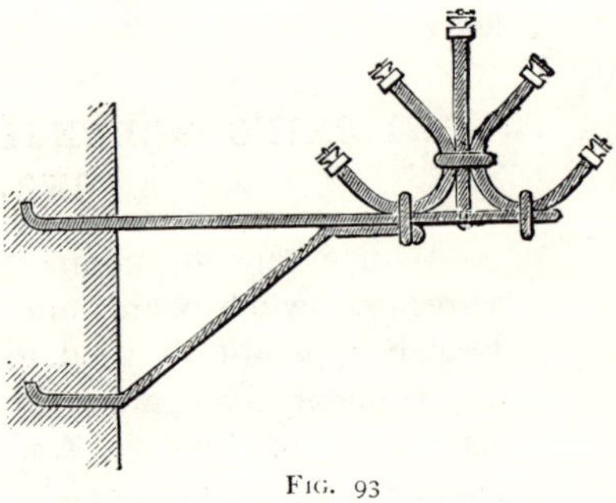

FIG. 93

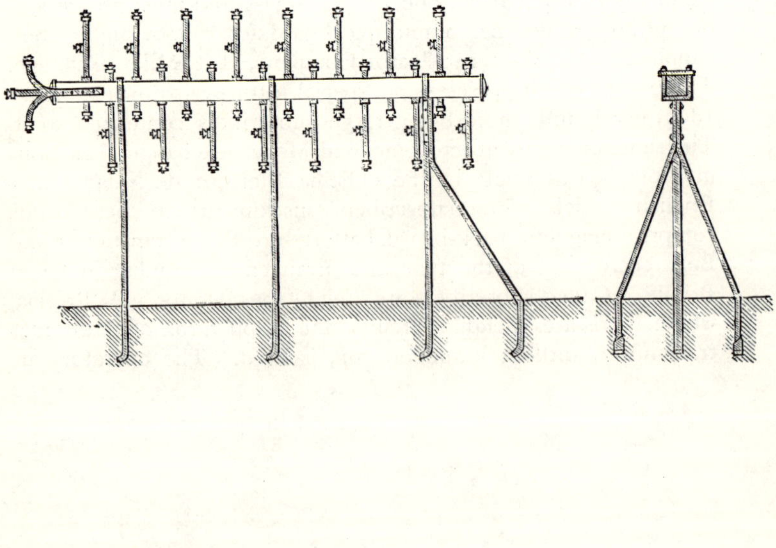

FIG. 91

usually girls by day, and men by night; but at Palermo, Catania, and Messina, males are exclusively employed.

HOURS OF SERVICE

The Telephone Company of Lombardy gives a perpetual service in all its exchanges, a good example which is followed in most of the larger towns. In the smaller, the hours vary from 7 or 8 A.M. to 8 or 9 P.M.

OUTSIDE WORK

The Telephone Company of Lombardy uses galvanised steel wire of 1·8 mm. diameter for its local, and galvanised iron wire of 3·17 mm. for its trunks to Monza and Legnano. Other companies follow the same practice, but bronze wire is nevertheless exclusively used in some places and partially in others. As its merits come to be better understood, bronze will doubtless oust iron and steel in Italy as it has already done in most other countries. The sole objection to bronze is its tendency, owing to the superior heat conductivity of the metal, to favour the formation of frost on its surface; but this should not weigh against it much in Italy. The wall-bracket form of construction is much in vogue, and it must be confessed, on the testimony of figs. 91 to 94, that the Italians have a pretty fancy in wall-brackets. Figs. 91 to 93 represent the practice of the Telephone Company of Lombardy. It will be perceived that the insulators are of a kind that would be altogether insufficient in our damp climate to prevent leakage overhearing between wire and wire, being merely short tubes of porcelain slipped over the bolt and fastened by a nut at the top. A strong, well-designed standard, built up of angle-iron on the Belgian plan, is shown in fig. 95 as an example of the Lombardy Company's roof work. The same company also employs tall iron-lattice ground poles very similar to those illustrated in the Belgian section. Two of its smaller poles are shown in figs. 96 and 97 as being of a more special design. They are formed of three parts, socketed one into the other, and, while providing a good carrying capacity, are far more ornamental than

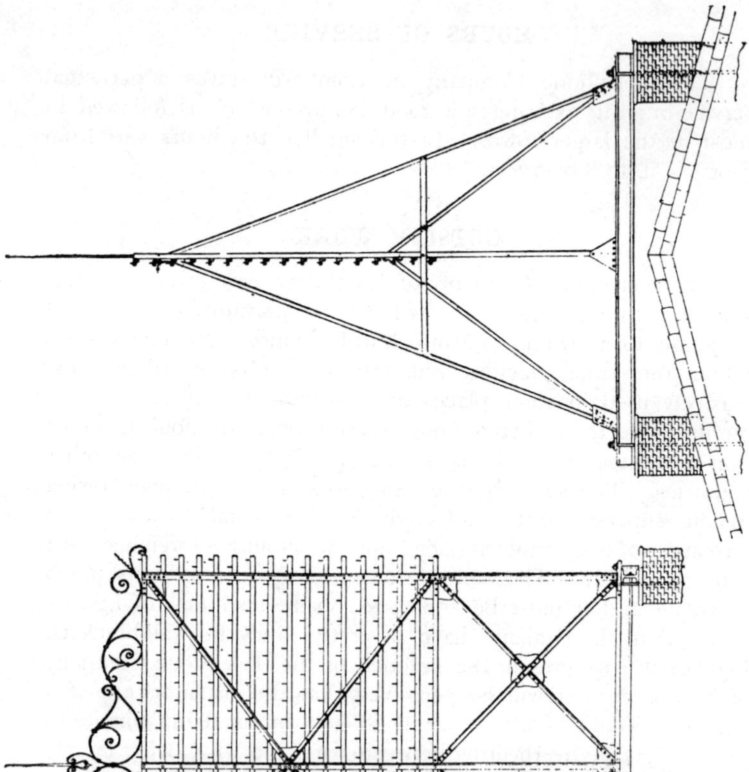

Fig. 95

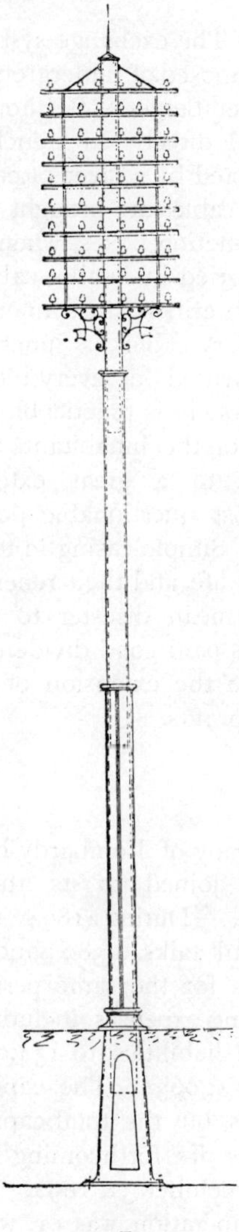

FIG. 96

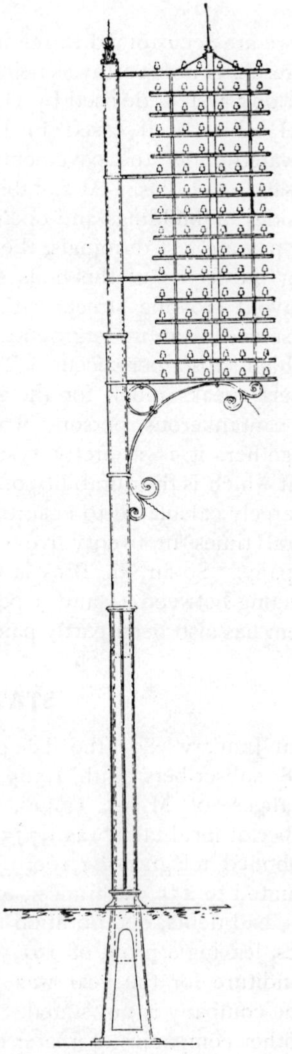

FIG. 97

any we are accustomed to see in England. The exchange system at Brescia is noteworthy as being largely composed of underground work on a system devised by Dr. von Wurstemberger. Berthoud-Borel cables, well cased in lead, are laid directly in trenches excavated under the pavements, and protected by a layer of coal-tar, sand, and tiles. At suitable points the cables are brought up the sides of buildings and opened out in junction boxes, whence, after passing test terminals, the wires are carried in smaller cables along the fronts of the houses to the subscribers' instruments. To avoid crossing streets with the secondary cables, a junction box served by an underground cable is provided for every block in which subscribers occur. That such a system is practicable in Brescia speaks much for the good nature of the inhabitants: a few cantankerous persons would spoil it to a great extent. Altogether, it is a pretty system, the most questionable point about which is the durability of the cables. Simple casing in lead is scarcely calculated to ensure them a long life, and their renewal several times in twenty-five years would mean disaster to the company. So far, the Brescia Company has paid good dividends, averaging between 4½ and 5 per cent., while the extension of its system has also been partly paid for out of profits.

STATISTICS

In January 1895 the Telephone Company of Lombardy had 1,518 subscribers, with 1,585 instruments joined to its three exchanges of Milan, Como, and Monza. During 1894 the number of local talks was 1,775,000; of trunk talks, 4,380; and of telephoned telegrams, 1,100. The receipts for the same period amounted to 255,598 francs; and the working expenses, including taxes, bad debts, deterioration fund, and all liabilities, to 154,017 francs, leaving a profit of 101,581 francs, or 4,063*l.* The capital expenditure for the year was 34,244 francs, but the total capital of the company is not stated. No statistics are forthcoming for the other companies of a later date than December 31, 1892. At that date the total number of systems in operation was 51, with 53 switch-rooms, 34 public stations, and 11,980 subscribers. The length of wire in use was 20,076 kilometers. The number of local talks for 1892 is returned at 17,748,559; of talks from public stations, 75,250; of telephoned telegrams, 2,022; and of trunk talks, 0. At the end of 1893 Rome had 2,350 subscribers divided between the Società Romana di Telefoni (1,750) and the Società Anonima Co-operativa dei Telefoni (600); Florence, 860; Genoa, 780; Turin, 762; Naples, 721; Palermo, 455; Leghorn, 370; Venice, 351; and Bologna, 300.

XIV. LUXEMBURG

HISTORY AND PRESENT POSITION

ONE of the smallest States of Europe, with an area (998 square miles) and a population (211,088 in 1891) practically the same as that of Dorsetshire, with a capital, too, counting only 18,187 souls, Luxemburg is nevertheless also one of the most telephonically active. In January 1895 the capital with its 18,000 people had 621 exchange instruments working, or 3·4 for each 100 inhabitants, while the whole Grand Duchy boasted 85 exchanges and 1,315 instruments, or ·62 of an instrument for each 100 inhabitants. Fancy Dorsetshire with 85 telephone switch-rooms within its borders!

By the law of December 17, 1884, the establishment of telephone exchanges was made a Government monopoly, and the existing regulations and charges were fixed by the law of March 9, 1887. The first exchange was opened in Luxemburg city in 1885.

The Luxemburg system differs from all others in Europe in one essential respect: there are no trunk rates. While all the villages (there is only one town, the capital) possess exchanges and are joined together by numerous trunk lines, the subscribers have nothing to pay beyond the subscription (a very moderate one as will presently appear) to their local exchange, and may call up any other subscriber within the limits of the Grand Duchy at will. That they are not backward in availing themselves of this privilege appears from the fact that in 1892 the inter-town talks numbered 671,937, considerably more than in the neighbouring republic of France for the corresponding period, while the local talks reached the total of 922,692, scarcely 50 per cent. more. This is a good traffic to develop within the area of one of the smaller English counties and amongst a population, scarcely equalling that of Edinburgh, chiefly employed in agriculture. It bears out the opinion so often reiterated by the author that the telephone possesses a sphere of usefulness all its own, which is at present but little understood in the United Kingdom—a sphere of usefulness that it will fill without artificial fostering, as it were spontaneously, whenever left to be introduced on its natural merits and at its legitimate price. The different methods of treatment pursued by the respective legislatures of the United Kingdom and Luxemburg produce the result that in London, the greatest commercial city in the world, there is about ·14 of a telephone to each hundred persons; and in Luxemburg, one of the poorest countries in Europe and possessed of no commercial importance whatever, the ratio is ·62. The British system would have been simply prohibitive in such a country, just as it has proved to be in many of the poorer British and Irish districts, which are to-day as innocent of telephones as they were in the reigns of Caractacus and Brian Boru.

SERVICES RENDERED TO THE PUBLIC

1. **Intercourse between the subscribers to the same exchange.**

2. **Intercourse between all the exchanges.**—Twenty of the chief villages have direct wires to Luxemburg; the remainder communicate through an intermediate switch-room.

3. **Telephoning of telegrams.**

4. **Telephoning of messages for local delivery or posting.**

5. **Public telephone stations.**—There are some sixty-five of these, which subscribers use without charge on producing a card of identity.

6. **Calling non-subscribers to the public stations.**—This facility is not confined as in other countries to the subscribers: a non-subscriber may go to one public station and have a non-subscribing client fetched to another.

7. **Parochial or communal stations.**—As in France and Switzerland, a local authority wanting a telephone station where the

Government is not disposed to establish one, at its own expense may arrange to contribute to the cost. In Luxemburg this is done by an annual subscription and by providing an office and operator at the charge of the commune. In January 1895 there were thirty-four such stations in operation.

TARIFFS

1. **Rates for local and (2) trunk intercourse.**—Within the limits of any town or village in which an exchange exists the annual subscription, which is payable half-yearly in advance, is 3*l.* 4*s.*

It is important to note that the State erects the lines, supplies the instruments, and maintains everything at its own expense. The 3*l.* 4*s.* per annum covers all charges and includes the right to communicate freely all over the Grand Duchy, which measures, roughly, 44 miles by 30.

When the subscriber is located at a distance from an exchange the tariff is modified. When his place lies not more than one and a half kilometers from an existing route of telephone wires the subscription is maintained at 3*l.* 4*s.*; for each additional kilometer it is increased by 2*l.* But the subscriber has, in any case, to reimburse the State the cost of his wire, at the rate of 4*l.* per kilometer, between its point of junction with the main route and the exchange. If he is located actually on an existing trunk route, but outside the radius of any exchange, the same system obtains: he bears the cost of so much of his line as lies outside the radius at the rate of 4*l.* per kilometer, and pays the usual local subscription of 3*l.* 4*s.* This rule, which, so far as the author is aware, has not its counterpart elsewhere, is by no means a bad one: it enables the distant subscriber, for one reasonable payment down, to bring out the exchange, as it were, to the nearest point on a main route to his dwelling, and puts him thenceforward on a par as regards annual subscription with his urban competitors.

Extra instruments are charged 1*l.* and extra bells 4*s.* per annum. In calculating distances the actual course of a wire is taken. Contracts are for three or five years, according to the subscriber's distance from the exchange. The use of instruments is restricted to the subscribers, their families, servants, and employees. Proprietors of hotels and other public places pay the ordinary rate and are allowed to place their instruments at the disposal of their customers, but are limited to 2,000 communications per annum. Any over that number are charged 3·36*d.* each, which charge the subscriber, if he likes, may collect from the person making the call.

In the event of a subscriber removing he must bear the cost of the labour, but not of the material, involved in shifting his instrument. Subscribers are entitled to a proportionate refund when an interruption lasts longer than thirty days.

3. **Rates for telephoning telegrams.**—For each telegram transmitted to, or received from, a telegraph office by telephone, a charge of ·98*d.* is made, irrespective of the number of words. The arrangement for ensuring payment of charges under this and the following heading is ingenious, and peculiar to Luxemburg. No deposit in advance is exacted, so that every subscriber can profit by the service without previous notice or agreement, but the subscription which he has paid in advance for his exchange line is debited with the costs of telegrams forwarded or received. At the end of the month a memorandum of the amount of this debit is presented, which the subscriber is expected to make good immediately : should he not do so, his exchange agreement is considered curtailed by the number of days represented by the amount of the debit, and his instrument may be taken out that number of days before the expiration of the period for which he had paid.

4. **Rates for telephoning local messages and mail matter.**—The charge is ·98*d.* per message, irrespective of length, plus the cost, 3·36*d.*, of the messenger employed to effect delivery, or of the postage, as the case may be.

5. **Rates at public telephone stations.**—The charge to non-subscribers is 3·36*d.* for five minutes' talk with any subscriber within the limits of the Grand Duchy. Two non-subscribers conversing together from different public stations are charged double fee. Subscribers, on showing a card furnished by the administration, use the public stations free.

6. **Rates for fetching non-subscribers to public stations.**—When called by a subscriber :—

3·36*d.* if resident within the telegram free delivery limits
4·8*d.* ,, 1½ kilometers beyond the limits
7·2*d.* ,, 3 ,, ,, ,,
9·6*d.* ,, 5 ,, ,, ,,
1·92*d.* for each kilometer above 5

When called by a non-subscriber :—

3·36*d.* in addition

7. **Rates applicable at parochial telephone stations.**—The local authority desiring the station pays the State 4*l.* per annum as rental for the line and instrument, and finds house room and attendance. The charge, which goes to the State, is, to all users, subscribers or non-subscribers, 3·36*d.* per five minutes.

WORK

Phosphor bronze wire of 1·4 mm. is used for the local; and of 2 mm. for the trunk lines, of which there are about seventy-six. Many of these are still single wires, but the more important are metallic circuits. The system is entirely aerial. There are, as yet, no multiple switch-boards employed. There is no night service, but any two or more subscribers who desire it are left plugged through during the close hours. Magneto instruments made by Messrs. Schäfer & Moutanus, Frankfort-on-Main, are used throughout the Duchy; the generator coils have to be cut in by pressing a button when ringing. Two receivers are provided to each instrument. Service is suspended during thunderstorms, and subscribers are required to earth their lines by means of a cord and plug attached to each instrument for the purpose.

STATISTICS

The latest available for telephones, apart from posts and telegraphs, are those for 1892. In that year Luxemburg possessed 50 exchanges, 54 kilometers of local routes, 531 kilometers of trunk routes, and 1,306 kilometers of trunk lines, used by 1,003 subscribers and 61 public stations. The local talks numbered 922,692; the trunk talks, 671,937; and the telegrams telephoned, 2,838. The capital expenditure amounted to 808,802 francs (32,352*l.*). The receipts for the year were:—

	Francs
Local subscriptions	60,989
Public stations and telegram service	3,505
Sundry receipts	4,717
Total	69,211

The working expenses amounted to 61,762 francs, leaving a profit of 7,449 francs (298*l.*) as evidence of the sufficiency of a 3*l.* 4*s.* rate.

Statistics for 1893, furnished to the author by M. F. Neuman, Director of Posts and Telegraphs, Luxemburg, give the following figures:—

Number of centres	52
,, subscribers	1,203
Length of routes, in kilometers	617
,, wire ,, ,, ...	2,333
Number of local talks	963,005
,, trunk ,, ...	765,929
,, public station talks	9,780
,, telegrams telephoned	2,661
Receipts for subscriptions, in francs	66,400
,, at public stations, ,, ...	3,816
,, sundries, in francs	2,517

Unfortunately the working expenses for 1893 are not shown separately from those of posts and telegraphs.

In January 1895 the exchanges and instruments connected throughout the Grand Duchy were as follow:—

	Ex-changes	Instru-ments
Luxemburg town .	2	621
Andorf	1	3
Aspelt	1	1
Bad-Mondorf . .	1	16
Bauschleiden . .	1	4
Carried forward .	6	645

	Ex-changes	Instru-ments
Brought forward .	6	645
Beckerich . . .	1	2
Befort	1	6
Beles	1	6
Berburg . . .	1	1
Carried forward .	10	660

	Exchanges	Instruments		Exchanges	Instruments
Brought forward	10	660	Brought forward	48	1,050
Bettborn	1	1	Medernach	1	1
Bettemburg	1	19	Mersch	1	24
Bettingen	1	5	Mertzig	1	1
Bissen	1	4	Mutfort	1	1
Bœgen	1	4	Niederanven	1	1
Bœvingen	1	1	Niederfeulen	1	2
Bourscheid	1	1	Niederkerschen	1	5
Clerf	1	21	Petingen	1	17
Consdorf	1	1	Rambruch	1	3
Consthum	1	1	Redingen	1	15
Cruchten	1	4	Reisdorf	1	1
Dalheim	1	3	Remich	1	28
Diekirch	1	66	Rodingen	1	16
Differdingen	1	15	Roodt	1	6
Dommeldingen	1	10	Rosport	1	3
Düdelingen	1	23	Rümelingen	1	24
Echternach	1	18	Sæul	1	5
Esch-on-Alzette	1	55	Sandweiler	1	6
Esch-on-Sauer	1	4	Schrondweiler	1	3
Ettelbrück	1	38	Simmern	1	1
Fels	1	15	Stegen	1	1
Frisingen	1	1	Steinfort	1	6
Garnich	1	1	Strassen	1	4
Grevenmacher	1	22	Tuntingen	1	1
Grosbous	1	5	Ulflingen	1	8
Harlingen	1	1	Useldingen	1	6
Heinerscheid	1	2	Vianden	1	9
Hellingen	1	2	Vichten	1	1
Hesperingen	1	3	Wahl	1	1
Hobscheid	1	1	Wasserbillig	1	6
Hoscheid	1	1	Wecker	1	6
Hosingen	1	13	Weiler (Pütscheid)	1	1
Itzig	1	1	Weiswampach	1	4
Junglinster	1	5	Wiltz	1	34
Kap	1	14	Wilwerwiltz	1	5
Kehlen	1	1	Wormeldingen	1	7
Kœrich	1	1	Walferdingen	1	2
Mamer	1	7		85	1,315
Carried forward	48	1,050			

XV. MONACO

THE Principality possesses a telephone exchange which in March 1895 numbers just seventy connections. It is conducted in every respect on the French plan, the instruments and mode of construction being French, and the tariff identical with that applicable to French towns of less than 25,000 inhabitants (see French section, p. 147). The list of subscribers is printed in Paris; the conditions of subscription, regulations, and instructions how to use the instruments are all copied verbatim from the French; so, when it has been stated that a trunk line gives Monaco communication with Antibes, Cannes, Grasse, Mentone, and Nice, there is nothing further to be said about the telephonic system of Albert I., Sovereign Prince of Monaco.

XVI. MONTENEGRO

No steps have yet been taken to provide this principality with telephonic exchange system.

XVII. NORWAY

HISTORY AND PRESENT POSITION

NORWAY, with a capital about the size of Dundee, half a dozen towns which may rank with Colchester, a multiplicity of villages, and a total population of 2,000,917, could not have presented itself to the imagination of the original pioneers of Bell's wonderful speaking trumpet precisely as a fountain of telephonic milk and honey. But it is rarely given to pioneers to realise the ultimate importance of their work ; and when the International Bell Telephone Company went to Christiania in 1880 intent on inducing the hardy Norseman to have his ears lengthened as it alone (as was then thought) could lengthen them, the task must have appeared (in view of the inertia exhibited in many far wealthier and more populous countries) an up-hill one indeed. It looked like sowing in ice with a prospect of reaping in snowballs ; but the event proved otherwise, for the Norse spirit of enterprise, which erstwhile discovered America, peopled Greenland and Iceland, and conquered Normandy and England, proved quite equal to the assimilation of the telephonic exchange idea. America may have discovered the telephone indeed, but had not Norway discovered America? So it came about that, within a year of the International Bell Company's start in Christiania, a local company was formed to oppose it, and oppose it it did in a hammer-and-anvil fashion that was all Norwegian. Indeed, so energetic was the battle—so frequent the encounters of legions of wiremen on the roofs—so exasperating the 'cross-talk' (both on the roofs and on the wires) to which it gave rise, that the Municipality intervened and threatened to cancel the concessions it had granted to the combatants unless peace could be successfully invoked. The subscribers, too, were tired of the incessant interruptions to which their wires were subjected, while the way-leave granters began to think that no telephone company was surely better than two which, usurping the time-honoured privileges of both proprietors and Tom cats, fought out their differences on the roofs. So in 1885, when the rival systems possessed 995 and 634 subscribers respectively, both were purchased by a new local association, the Christiania Telephone Company, which has since carried on the business, under Mr. Knud Bryn's able management, with marked satisfaction to both its subscribers and shareholders. Starting with 1,493 subscribers in its first working year, it had increased to 3,150 in 1890, 4,210 in 1892, and 4,624 in October 1894. The capital cost has been just 50,000*l.*, or nearly 11*l.* per subscriber—practically the same as that of a similar system in England. The rate is 4*l.* 8*s.* 11*d.* per annum, everything included, which has sufficed to pay dividends of 5 and 5½ per cent. (the company's concession limiting dividends to 6 per cent.) per annum. The company possesses no special way-leave privileges, and its construction work has been superior, as a rule, to that of the United Telephone Company and its subsidiaries in England.

The International Bell Company started in only one other Norwegian town—Drammen—which it continued to work until 1889, when the business, then comprehending 147 connected instruments, was transferred to the Drammen Telephone Company. In February 1895 the number of instruments connected had risen to 401. The population of Drammen being only 20,000, the development here, on the same inclusive rate (4*l.* 8*s.* 11*d.*) as in Christiania, must be considered satisfactory. It covers connections up to two kilometers in length. The Drammen Company has paid good dividends. At the end of 1894 the capital expended was 4,011*l.* The receipts amounted to 1,483*l.*, and the management and maintenance to 659*l.*, leaving a profit which enables a dividend of 7 per cent. to be paid after placing a substantial amount to the reserve fund. The dividends have always ranged from 5 to 7 per cent. At December 31, 1894, the company's system comprised 507 kilometers of line, of which 479 kilometers were single wire.

The Drammen Telephone Company declined to extend its

lines beyond the immediate precincts of the town, a policy which gave umbrage to the country folk, who wanted to share in the benefits flowing from telephonic communication, and ultimately led to the formation of the Drammen Uplands Telephone Company, which obtained a concession for a tract of country around Drammen measuring 230 kilometers from north to south and extending over five counties, forming the largest concessionary tract in Norway. It began business in June 1890, and at December 31, 1894, owned 24 switch-rooms, 2,500 kilometers of routes, comprising 770 kilometers of poles and 1,080 kilometers of metallic circuits, all for the benefit of 292 subscribers. The principal places within its area are the townlets of Kongsberg and Hönefros. The annual subscription, which is inclusive, and covers lines not exceeding two kilometers in length, is 5*l.* 11*s.* 1*d.*, for which sum free communication over the whole of the company's area is allowed. Up to December 31, 1894, the system had cost 9,815*l.*, and the receipts for 1894 amounted to 2,332*l.*, the repairs to 373*l.*, and the net profit to 752*l.* Since its commencement the company has regularly paid a dividend of 6 per cent. Last year 900 kilometers of new line were run. The success of this Drammen Uplands Telephone Company is most interesting, and most creditable to the managers. The company has shown how a large tract of sparsely populated country, containing nothing larger than a village, can be telephoned and maintained year after year at a handsome profit. It is a lesson which the author fears will nevertheless be quite without effect on the British Post Office. It should be added that the whole of the Uplands system is in trunk communication with Drammen town, Christiania, and the network of lines radiating therefrom.

The third exchange established in Norway was that of Trondhjem (population 30,000), commenced in 1881 by a private concessionary, and worked by him until 1889, when it was purchased by the Town Council for 1,650*l.* At that time it numbered 315 subscribers; in October 1894 these had increased to 700. The rate, which is an inclusive one and represents all the expense for which a subscriber is liable, is only 2*l.* 10*s.* per annum for business connections, and 1*l.* 5*s.* for private houses within a radius of one and a half kilometers. At the end of 1892 the total capital expenditure was 6,000*l.*; the annual income, 1,240*l.*; the working expenses and maintenance, 1,000*l.*; and the profits 240*l.*, equal to 4 per cent. on the capital. This is a good specimen of what may be done by a municipality owning its own telephones. It cannot be said that the cheapness is due to indifferent work, because the lines are well constructed of bronze wire strung on substantial wooden and iron poles and standards, while the switch-boards are Ericsson's make, as are also most of the subscribers' instruments. The municipality, as controlling the roads and streets, may have some advantage over a company in the matter of way-leaves, but it possesses no rights over private property. It must be remembered, too, that, unlike the practice in many of the Norwegian systems, the Trondhjem subscribers' instruments are provided by the exchange and are included in the subscription.

The Bergen (population 53,000) exchange was begun in 1882 by a local company, and is noteworthy as being the first in Norway in connection with which the subscribers were required to pay for their own instruments. These are sold to them by the company, and their purchase amounts in effect to an entrance fee, similar to that payable in Sweden, of some 2*l.* 10*s.* The maintenance of the instruments after erection is included in the annual subscription. In Bergen itself the company finds and maintains the lines, but subscribers located outside the town have to pay the first cost of their wires according to a distance scale. The annual subscription, which, with the above-noted exceptions, is an inclusive one, is 3*l.* 8*s.* 10½*d.* per annum, both in town and country. For this, day and night service is given, and the company can afford to assign its girl operators a maximum duty of six hours daily. At December 31, 1894, the number of subscribers was 1,439, renting 1,516 instruments, of which 35 were connected to country branch switch-rooms. The capital expended on construction to the same date was 10,460*l.* In 1893 the total income was 4,244*l.*; in 1894, 4,621*l.*, out of which, after paying all expenses and providing for the maintenance of the system, the usual dividend of 6 per cent. per annum was paid to the shareholders. The instruments used are magnetos of the best type, and the equipment generally is creditable. A considerable extension, notwithstanding the present ratio of 2·9 instruments to every 100 souls, is looked for, and an order has been placed with Messrs. Ericsson & Co. for a multiple switch-board comprising the latest improvements. It is usual for partisans to asseverate that, even if very low subscriptions do exist, they are applicable to very small exchanges only. Here, however, is an instance of a system, surpassing in size the vast majority of those belonging to the National Telephone Company, paying a 6 per cent. dividend year after year on a subscription of 3*l.* 8*s.* 10½*d.*! The fact speaks eloquently of the competency and conscientiousness of the Bergen managers, as well as of the enterprise of the population.

Besides these five chief exchanges, there are about one hundred and seventy others in Norway, mostly worked quite independently (although many of them are joined by trunk lines) by concessionary companies, co-operative societies, or individuals, but occasionally by municipalities or rural authorities. The rates charged are, from a British point of view, absurdly small; but two facts cannot be gainsaid: that this system of concessions enables the Norwegian citizens and even peasantry to enjoy facilities which are denied to the English public; and that, low as the rates are, the companies succeed in more than making ends meet. The following statistical table, which the author has been enabled by the courtesy of the companies and gentlemen named therein to compile, abundantly demonstrates these facts, and also gives some idea of the constitution and mode of working of enterprises which contrive to do so much for so little. One of them, that of Hammerfest, is well within the Arctic circle, being situated in latitude 71·6° and within a few miles of the North Cape. The population of Hammerfest is some 2,500, yet telephonically it is far in advance of some of the London suburbs with populations counted by the fifty thousand, and of a greater number of British towns than could be tabulated in a day's work. There is another small exchange, that of Tromsö, within the Arctic circle. With these exceptions, nowhere does the midnight sun obtain a glimpse of a telephonic switch-board. Two of the towns named in the list—Christianssand and Hammerfest—were destroyed, together with their telephone exchanges, by fire a few years since; but the

STATISTICS OF SOME PROVINCIAL NORWEGIAN

—	Town	Population	Exchange opened	By whom owned	Number of switch-rooms	Number of subscribers	Do subscribers pay for their instruments?	Do subscribers pay for their lines?	Annual subscription
1.	Christianssand *a*	12,813	Oct. 1883	Company	1 Central 4 Branch	230	No	No	In town, 2*l.* 15*s.* 7*d.*; in suburbs, 2*l.* 4*s.* 1*d.*
2.	Christianssund	10,381	1888	Co-operative Society	1	100	Yes	No	2*l.* 9*s.* 7*d.* (*b*) 3*l.* 1*s.* 1*d.* 3*l.* 12*s.* 2*d.*
3.	Flekkefjord .	—	Sept. 1894	Co-operative Society	2	36	Yes	Yes	1*l.* 13*s.* 2*d.* shareholders; 2*l.* 4*s.* 1*d.* others
4.	Fredrikstad .	11,217	May 1883	Company	2	277	No	No	3*l.* 6*s.* 8*d.* business place; 2*l.* 15*s.* 7*d.* residence
5.	Grimstad .	3,000	Nov. 1891	Co-operative Society	3	119	No	Not, in town; outside subscribers pay a proportion of cost	1*l.* 13*s.* 2*d.* for one, 2*l.* 9*s.* 7*d.* for two, and 3*l.* 6*s.* 8*d.* for three instruments
6.	Hammerfest .	2,500	1887	Mr. H. Wingaard Früs	1	23	Yes	Yes	2*l.* 4*s.* 1*d.*
7.	Haugesund .	—	Oct. 1888	Company	1	104	Yes	Yes	Entrance fee 4*l.* 8*s.* 11*d.*; annual subscription, 1*l.* 7*s.* 10*d.*
8.	Hortens . .	—	Feb. 1889	Company	1	120	No	Not, in town; in country, yes. Country members also pay maintenance of their lines outside town	Entrance fee 2*l.* 15*s.* 7*d.*; annual subscription, 1*l.* 13*s.* 2*d.*
9.	Mandals . .	—	June 1892	Company	5	67	Yes	Yes	2*l.* 4*s.* 1*d.*
10.	Röros . .	—	Oct. 1894	Company	1 Central 5 Branch	20	Yes	Yes	1*l.* 13*s.* 2*d.*
11.	Skien . .	—	June 1883	Co-operative Society	1	180	Yes	Yes	2*l.* 4*s.* 1*d.*
12.	Stavanger .	24,000	Oct. 1881	Messrs. Grene & Egends	1	304	Yes	Yes	1*l.* 8*s.* 0½*d.*
13.	Tromsö . .	5,409	April 1886	Mr. Andr. Risock	1	76	Yes	Yes	2*l.* 15*s.* 7*d.*

(*a*) Shareholders mostly subscribers. Lines measure 250 kilometers. Central station was destroyed by fire in 1892. (*b*) According to island on which subscriber lives, town being built on three islands.

EXCHANGES FOR YEAR ENDING DECEMBER 31, 1894

—	Radius or distance to which it applies	Hours of daily service	Total expenditure on system	Amount of annual revenue	Amount of annual working expenses, repairs, &c.	Amount of profit for 1894	How profit is disposed of	Number of connections per month	Nature of instruments used by subscribers
			£	£	£	£			
1.	Town and vicinity	14 hours	1,923	549	384	165	One-third to 'extension' fund; one-twentieth to shareholders; balance to reserve	25,000	Not stated
2.	2½ kilometers	8 till 9	440	305	Not yet ascertained for 1894	Not yet ascertained for 1894	5 per cent. of profits to reserve; subsequently, not exceeding 8 per cent. to subscribers	8,000	Magnetos
3.	5 kilometers	11 hours; but exchange can be called all night for extra payment	110	(*a*)	(*a*)	(*a*)	Reserve fund	1,300	Magnetos
4.	3 kilometers	Day and night	2,692	846	Not yet ascertained for 1894	Not yet ascertained for 1894	5 per cent. to shareholders; balance to reserve	42,000	Magnetos
5.	District extends to about 30 kilometers	13 hours	1,044	143	143	(*b*)	(*b*)	10,500	Battery calls
6.	1 kilometer; one subscriber 2 kilometers off, pays 3*l.* 6*s.* 1*d.*	10 hours	192	(*c*)	(*c*)	(*c*)	—	1,400	Magnetos
7.	6 square kilometers	14 hours	193*l.* in 1888; since, entrance fees have paid for construction	230	55	175	At discretion of shareholders	4,500	Magnetos
8.	Town and vicinity	Day and night	935	248	170	78	General meeting decides	11,500	Magnetos
9.	—	12 hours	823	256	100	137	General meeting decides	6,000	Magnetos
10.	3 kilometers	9 till 5	329	(*d*)	(*d*)	(*d*)	5 per cent. to shareholders; rest to reserve	1,200	Battery calls
11.	Town and vicinity	14 hours	395	395	(*b*)	(*b*)	(*b*)	65,100	Not stated
12.	1 kilometer	Day and night	1,980	494	340	154	—	Not stated	Some magnetos, some battery calls
13.	1 kilometer	12½ hours	Not stated	164	99	66	Owner	6,445	Magnetos

(*a*) New society. Uses bronze wire. (*b*) Subscriptions adjusted to cover all expenses, leaving no profit to divide. (*c*) Receipts are made to balance expenses. Instruments and lines are bought from Mr. Früs. (*d*) New company.

exchanges have been re-established, and are worked at a profit. At the end of this section will be found the accounts and balance-sheet of the Christiania Telephone Company for 1893, which cannot fail to be instructive to telephone managers who doubt the vitality of a 4*l.* 8*s.* 11*d.* rate.

The Norwegian Government held aloof from matters telephonic until a proposal was made to connect Christiania with Drammen, when, in 1881, it passed a law conferring on the State the exclusive right to establish inter-town communication. This effectually put a stop to the projected trunk lines, as the State had no funds available wherewith to undertake the construction itself, and, influenced by the usual bogle of competition with the Government telegraphs, refused to license the companies to do the work. It granted permission for each to operate within a radius of eleven kilometers of its central office, and so secured reasonable facilities for communication between a town and its suburbs and immediate surroundings, but no two such radii were allowed to be joined; and if two eleven-kilometer radii each containing a telegraph office chanced to overlap, the radius of each was to be restricted on the overlapping side in such a manner that two kilometers of neutral ground were to intervene between them. However, by 1885 the local telephonic systems had multiplied and grown to such an extent that the Government was no longer able to escape from the necessity of either constructing or licensing, and in that year it allowed the local telephone companies of Skien and Porsgrund to join their systems by a trunk line conditionally on their paying to the State an annual sum of 25*l.*, the estimated loss of telegraphic revenue between the two places. Other trunks soon followed, and it was not long before Christiania had joined ears with Drammen, Gjövik, and twenty other towns in its vicinity. To show how groundless was the fear for the telegraphic revenue, it may be mentioned that in 1891 the amount payable to the Government (on its own valuation be it remembered, as the companies had to pay whatever the State demanded) was only 489*l.* for the twenty-two trunk lines radiating from Christiania, some of which extended to a distance of 120 kilometers. On only one of these trunks, that to Drammen, the conversations had averaged 100 per working day, which, at the tariff of 6½*d.*, meant a telephonic revenue of 847*l.* per annum. It consequently became obvious that the two systems of communication could exist side by side. It should, however, be noted that the Norwegian Government had acted wisely from the first in availing itself of the telephonic exchanges as feeders of the telegraph; and had even inserted a clause in the companies' concessions binding them to allow their lines to be used for the transmission of telegrams. The British Post Office, on the other hand, moved heaven and earth to prevent the English telephone companies doing anything of the kind; thereby proving itself far less enlightened, and appreciative of the new state of affairs that had arisen, than that of Norway. In 1894 the Norwegian trunk system has grown to such an extent that space cannot be spared in the present work for a mere enumeration of the lines.

In Norway, particularly in the north, many telegraph lines exist which, prior to the advent of the telephone, were used only during the fishing season, the traffic during the rest of the year not sufficing to pay the cost of skilled operators, lighting, warming, &c. As the towns and villages concerned were nevertheless desirous of enjoying a service all the year round, the Government determined to utilise the telephone—the employment of which does not call for any special skill—for this purpose; and, on the towns agreeing to bear the cost of warming and lighting and to find persons satisfactory to themselves to act as operators, some of these fishing wires were brought into acceptable use during the winter. Others, which happened to connect towns or districts in which local telephone exchanges had been established, were handed over to the telephone companies to serve as trunk lines conditionally on the companies agreeing to transmit telegrams for non-subscribers, when required, at the State tariff rates. In this matter again the Norwegian Government showed a happy adaptation to special circumstances and a freedom from red-tapeism which cannot be too highly commended.

The latest telephonic development in Norway is the international trunk line to Stockholm.

The subscribers who use this are already on metallic circuits; all others throughout Norway are as yet connected with their exchanges by single wires, but the Christiania Company has definitely resolved to convert its system to metallic circuit, and the alteration will be commenced as soon as the new switch-board has been installed.

As being by far the most important and at the same time typical of all, the system of the Christiania Telephone Company is particularly referred to (unless otherwise stated) in the following description. The concessions of all the companies are much on the same lines, and the services rendered to the public, except when modified by special local conditions (as the fishing wires already mentioned), are essentially of the same nature. They all have the right to telephone telegrams, to open public telephone stations, and to use trunk lines; but the international line to Sweden is at present only available from Christiania and towns which, like Drammen, are joined to it by metallic circuit trunks; and Kongsvinger, where the Norwegian Post Office has opened a public station.

SERVICES RENDERED TO THE PUBLIC BY THE CHRISTIANIA TELEPHONE COMPANY

1. **Intercourse between the subscribers and public tel:phone stations of the same town or district.**

2. **Internal trunk communication.**—There are several groups of trunk lines, at present unconnected with each other, but the only one of importance is that having Christiania for its centre. This is, however, very extensive. Not a town, and scarcely a village, on both coasts of the Christiania fjord, down to Sarpsborg and Fredrikshald on the one side and to Skien and Fredriksværn on the other, but has its trunk; while to the north of the capital five main routes exist, embracing Gjövik, Hamar, Elverum, and Lillehammer, with every place of importance for some 400 kilometers, making a total distance of about 500 kilometers (284 miles) that may be spoken over from south to north. The trunks are erected under agreement between the companies concerned, each association sharing in the traffic of a particular trunk contributing equal proportions to the cost of erecting and maintaining it, irrespective of the mileage within its own specific area. Each company retains the whole of its receipts for trunk talks, but may not demand two consecutive connections if another partner com-

pany wants the wire. The payment to the State to compensate for loss of telegraph traffic is borne by the different companies proportionately to the number of messages originating with each. Trunk talks may be booked several hours in advance, and this plan is in common use. If a called subscriber proves not to be in, the caller has to pay the unit trunk charge all the same, but is allowed a second inquiry later in the day, when, if his man is then in, he obtains a connection without further payment.

3. **International trunk line communication.**—This is at present, and, owing to the geographical situation of Norway, is likely to be for a long time, restricted to the metallic circuit trunk line to Stockholm. The length of the line is about 325 miles, and the tariff 1*s.* 8*d.* per three minutes, a rate which is found to produce a satisfactory traffic. The line has been erected by the Norwegian and Swedish State telegraph departments within their respective territories, but on the Norwegian side it is worked by the Christiania Telephone Company. By agreement with the State, only those subscribers who have special metallic circuits are allowed to be connected to the trunk. There are already seventy such metallic circuits (for which an additional subscription of 3*l.* 6*s.* 9*d.* per annum is charged) in Christiania. To cover operating and administrative expenses the State pays the company 1·9*d.* on each international trunk talk originating in Norway; but, on the other hand, the company must make all connections demanded from Sweden gratis.

4. **Telephoning of telegrams to the State telegraph office.**—This is practised very largely, and is conducted by the company's employees, who attend at the State telegraph office for the purpose and who are sworn to observe secrecy. They write down and hand to the Government clerks messages dictated to them through the telephone, and receive from the Government clerks, and telephone on, messages destined for the subscribers, copies of which are afterwards delivered by messenger. Non-subscribers may forward telegrams in this manner from the public telephone stations, which thus become branch telegraph offices, the use of which, however, entails payment of the company's charges in addition to the ordinary telegram tariff. The State pays the Christiania Company an annual subsidy of 27*l.* 15*s.* 5*d.* in respect

to this service ; but in all other places the proprietors of the telephone exchanges have to rely entirely on the charges they impose on their subscribers, although they too, as a rule, have to furnish the necessary attendants at the telegraph offices.

5. **Telephoning of messages for local delivery.**—Subscribers (and non-subscribers using public telephone stations) may ring up the central office and dictate messages to be delivered direct by company's messenger without the intervention of the State.

6. **Public telephone stations.**—There are 71 of these in Christiania and 45 in the suburbs, making 116 in all. Many are at subscribers' offices. In this case the keepers pay the ordinary tariff for their connections and are permitted to retain 30 per cent. of the receipts. In a good many instances automatic slot boxes are employed to receive the initial payment of ten öre (1·3*d.*), without which no service is rendered ; in others, a simple box is hung up into which the user drops the coin. The charges for trunk talks and telegrams are paid to the keeper, these being too variable and complicated to be dealt with by automatic boxes. The slot machine favoured, after several years' experimental trial of many different patterns, is that of Mr. Jakobsen, of Christiania. Subscribers pay the same as strangers when using the public stations ; but quarterly, half-yearly, and yearly tickets, covering the use of one or more stations, are issued.

7. **Messenger service.**—Messengers are kept at, or within call of, the central station and some of the public telephone stations, who, on demand, are sent round to subscribers' offices or houses, or utilised to summon to a public station non-subscribers with whom subscribers wish to speak.

TARIFFS

1. **Rates for local exchange communication** :

	Per annum £	s.	d.
For one instrument on a direct line not exceeding 1,500 meters in length	4	8	11
For each additional 500 meters	0	8	4

For additional instruments on the same line and in the same building:—

	Per annum £	s.	d.
Per instrument, if to the same subscriber . . .	1	2	3
,, ,, other persons	1	13	4

For additional instruments on the same line but in different buildings:—

	£	s.	d.
Per instrument, if to the same subscriber . . .	1	13	4
,, ,, other persons	2	4	5
For an extra bell or extra microphone	0	5	7
,, receiver	0	3	4
A second person, unconnected with a subscriber in business, may use his instrument and have his name printed in the subscribers' list for	0	11	3

When one person or firm takes more than one connection the tariff rate of *each* is reduced by 11*s.* 1*d.* Thus a subscriber can have his private house joined up for 3*l.* 6*s.* 8*d.*, being 22*s.* 3*d.* less than the tariff for his two connections. Three exchange lines would cost such a person 11*l.* 13*s.* 6*d.*, and four 15*l.* 11*s.* 4*d.*, per annum. Lines that are only used six months out of the twelve are charged 3*l.* 6*s.* 8*d.* per annum.

Contracts for one year only. There is no payment down on connection as practised in Sweden. The subscribers do not find their own instruments, and the rates are inclusive of all expenses of installation, maintenance, and service.

2. **Rates for internal trunk communication.**—The time unit is five minutes.

Rates from Christiania to Gjövik, Toten, or Grän . .	3·3*d.*
,, ,, any other town connected . .	6·5*d.*

3. **Rates for international trunk communication.**—The time unit is three minutes. Rate between Christiania and Stockholm, 1*s.* 8*d.*; Drammen and Stockholm, 1*s.* 10*d.*

4. **Rates for the telephoning of telegrams.**—This important traffic may be paid for per single message, or by annual subscription.

For each message telephoned to the State telegraph office from a subscriber or a public telephone station, or telephoned from the State telegraph office to a subscriber :—

If not exceeding 20 words	2·6*d.*
For each additional 10 words	·66*d.*

Telegrams may likewise be telephoned to a public telephone station in the neighbourhood of which there is no telegraph office and delivered by messenger on payment by the sender of 2·6*d.* per telegram, without regard to the number of words ; if the addressee is not a telephone subscriber, a similar amount is collected from him also.

Subscribers who telegraph often, obtain a decided advantage by paying annually as follows :—

	Per annum £	*s.*	*d.*
For 100 telegrams	0	11	1
,, 101 to 300	0	16	8
,, 301 ,, 600	1	2	3
,, 601 ,, 1,000	1	7	10
,, each additional 500	0	5	7

Messages containing over twenty but under forty words are counted as two ; over forty but under sixty words, as three telegrams ; and so on.

Deposits to cover the cost of despatched telegrams are not obligatory, but the company can demand them if not satisfied with the standing of subscribers ; usually the company pays for the messages, and charges the subscribers 2 per cent. on the cost for the accommodation. Accounts are rendered once a month to subscribers of acknowledged position, or oftener, at the company's discretion. Accounts for telegrams emanating from hotels are rendered the same day.

The foregoing particulars apply to Christiania only ; in the provinces the charges for telephoning telegrams vary very much between limits of 2*d.* to 6·5*d.* per message.

5. **Rates for messages telephoned for delivery by the company :**

If addressee is located within 1 kilometer of central station .	4*d.*
,, ,, ,, 2 kilometers ,, ,, .	5·3*d.*
,, ,, ,, 3 ,, ,, ,, .	6·5*d.*

These charges cover thirty words, exclusive of address and signature, and are increased, irrespective of distance, by ·132*d.* for each extra word. A person receiving such a message may send back by the messenger a written reply at half-price if not exceeding thirty words, with ·132*d.* for each extra word.

6. **Rates applicable at public telephone stations.**—Persons using a public station must pay a first charge of ten ore, or 1·3*d.* This covers a five-minute talk with a subscriber within Christiania. If any other service is taken advantage of, the following *additional* charges are made :—

For a five-minute talk to a suburban subscriber (according to distance)	2*d.* to 3·3*d.*
For a five-minute trunk talk	6·5*d.*
For a telegram to the State telegraph office, 2·6*d.* for 20 words, with ·66*d.* for each additional 10 words .	
For a message for local delivery by the company (according to distance of addressee from point of delivery)	4*d.*, 5·3*d.*, and 6·5*d.*

Habitual users of public stations may obtain some reduction on the tariff charges by subscribing for quarterly, half-yearly, or yearly tickets.

7. **Rates for messenger service :**

For fetching a non-subscriber to a public station (payable by person called)	2·6*d.*
For sending a messenger to a subscriber's premises . .	1·3*d.*

WAY-LEAVES

None of the companies possesses any compulsory powers, and way-leaves have to be arranged by negotiation with the proprietors and local authorities concerned. In Christiania facilities have, as a rule, been obtained on favourable terms, the maximum consideration given being a free telephone connection, corresponding to 4*l.* 8*s.* 11*d.* per annum. Many buildings are roofed with iron, which is not nearly so susceptible to damage as slates or tiles, a fact which has helped the company to obtain and keep its way-leaves.

SWITCHING ARRANGEMENTS

The existing switch-board at Christiania is an ordinary Western Electric Company's single-wire, double-cord, series multiple, with an ultimate capacity of 6,400. The test employed differs, however, from the usual one, inasmuch as the testing-cord includes a make-and-break, the effect of which is to give the operator a vibrating signal instead of a single click when a line proves to be engaged. The number of connections averages about nine per subscriber per diem, and each operator attends to 100 lines. The arrangements for trunk-line switching comprise a special section to which each operator has a sufficient number of junction lines to meet the requirements of her own set of subscribers, these junction lines being used indiscriminately for up and down traffic. In case of need, an operator can borrow additional junctions from the sections to her left and right. At the trunk section four lines are allotted to each girl, who, in addition to the actual switching, has to make the necessary notes for the subscribers' accounts. The testing, lightning-guard, and cross-connecting boards are of an ordinary pattern, and call for no remark. The present switching arrangements are to give way during 1895 to a new switch-board by the Bell Manufacturing Company of Antwerp, comprising parallel jacks, self-restoring drops, and accommodation for 9,000 metallic circuits. The new installation is to cost some 10,000*l.*, a fact which does not seem to augur any lack of confidence or of expectation on the part of the Christiania Telephone Company in the sufficiency or possibilities of a 4*l.* 8*s.* 11*d.* rate.

Called subscribers are rung by the operator, and much of the confusion attendant on the frequent dropping of the ring-off shutters avoided. The service is smartly performed, and the speaking generally very good.

In the neighbourhood of Christiania there are a few groups of subscribers working by means of automatic commutators (Cedergren and Ericsson's patent), placed generally at or near a railway station. These groups, and some others who subscribe amongst themselves for the housing and operating of an ordinary switch-board, communicate with the capital by a single junction wire, and are admitted at very low rates of subscription.

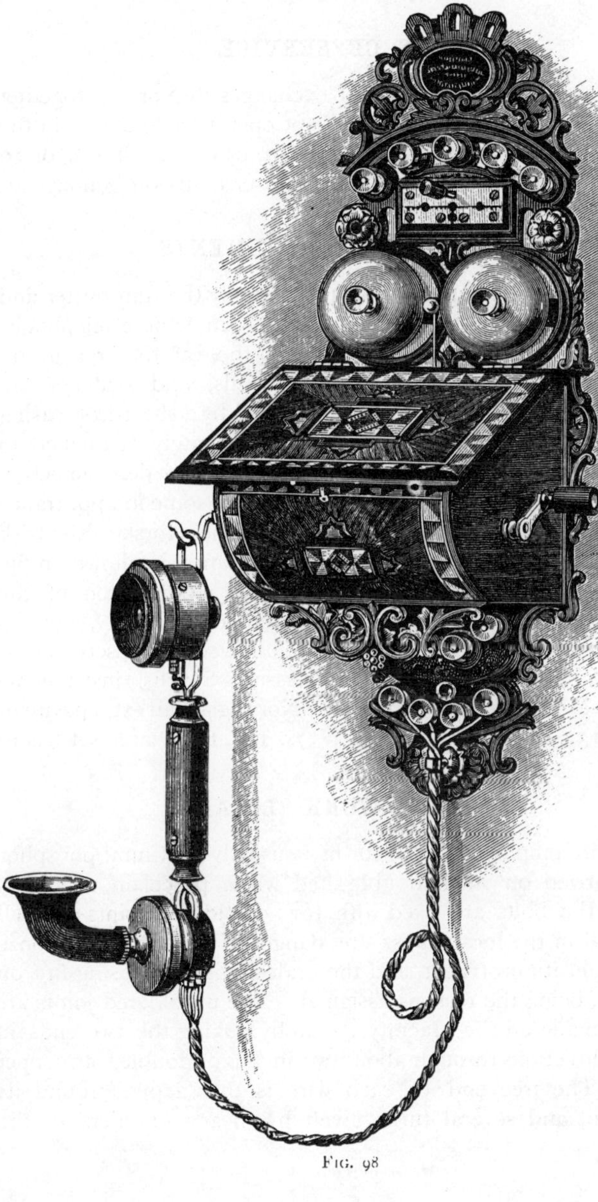

Fig. 98

HOURS OF SERVICE

Christiania and the other chief exchanges in Norway, together with many of the smaller ones, are kept open permanently ; in the others the hours of service vary from 6, 7, or 8 A.M. till 8, 9, or 10 P.M. on week days, with, sometimes, shorter hours on Sundays.

SUBSCRIBERS' INSTRUMENTS

These comprise magneto ringers. Usually the transmitter and receiver are attached to the same handle, in 'micro-telephone' form, so that when the receiver is held to the ear the transmitter is before the mouth. The magneto, bells, and switches are mounted on a cast-iron frame and protected by a sheet-iron casing which forms a writing-desk, and is so elaborately enamelled in imitation of ornamental and inlaid woods as to defy detection by the eye. The instruments, which are handsome in appearance and of good workmanship, are made by the Norske Elektrisk Bureau, Christiania. Their general appearance is shown in fig. 98. The transmitter is usually the Oyan modification of the Hunnings. When transmitters are mounted separately, double-pole receivers of the Bell type are employed. Subscribers in Christiania may, on demand, have Ericsson's Swedish instruments fitted ; but as these are dearer than the Norwegian, an extra payment down of 13*s*. 4*d*. for a wall-, and 1*l*. 7*s*. 10*d*. for a table-set has to be made.

OUTSIDE WORK (LOCAL)

The wire employed in the towns is usually 1·25 mm. phosphor bronze, carried on small double-shed white porcelain insulators in which the bolts are fixed with tow. Soldered joints are still exceptional in the local wires—the danger of softening the bronze by the application of heat, and the undesirability of using fire on housetops, being the reasons assigned. The unsoldered joints are generally made on the Macintyre plan by poking the two ends in opposite directions through about two inches of double soft copper tubing. The free end of each wire is then lapped round its companion, and several turns given by means of pliers to the whole joint, the effect being to twist the soft copper tubes into reversed spirals, within which the line wires are so tightly grasped that the parts in contact are permanently protected from the

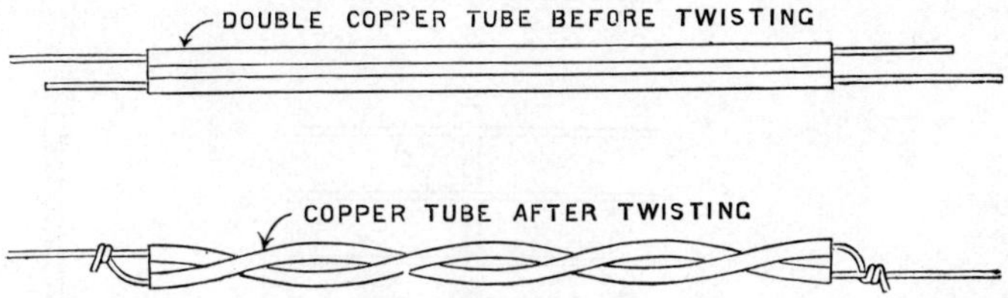

FIG. 99

weather. This joint is shown in fig. 99. When it is considered desirable to solder, the form of joint shown in fig. 100 is used : the heat being applied at the point A, can have no effect on the

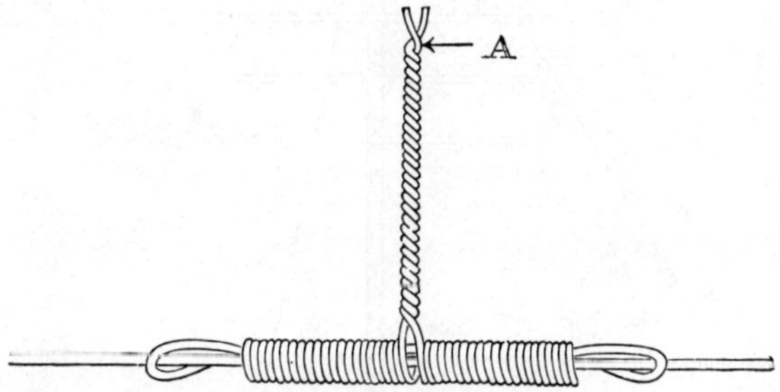

FIG. 100

running wire. When too much vibration is set up in the houses it is damped by placing several inches of wire on each side of the insulator (fig. 101) in a split vulcanised rubber tube, and then

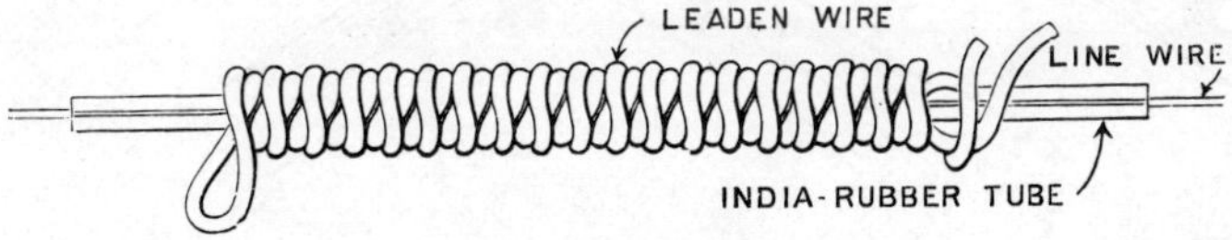

FIG. 101

tightly twisting over all two spiral layers of heavy leaden wire or strip. The central station fixture is a large and substantial structure built up of channel and angle iron, but devoid of decorative pretensions. From this fixture a great many of the subscribers' wires are carried in aerial cables, each containing twenty single wires

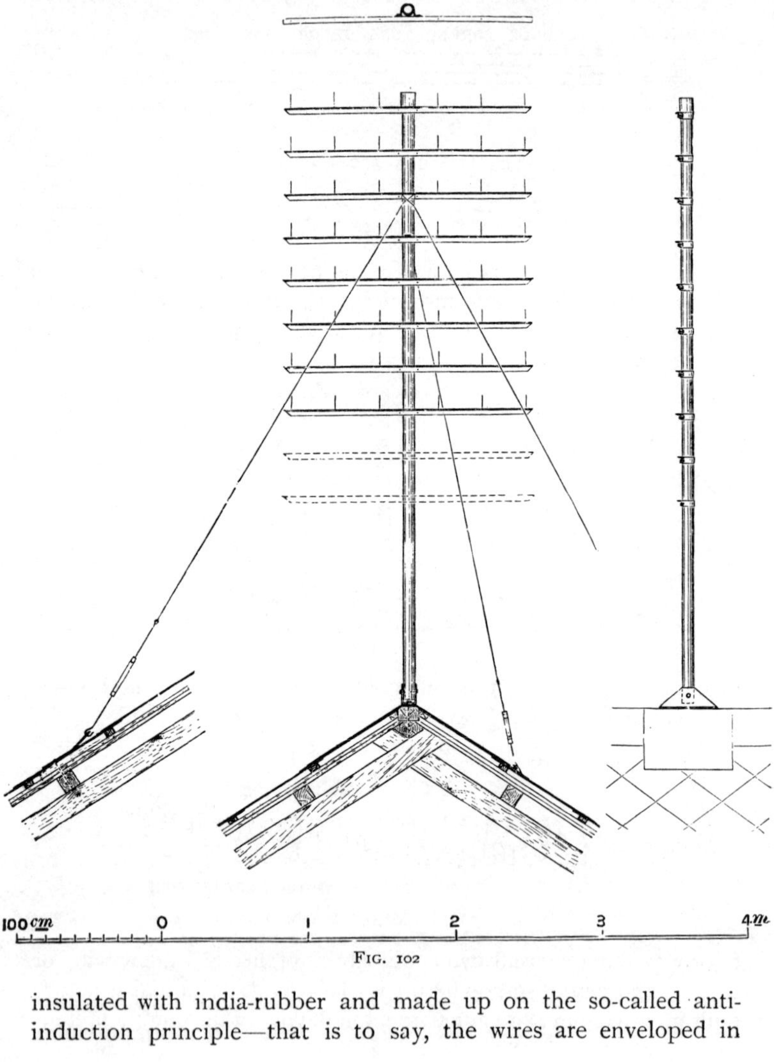

Fig. 102

insulated with india-rubber and made up on the so-called anti-induction principle—that is to say, the wires are enveloped in metal foil connected to earth. The cables are slung by galvanised iron hangers from stranded steel suspenders, and at their junctio with the open wires, which takes place as soon as the crowded vicinity of the central station is cleared, are passed through joint boxes fitted with lightning arrestors. The standards are well designed, and carefully erected with due regard to safety in the

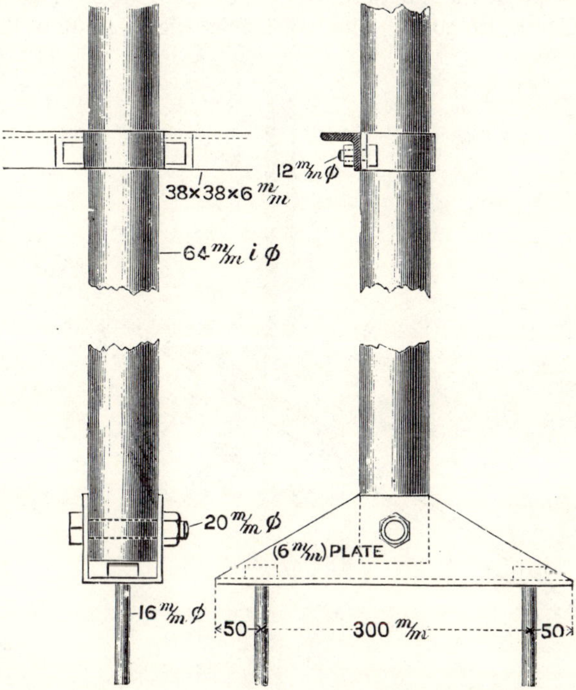

FIG. 102A

face of untoward fires or storms. The single form (figs. 102 and 102A) consists of a wrought-iron tube fitted with English angle-iron cross-arms. The foot-plates and fastenings are all of Swedish iron. The standard is sometimes placed on the slope of a roof (fig. 102B) instead of on the ridge. A triple standard for 300 wires, with its details, is shown in figs. 103 and 103A. In this case the uprights are each composed of two pieces of Belgian channel-iron bolted together, and the cross-arms are also of channel-iron arranged so as to form a shelter for any insulated wires that may be used for cable or cross connections. The uprights, when extra strength is called for, are strutted on one or both sides with riveted channel- or angle-iron. The uprights are riveted to iron foot-plates adapted to the slope of the rafters to which they are bolted. The ground pole work is also good. The larger poles (fig. 104) are of the best

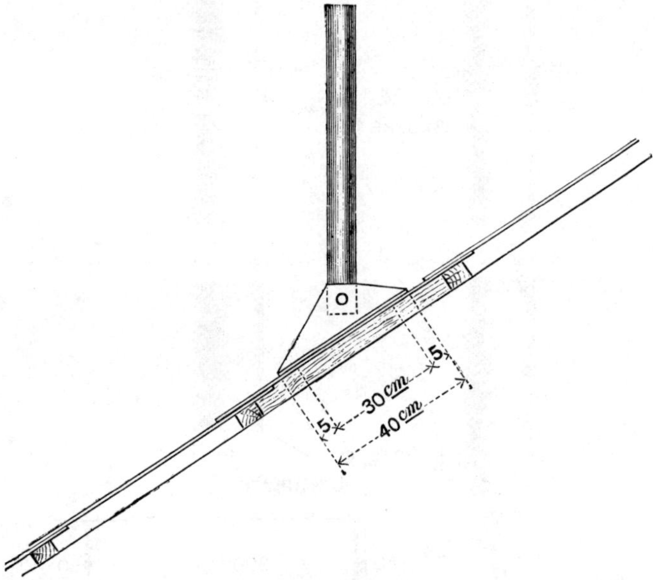

FIG. 102E

fir; their butts are usually soaked in boiling creosote to above the ground line, and the weather is excluded by roofs of the English pattern. The arms are of angle-iron (wooden arms are quite exceptional in Norway) made into a frame by riveting to four vertical bars, the frame being fastened to the wood at three points by strong straps and wood screws. This plan secures a neat job, since the frame is constructed before attachment to the pole, and it is easy to make the arms truly parallel; on the English plan it is difficult to secure parallelism when so many long arms have to be notched for and attached individually, perhaps at different times and by different men. On the other hand, the English method permits of arms being added exactly as they are wanted; while a frame must contain a certain number of spare arms, representing unremunerative capital, to allow for developments.

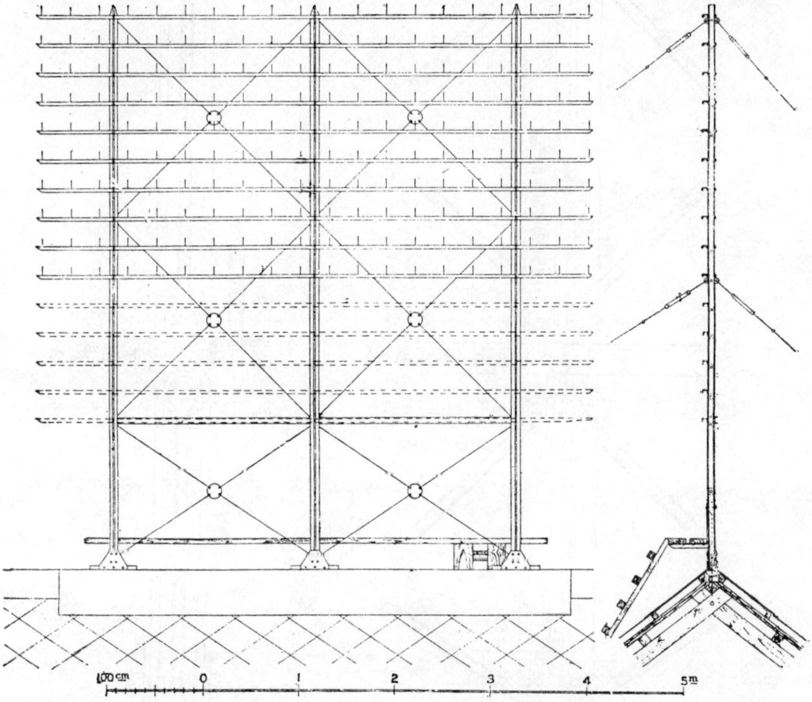

FIG. 103

But if the Norwegians with microscopic tariffs can afford to invest capital in neat and pleasing workmanship, such poles should not be absolutely beyond our own reach. There is at present no underground work in existence in Norway, but it is proposed to make a beginning with it in connection with the change to metallic circuits shortly to be commenced in Christiania.

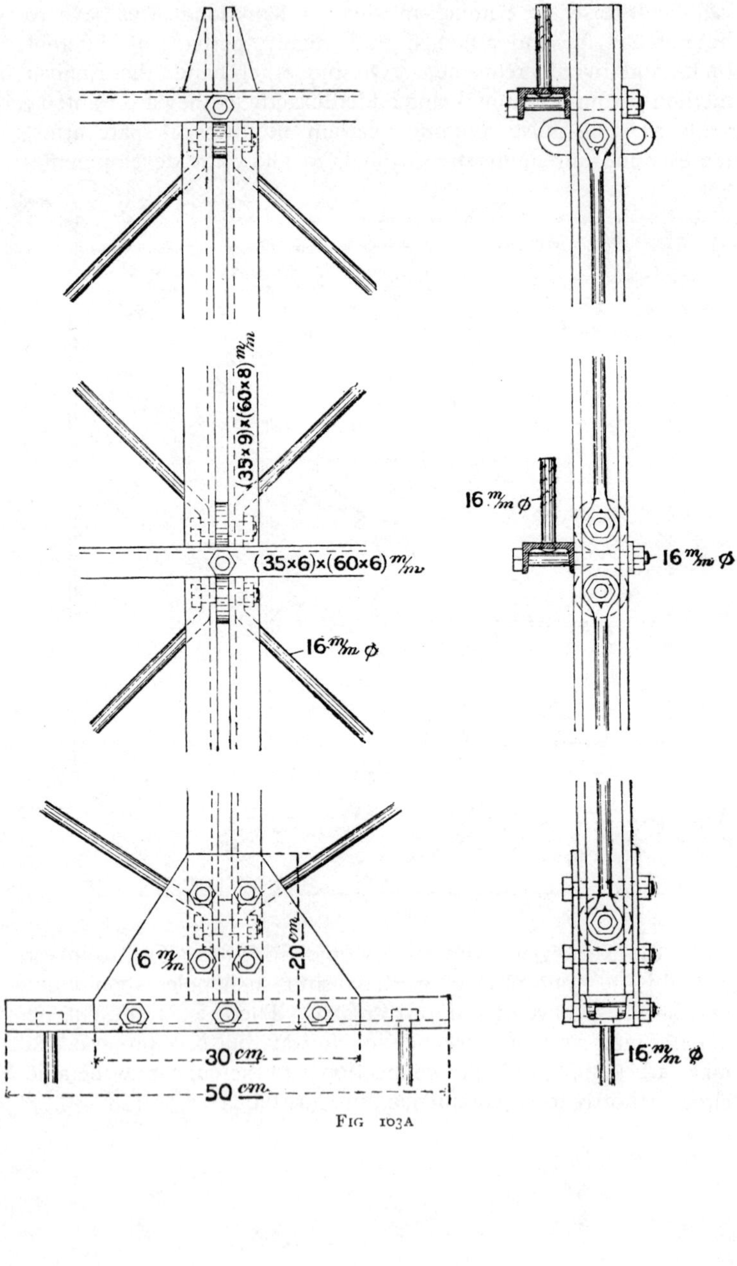

(35×9)×(60×8) m/m
(35×6)×(60×6) m/m
16 m/m ϕ
16 m/m ϕ
16 m/m ϕ
(6 m/m
20 cm
30 cm
50 cm
16 m/m ϕ

FIG 103A

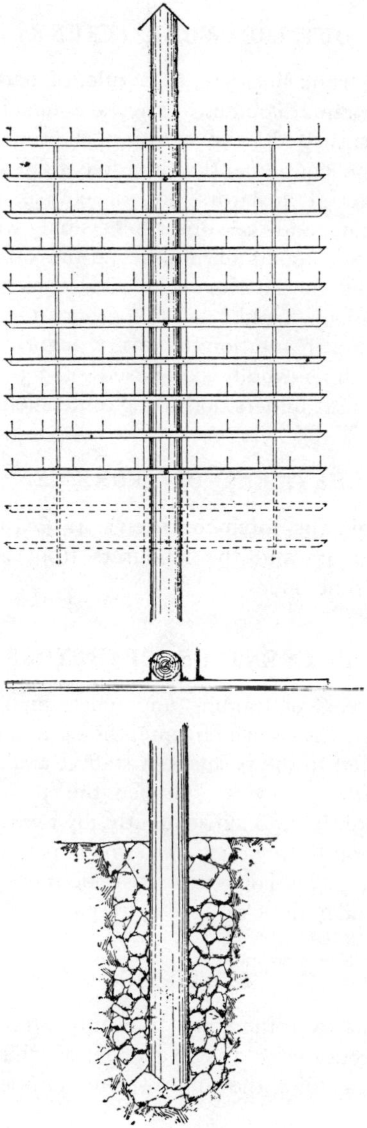

FIG 10

OUTSIDE WORK (TRUNK)

The internal trunk lines are, as a rule, of hard-drawn copper of from 2 to 2·5 mm. diameter. They are metallic circuits, and are crossed, not twisted: the crossing is properly carried out, and the lines are consequently quite free from overhearing and inductive noises. Translators are, of course, placed between the metallic circuits and the subscribers' single wires. The international trunk to Stockholm, unlike all the others, was erected and is maintained by the Norwegian Government on the Norse side of the frontier. Unlike all the others, too, it is twisted so as to complete a revolution at every eighth pole. On the Norwegian side it is wholly composed of 3·3 mm. hard-drawn copper. The line is understood to be quite silent and the speaking very good.

PAYMENT OF WORKMEN

In Christiania the foremen receive 4*s.* 6*d.*, the skilled wiremen from 3*s.* to 4*s.*, and the labourers from 2*s.* 6*d.* to 3*s.* per working day of nine hours.

PAYMENT OF OPERATORS

After two years of training and occasional employment as reserve operators, during which time they are paid by the hour, girls are appointed to the permanent staff at a salary of 2*l.* 15*s.* 7*d.* per month; after two years' service the pay is advanced to 3*l.* 1*s.* 1*d.* per month; and subsequently, by two-yearly increments of 5*s.* 6*d.* per month, to 3*l.* 17*s.* 8*d.*, which is the maximum for a simple operator. The daily work is six hours. They take turns at night and Sunday duty without extra pay.

STATISTICS

In Christiania town there are (November 1894) 4,174 instruments in connection with the exchange, of which 3,786 are on direct wires. Including the suburbs, the number of instruments is 4,627.

No statistics for the whole of Norway of later date than 1892 are forthcoming. In that year the total number of subscribers was returned as 9,490, making use of 10,437 instruments. The total length of their wires was 11,878 kilometers, of which Christiania possessed 4,210, Bergen 1,322, Drammen 355, and Trondhjem 350 kilometers; and of the trunk lines, 4,908 kilometers. The number of exchanges was 175; of public telephone stations, 546. The trunk talks numbered 391,966; and the telegrams telephoned 78,323, of which 43,594 were credited to Christiania. The total amount of receipts was 31,136*l.*; of working expenses and repairs, 19,762*l.*; and of capital expended in construction, 118,790*l.* The cost of connecting each subscriber, even adding in the cost of the trunks, which we must do as it is not returned separately, was consequently only a little over 12*l.*, truly a marvellous result when it is borne in mind that most of the material and apparatus used had to be imported and to pay duty at the Norwegian Custom House. The figure of 12*l.* per subscriber, however, tallies well with experience in Great Britain when results have not been vitiated by incompetence and mismanagement.

In order to show how a 4*l.* 8*s.* 11*d.* inclusive rate can be made to pay in a capital city, the accounts of the Christiania Telephone Company for 1893 are, with the kind permission of Mr. Knud Bryn, annexed.

CHRISTIANIA TELEPHONE COMPANY'S ACCOUNTS, 1893

1 krone = 1*s.* 1¼*d.* 1*l.* = 18·2 kronor

Revenue Account

Dr.

	Kronor	Kronor
Management		
Salaries . . .	18,350·11	
General office expenses and advertising .	5,474·12	
Messengers' wages and uniforms . . .	4,614·46	
		28,438·69
Central stations working expenses		
Salaries; watchmen; central station messengers . . .	61,934·10	
Contribution to the lady operators' benevolent fund . .	2,342·06	
Subscribers' lists and supplements, including distribution .	4,321·82	
		68,597·98
Working expenses outside system		
Engineers' salaries .	7,409·74	
Inspection and improvements . .	21,311·60	
Maintenance . .	61,628·58	
Roof repairs and wayleaves . . .	9,171·21	
Tools and instruments	1,688·51	
Contribution to the workmen's benevolent fund . . .	1,640·69	
		102,850·33
Sundry expenses		
Rent, central station, offices and stores .	6,000·00	
Wharfage, lighting, firing, and cleaning .	7,046·78	
Building account .	876·18	
Royalties . . .	5,463·29	
Sundry expenses in the suburbs . . .	3,689·31	
Insurance and taxes .	8,113·17	
Bad debt reserve fund .	1,500·00	
Sundry expenses .	4,066·34	
Interest . . .	1,407·51	
Directors' fees . .	5,000·00	
		43,162·58
Central station renewals . . .	25,000·00	25,000·00
Written off		
Fixtures account, 10 per cent. of value . .	1,005·08	
Building capital account . . .	10,000·00	
Telephone system account . . .	5,396·41	
		16,401·49
Dividend at rate of 5½ per cent. (2,847*l.*) .	51,821·00	51,821·00
		Kr. 336 272·07

Cr.

	Kronor	Kronor
Subscriptions collected during the year .	333,781·65	
Less proportion carried forward of unearned rentals . . .	30,999·07	
		302,782·58
Receipts at public telephone stations . .	5,644·05	
Receipts, messenger service . . .	394·49	
Receipts for telephoning and delivering local telegrams .	2,912·27	
Receipts, trunk lines .	24,008·68	
Sale of shares . .	530·00	
		33,489·4
		Kr. 336,272·07

BALANCE SHEET

Liabilities	Kronor	Kronor
Share account . .		942,200·00
Last year's dividends unclaimed . .	3,165·00	
Dividend for 1893 .	51,821·00	
		54,986·00
Set aside for central station renewals .		25,000·00
Christiania Savings Bank loan . .		100,000·00
Mortgage on building No. 12 Slotsgade .		40,000·00
Lady operators' benevolent fund . .		10,644·39
Workmen's benevolent fund . .		1,594·47
Sundry account		
Sundry creditors .	19,057·49	
Less for sundry debtors . . .	6,507·65	
		12,549·84
Proportion of rentals for 1894 paid in advance . . .		30,999·07
		Kr. 1,217,973·77

Assets	Kronor
Construction account	949,387·91
Trunk line account . . .	56,010·00
Buildings capital account . .	125,000·00
Stock of instruments and material	33,296·85
Sundry debtors, arrears of subscriptions	29,536·76
Cash in hand	1,286·34
Christiania Bank, cash balance .	23,455·91
	Kr. 1,217,973·77

Signed by the Directors of the Christiania Telephone Company.

EVALD RYGH. N. A. EGER. A. M. LUND. E. SUNDE.

KNUD BRYN, General Manager.

February 10, 1894.

I hereby certify that the above balance sheet is in conformity with the company's books.

TH. HAMMOND, Auditor.

CHRISTIANIA: *February* 20, 1894.

Note.—Since going to press, the accounts for 1894 have been received. They show subscriptions collected for the year Kr. 337,564, and the amount available for dividend Kr. 56,325 (3,095*l.*); the assets having increased to Kr. 1,368,703, and the share capital to Kr. 1,125,000. The usual dividend of 5½ per cent. was paid.

XVIII. PORTUGAL

HISTORY AND PRESENT POSITION

THAN Portugal few European countries possess shorter or more uneventful telephonic histories. In 1882 a concession was granted by the Government to the Edison-Gower-Bell Telephone Company of Europe, in virtue of which exchanges were soon established in Lisbon and Oporto. In 1887 the business was taken over by an English company formed for the purpose, the Anglo-Portuguese Telephone Company, Limited, of 53 New Broad Street, London, E.C. On this occasion the concession was renewed to the new company for a period of thirty years. The two exchanges have thrown out branches to the suburban towns in their immediate neighbourhood, but are not yet themselves in connection; while the remainder of Portugal remains, so far, an unexplored territory. Rumours have been heard of an international trunk line to Madrid, but the scheme has not yet assumed any solidity. All lines in Portugal are still single. Practically the only service rendered to the public is the local exchange connection, since there is no telephoning of telegrams, no telephonograms, no trunk lines, and no public telephone stations. There are call offices for the use of subscribers only on the production of a ticket of identity, but this can scarcely be considered a public convenience. In January 1895 the number of subscribers was returned at 763 for Lisbon and 720 for Oporto, including the suburban exchanges in each case. Lisbon has three, and Oporto five suburban switch-rooms.

SERVICES RENDERED AND TARIFFS

1. **Local exchange intercommunication between the subscribers in Lisbon and Oporto and their respective suburbs.**—The tariff depends on the length of line and nature of the connection, as follows :—

Distance	Business places		Doctors and private houses
	First connection	Subsequent connections	
	£ s. d.	£ s. d.	£ s. d.
1 kilometer, per annum . .	7 10 0	5 12 6	5 12 6
1½ " " . .	9 0 0	5 12 6	5 12 6
2 " " . .	10 5 0	6 7 2	6 0 0
2½ " " . .	12 0 0	7 4 6	6 7 2
3 " " . .	12 15 0	7 17 6	7 4 6

When private houses are joined as an extension from a business place considerable reductions are made, as follow :—

Extension Instrument	£ s. d.
In the same building, per annum	2 5 9
500 meters distant "	3 15 0
1 kilometer " "	4 10 0
1½ kilometers " "	5 12 6
2 " " "	6 0 0
2½ " " "	6 7 2
3 " " "	7 4 6

In considering these tariffs it must be borne in mind that in the terms of its concession the company pays three per cent. of its gross revenue to the State, and that they are considerably below the maximum permitted to the company by the terms of its concession.

WAY-LEAVES

Considerable difficulty was at first experienced in obtaining attachments, but this has now in great measure been happily overcome. The company possesses no exceptional privileges, and is wholly dependent on the good will of the proprietors.

SWITCHING ARRANGEMENTS AND SUBSCRIBERS' INSTRUMENTS

The necessity for multiple boards has not yet been felt [traffic is small in Portugal, the calls averaging only two per line per day at Lisbon, and four at Oporto], and the service is carried on by simple boards manufactured by the Consolidated Telephone Construction and Maintenance Company, London. These are of three types: (1) a 50-line modified cross-bar, peg commutator, mounted vertically, with the indicators above; (2) a

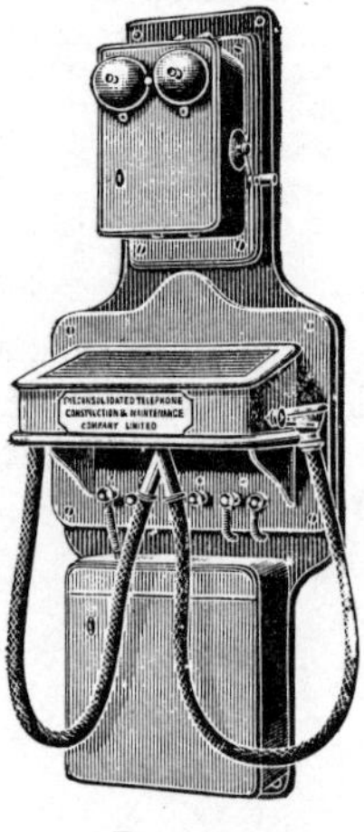

FIG. 105

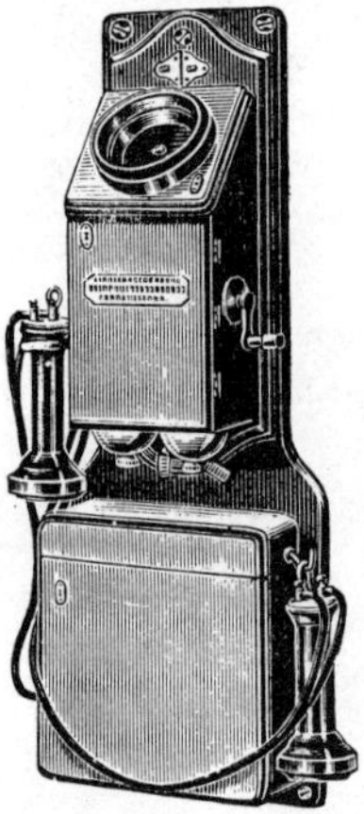

FIG. 106

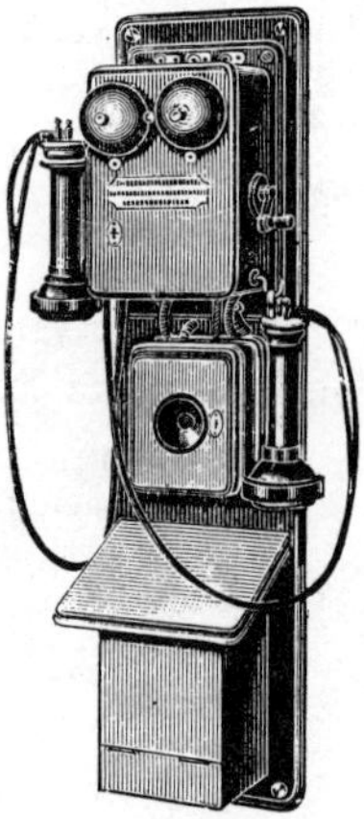

FIG. 107

100-line board of the same nature, but mounted horizontally, with the indicators on a vertical board at the back; (3) a 100-line spring-jack and plug and cord board, the general arrangement of which resembles that of the Western Electric Standard board, but without ring-off indicators. The first and third are used at Lisbon, the second at Oporto. In all three speaking has to be carried on through two indicators, which both fall when a ring through or ring off is given. Each operator usually attends to seventy-five subscribers, but in Oporto during the busy hours this number is reduced to fifty. Magneto ringers are exclusively used.

Originally the subscribers' sets consisted of Gower-Bell instruments, combined with magnetos and battery-boxes on a common back-board (fig. 105); later, the Gower-Bell receiver and tubes were replaced by Bell receivers, the transmitter being retained (fig. 106); later still, the Blake transmitter succeeded the Gower, and the instrument assumed the aspect which was so long familiar in Great Britain (fig. 107). All the instruments have been manufactured by the Consolidated Telephone Construction and Maintenance Company. The most recent form of transmitter supplied by this company is a Hunnings of the construction shown in fig. 108, in which A is an ebonite mouthpiece, which directs the sound waves to a ferrotype diaphragm F, having

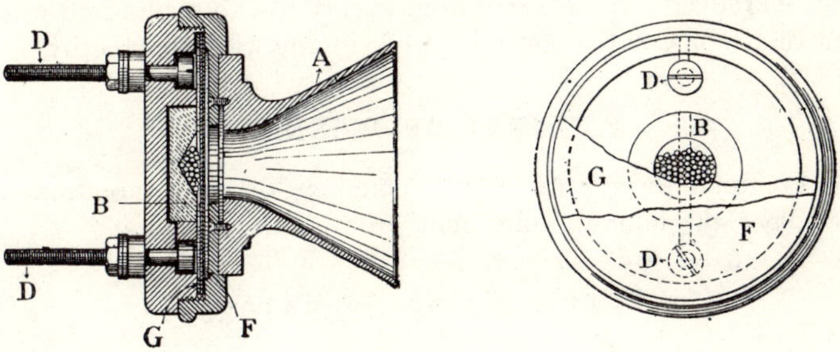

FIG. 108

behind, and in true contact with it, a thin carbon disc G. B is a rigidly fixed carbon block furnished with a conical pocket, which is nearly filled with truly-spherical carbon balls. The electrodes are the carbon disc and carbon block respectively, and form their connections through the screws D, which also serve to clamp the transmitter to the magneto. When used with two or three cells and a good receiver, the loudness of the transmission obtained is very remarkable, while its clearness leaves nothing to be desired.

HOURS OF SERVICE

The Portuguese exchanges enjoy a perpetual day and night service.

OUTSIDE WORK

The wire generally employed is 1·25 mm. phosphor bronze, although there is also some 1·5 mm. bronze and 2 mm. galvanised iron. With the exception of a few short lengths of aerial cable, used where way-leaves were difficult to obtain, the whole of the system is open wire. No commencement has been made with underground work. The Lisbon roofs are not well adapted for the erection of standards, and the fixtures are mostly of the wall-bracket kind illustrated in the French, Italian, and Austrian sections. These fixtures and their wires are erected and attended to by the aid of telescopic fire-escape ladders. In Oporto single iron standards on the roofs and ground poles are also employed ; these carry from twenty-five to a hundred wires, but their constructive details are not in any wise noteworthy.

PAYMENT OF WORKMEN

Foremen receive 6*s.* 8*d.*, wiremen 3*s.* 6*d.*, and labourers 2*s.* per day ; the hours of duty being from 7 A.M. till 6 P.M.

PAYMENT OF OPERATORS

Girls commence as probationers, and give their services gratis until competent. Thereafter they receive 30*s.*, rising to 2*l.*, and finally to 2*l.* 10*s.* per month. Lady superintendents receive 3*l.* The hours of daily duty are eight. The service between 6 P.M. and 8 A.M. is performed by men.

XIX. ROUMANIA

THE Government has assumed the exclusive care of telephone exchanges in Roumania, and has most wisely determined to adopt the metallic circuit throughout. Exchanges have been opened in Bucharest, Braïla, Galatz, and Crajowa ; but development halts, there being only some 100 members at Bucharest after nearly two years' working. This disappointing result may perhaps be most reasonably ascribed to the tariff in operation, which, everything considered, is probably the most illiberal in Europe. Three trunk lines have been put into use between Braïla and Galatz, one being an exclusively telephonic metallic loop, and the other two adaptations of Van Rysselberghe's system to existing telegraph wires. Braïla and Galatz also speak to Bucharest on a duplex line. The capital has likewise communication with Ploesti and Sinaïa, in which towns there are public telephone stations, but no exchanges. The subscribers' sets of instruments comprise transmitter, two receivers, bell, and lightning protector.

SERVICES AND TARIFFS

1. **Local exchange communication.**—Payments come under three headings : (*a*) contribution to the cost of the line and instrument ; (*b*) annual subscription ; (*c*) charge for conversations originated exceeding 1,000 per annum. These again vary with the location, inside or out of the fortifications, of the subscriber.

The contribution on joining amounts to 6*l.*, which is payable in four quarterly sums of 1*l.* 10*s.*

The annual subscription, which franks only 1,000 communications not exceeding five minutes in duration each, per annum, is :—

Within the fortifications	8*l.*
Without the fortifications, but within three kilometers of the exchange	20*l.*

When more than 1,000 conversations per annum are originated by any subscriber the excess must be paid for at the rate of 16*s.* per 100 or fraction thereof if he is located within the fortifications, and of 40*s.* if without. Contracts are accepted for three years only on first joining, which are subsequently renewable from year to year

2. **Internal trunk line communication.**—The time unit is three minutes.

For the first 100 kilometers or less	14·4*d.*
Each additional 100 kilometers	9·6*d.*

A considerable reduction may be had by paying for fifty talks in advance, thus—

	£	*s.*	*d.*
100 kilometers or less, 50 three-minute talks . .	2	0	0
Each additional 100 kilometers, extra	1	10	0

3. **Public telephone stations.**—The time unit for local talks is five minutes.

Non-subscribers	9·6*d.*
Subscribers to local exchanges, or persons who have paid for 50 trunk talks in advance	4·8*d.*

Trunk talks are charged as from subscribers' offices.

4. **Telephoning of telegrams** :

Per telegram forwarded or received by a subscriber . .	·96*d.*
In addition, for each five words contained in the telegram .	·48*d.*

Messages must be in a language understood by the telegraph clerk who receives or dictates them by telephone. Copies of telegrams telephoned to subscribers are subsequently delivered by messenger without charge.

5. **Messages telephoned for local delivery.**—For a message containing twenty words telephoned by a subscriber located within the fortifications to the central office for delivery locally to a non-subscriber, the charge is 4·8*d.* plus 1·92*d.* for each twenty words in excess. For a subscriber situated beyond the fortifications, or for a non-subscriber telephoning from a public station, these charges are doubled.

XX. RUSSIA

THE first exchanges in Russia—those of St. Petersburg and Moscow, opened in 1881—were due to the enterprise of the International Bell Telephone Company, which subsequently obtained concessions for, and commenced business in, Lodz, Odessa, Riga, and Warsaw. The rates charged by this company in the two first-named towns (in which it is secured by the terms of its concession from competition for a long term of years) have the distinction of being the highest in Europe—25*l.* per annum, out of which it has, in common with all other concessionaries, to pay 10 per cent. to the Government. In other towns, however, rates are much more reasonable. In Rostoff-on-Don (680 subscribers) and Reval (110 subscribers), for instance, for which places Mr. C. Siegel of St. Petersburg holds the concessions, the annual subscriptions are 12*l.* 10*s.* and 10*l.* respectively. The radius allowed is, however, liberal, extending to 3 versts ($2\frac{1}{4}$ miles) from the exchange, within which area no extra charge is made. For many persons 12*l.* or 10*l.* applied in this manner may mean a better bargain than a 5*l.* rate restricted to one mile. The State has also opened a good many exchanges, and contemplates the construction of an extensive system of trunk lines.

At date of writing (January 1895) the Odessa–Nicholaieff is the only one of importance reported finished, although Sebastopol is connected with Simferopol by railway wire, and other inter-town lines have been established for military purposes.

The system of construction adopted is the single wire, run overhead on roof standards and poles. Way-leaves are reported to be readily obtained on reasonable terms, but no information is forthcoming as to the status of the Government in this connection in the towns it itself exploits.

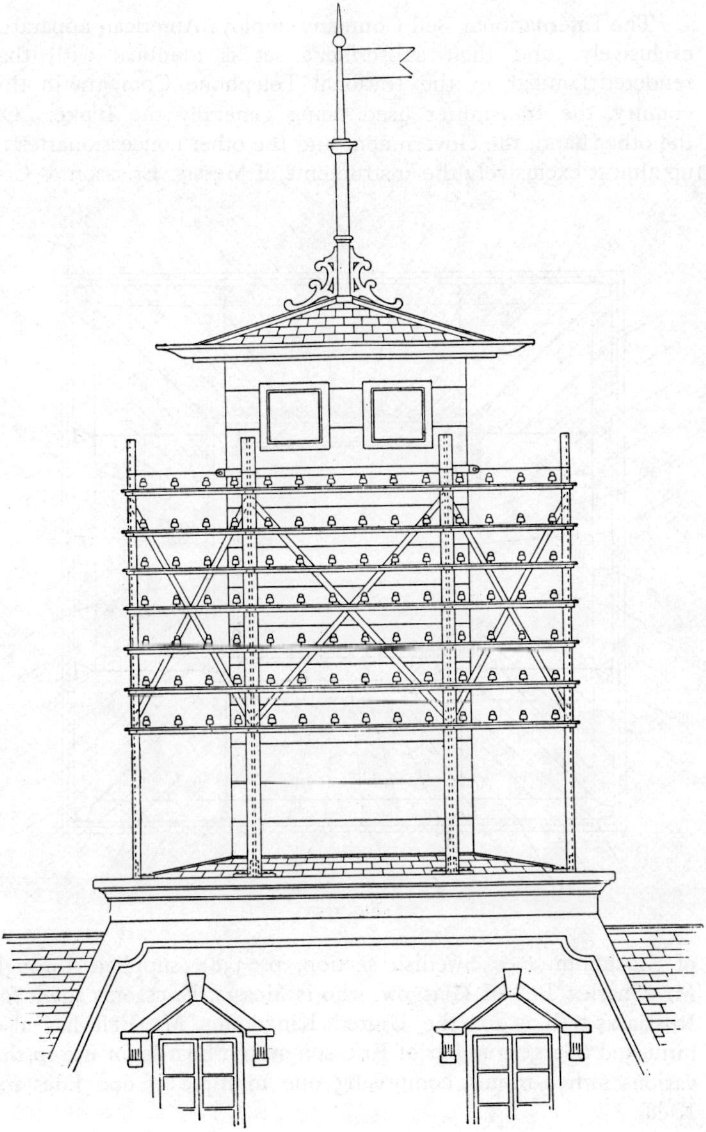

Fig 109

The International Bell Company employs American apparatus exclusively, and their subscriber's set is identical with that rendered familiar by the National Telephone Company in this country, the transmitter used being generally the Blake. On the other hand, the Government and the other concessionaries fit up almost exclusively the instruments of Messrs. Ericsson & Co.,

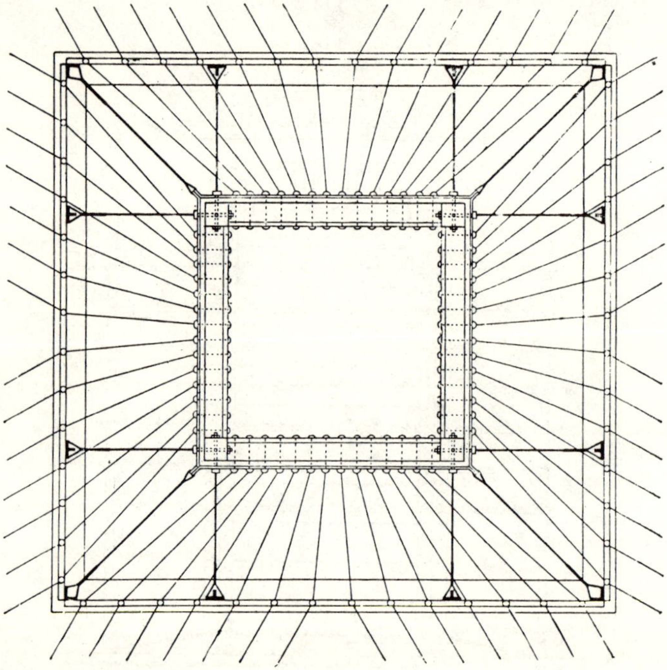

FIG. 109A

of Stockholm (see Swedish section, p. 358), supplied through Mr. Charles Bell of Glasgow, who is Messrs. Ericsson's agent for Russia as well as for the United Kingdom. Mr. Bell has also furnished a large number of Ericsson switch-boards for use in the various switch-rooms, comprising one multiple of 600 lines for Kieff.

The International Bell Company have Western Electric multiples at St. Petersburg (1,400 lines), Moscow (1,400 lines), and Warsaw (800 lines), with Gilliland boards at Lodz, Odessa, and Riga.

The principal Government exchanges are at Charkoff, Gatschina,

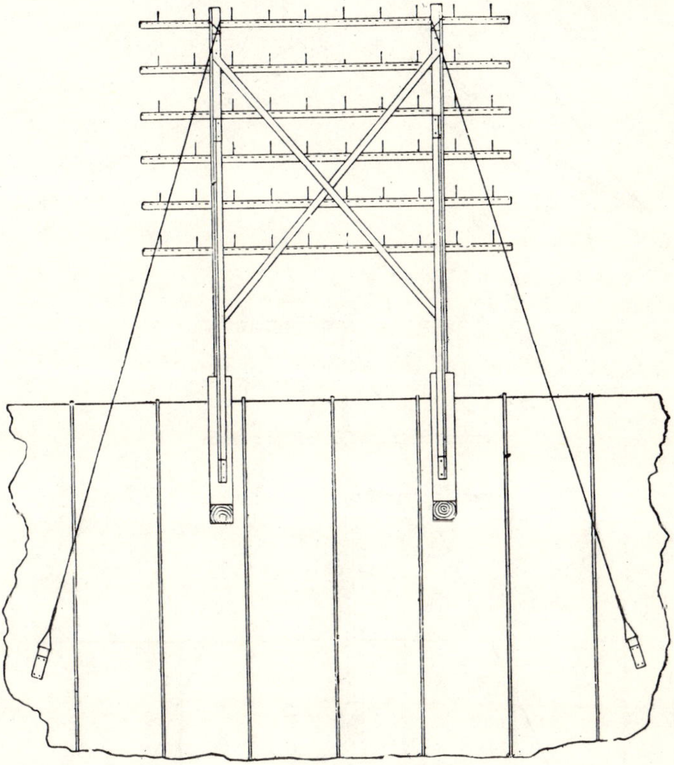

Fig. 110

Kazan, Libau, Nicholaieff, Nijni-Novgorod, Novorosisk, Novotcherkask, Pavlovsk, Selo, Taganrog, and Zarsko.

Through the kindness of Mr. C. Siegel the author is enabled to give some idea of the designs of Russian wire supports. Figs. 109 and 109A represent, in elevation and plan, the central station

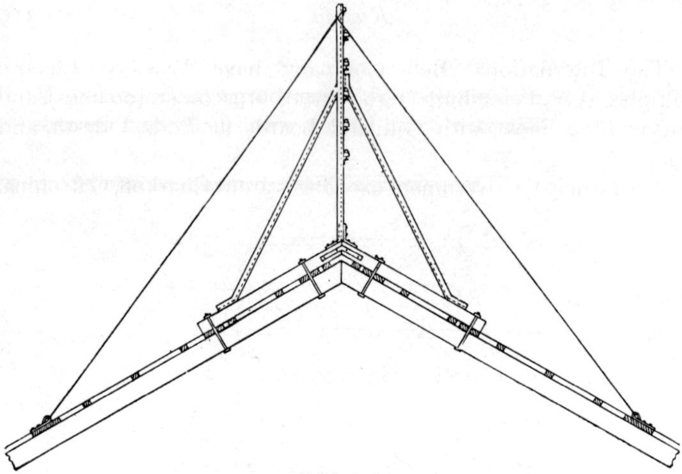

FIG 110A

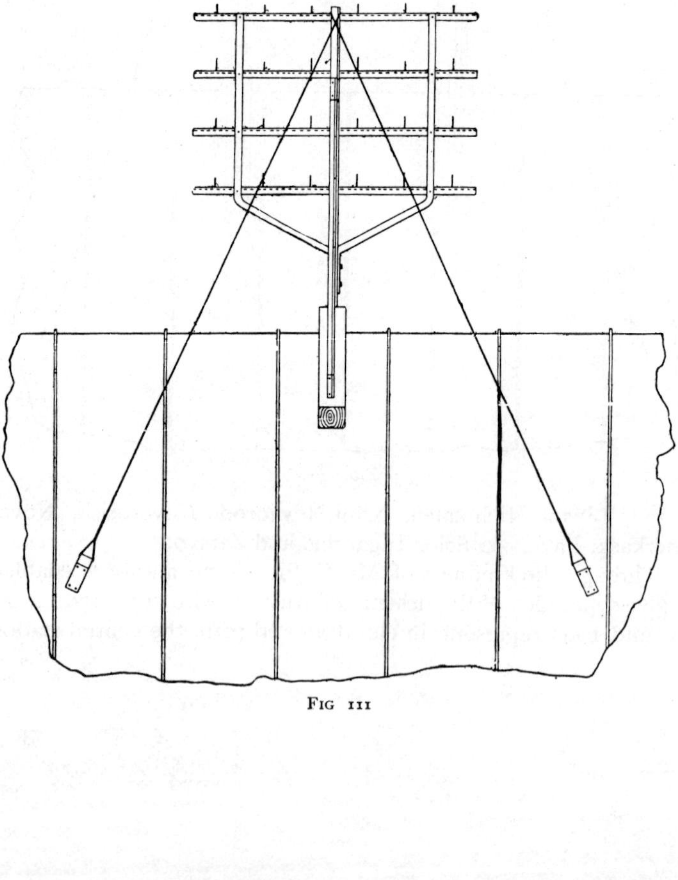

FIG 111

standard at Rostoff-on Don. Figs. 110 and 110A show a strongly constructed double standard, and figs. 111 and 112 respectively a single standard and a wall-bracket.

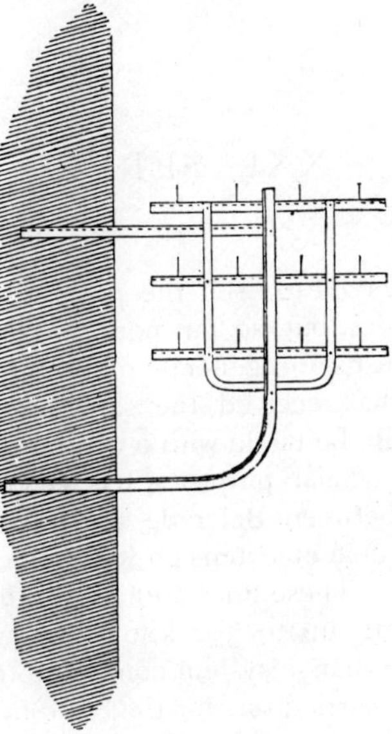

FIG. 112

STATISTICS

No returns later than 1892 are available. The following figures relate to that year.

	State	Companies
Number of exchanges	18	11
,, ,, switch-rooms	18	14
,, ,, public stations	19	—
,, ,, subscribers' lines	2,216	5,148
Length of wire in use, in kilometers	5,568	15,436
Number of conversations between subscribers, for year	3,033,139	7,631,016
Number of conversations from public stations, for year	2,744	—
Number of telegrams telephoned, for year	19,106	80,061
Receipts, for year	32,317*l.*	Not stated

XXI. SERVIA

THE State has reserved to itself the power to establish and work telephone exchanges, but, so far, none has been opened. Still a law regulating tariffs and general conditions has been prepared, and as soon as it has received the sanction of the legislature a commencement will be made with a central station at Belgrade. But the State, for official purposes, has erected metallic circuits of 3 mm. bronze between Belgrade and Nisch (250 kilometers), which line is in course of extension to Sibervcz, a further distance of 125 kilometers. These lines and a few others, which bring the total existing length up to 322 kilometers, will be available as trunks when the exchange system comes into operation. Most of the work has been carried out for the State by Mr. J. Berliner, of Hanover, through his agent in Vienna, Mr. Hax Hahn ; and the instruments used are the Berliner transmitter with double-pole Bell receivers and magnetos. There are also twenty-four kilometers of private lines in Servia. Such lines require a licence from the Minister of Commerce, and have to pay an annual tax of 1*l.* 12*s.* for every three kilometers, or less, of wire erected, and of 2*l.* 8*s.* if the length exceeds three and is not more than eight kilometers.

XXII. SPAIN

HISTORY AND PRESENT POSITION

A ROYAL decree, dated August 11, 1884, made telephonic exchange communication a Government monopoly ; but the experience gained during the next two years was so little to the taste of the officials that in June 1886 another decree entirely reversed the first one and provided that the exploitation of telephones in Spain should henceforth be left to private enterprise. In explanation of this change of front the decree said, 'So long as the telephonic service is administered by the State it can never develop and attain the proportions demanded by the necessities of modern life. Private enterprise, on the other hand, while adapting itself to public requirements, will find in this novel means of communication a vast field for activity in which apt initiative will be repaid by satisfactory development.'

While it is undoubtedly rather amusing to find the Spanish Government naïvely confessing itself so much behind the age as to be impotent to deal with the exigencies of modern life, there was certainly a strain of good sense in its argument. Government departments are generally very inelastic affairs, averse to innovation and desirous of running on in the grooves to which they have been accustomed. Such exceptions as may be cited are explainable by the unquestionable fact that good and energetic men—wishful to earn laurels for themselves and their country, and of force of character sufficient to overcome the inertia which pertains to Governments—must now and then come to the surface, even in a Government department and in defiance of its humdrum traditions and training. But such a good man, after having animated the mummy for a series of years, with results creditable to himself and beneficial to the public and all concerned, may be succeeded by one of quite another stamp, desirous only of pursuing as uneventful a career as is compatible with the retention of his office ; or, worse still, by one who expends his energy in combating instead of fostering the requirements of the community. It is not conceivable that the principles of promotion by seniority, or seniority tempered by patronage, which prevail in so many countries can produce any other result, for they open the doors to dullards, routine-worshippers, red-tape and sealing-wax champions, and others who, good enough men in their own small way, are not designed by nature to lead or initiate. On the other hand, commercial companies—which have to contend with competition, which can only exist by earning dividends, and which have a day of reckoning at least once a year—cannot afford to tolerate triflers or idlers. Promotion with them should be, and generally is, by seniority tempered by proved ability to keep abreast of the times : if inadvertently a round peg gets into a square hole he cannot catch on, and is soon shunted by the force of the circumstances which he cannot control. The smart official looks after the shareholders' dividends, and the competition looks after the public. The only exception is when a company has a rich monopoly which cannot be spoiled even by bad management. Such a company, by force of its monopoly, may do well for itself ; but it will not, unless directed by an enlightened and superior man, do well for its customers, whom, as likely as not, it will regard in the light of enemies, to be snubbed and repressed on every occasion : it is, in fact, liable to all the abuses and drawbacks of a Government department.

When a commencement was made with the new order of things in Spain it became apparent that the Government's idea of how to foster a telephonic development commensurate with the exigencies of modern life was to put the various towns up to auction and knock them down to the company or person willing to part with the greatest proportion of the gross receipts to the State, no offer of less than 10 per cent. being entertained under any circumstances. At the same time, to safeguard the interests of the public (so it was said), a scale of maximum charges was prepared, and regulations for the conduct of the traffic—some of which were distinctly worthy of commendation—drawn up, to which the concessionaries had to undertake to conform.

Under this decree concessions for thirty-five exchange systems were granted, the State proportion of the gross receipts varying from 10 per cent. in Valladolid, Seville, Granada, and Alicante, to 20 per cent. in Madrid and Saragossa, $31\frac{1}{2}$ per cent. in Valencia, $33\frac{3}{4}$ per cent. in Barcelona, and 34 per cent. in Bilbao; and averaging 20·66 per cent. all round.

The principal maximum rates, payable quarterly in advance, as fixed by law were as follow:—

	Per annum £	s.	d.
Subscriber to a local exchange located within the municipal boundary, with the instrument in his private office or house	12	0	0
For a telephone connected to the local exchange, but fixed in a casino, club, hotel, café, theatre, railway station, or other place where it could be used by strangers	40	0	0
Three-minute local talk from a public telephone station .	0	0	1·44

This local rate of 12*l.* payable by the subscriber meant that the concessionary companies had to earn dividends

	Per annum £	s.	d.
In Madrid on 12*l.*—20 per cent.	9	12	0
,, Bilbao on 12*l.*—34 per cent.	7	18	5
,, Barcelona on 12*l.*—33·75 per cent.	7	19	0
,, Valencia on 12*l.*—31·5 per cent.	8	4	5
,, Valladolid, &c., on 12*l.*—10 per cent.	10	16	0

and so, providing subscribers were forthcoming in any decent number, were in clover, even the lowest net rates being ample for the purpose. How Swedish, Swiss, and Dutch telephone managers, accustomed to work on 5*l.*, 4*l.*, and even 2*l.* 9*s.* 7*d.* rates, would revel in such exuberant figures! Imagine Mynheer Jan Sōt (see Dutch section, p. 220) established on the banks of classical Guadalquivir with net rates ranging from 8*l.* to 10*l.* 16*s.*! Spain, to him, would be a telephonic El Dorado indeed.

But a telephone company without subscribers gets on but indifferently well. The anticipated rush of hotel and casino keepers and railway managers anxious to pay 40*l.* a year for a local telephone connection developed but slowly, and even the stream of ordinary twelve-pounders who did not keep casinos bore more resemblance to a Ravensbourne than to a Mississippi. A concession was granted for Felanitx, the inhabitants of which town were assumed to be eager to get level with the necessities of modern life on the 12*l.* a year terms. An exchange was built and declared opened in October 1888, at which time the only connection to it was a public telephone station. At December 31, 1891, the date of the last report, the system had neither grown nor decreased—but was still open. At the end of 1890, after more than four years' development, the exchanges in Madrid and Barcelona having been opened in 1886, the number of subscribers in all Spain was 8,680, connected to thirty-two exchanges, giving an average of 271 subscribers per exchange. The total annual subscriptions actually collected in 1890 amounted to 1,726,284 francs, or 198 francs (7*l.* 18*s.* 5*d.*) per subscriber. These results were, rightly enough, considered unsatisfactory, and a third royal decree made its appearance in November 1890 and came into operation on January 2, 1891. The decree set forth that the State, instead of being, as the royal decree of 1886 had alleged, a perpetual obstacle to the development of telephoning, had, in Spain, proved its greatest supporter. That opinions were now divided as to the better method of control, State or company, so that it was deemed judicious to recall the decree of 1886 in order that the State might again be free to undertake exchange work where expedient. At the same time, it was proposed to give future companies a greater degree of freedom. This it certainly did in various ways. The auction system was abandoned, and the royalty reduced from as much as could be screwed out of the concessionaries to 10 per cent. on the *net* earnings, with a minimum payment for each town based on the number of inhabitants. Thus a town of 10,000 inhabitants or less must pay a minimum royalty of 40*l.* per annum ; 10,001 to 20,000, 80*l.* ; 20,001 to 50,000, 200*l.* ; increasing by steps to 2,000*l.* for a town of 200,001 or more inhabitants.

The rates were generally reduced, even the unhappy casino keepers being remembered, and new regulations issued. As these rates and regulations represent the conditions under which the telephonic industry in Spain is now pursued, they are given rather fully below. With regard to trunk lines, concessions have been granted for connecting Madrid to Saragossa, Barcelona, Pampeluna, St. Sebastian, Vittoria, Bilbao, Valencia, Tarrasa, and Sabadell. Of these, only the Madrid–Barcelona and the Bilbao–Vittoria are at the date of writing (February 1895) reported finished.

The new policy has, it is understood, been attended by considerable development. The last official report only extends to the end of 1892, when the number of exchanges was forty-six, and of subscribers 10,984. Practically the whole of the increase over 1890 had been won by the companies, for although the State had, in pursuance of the new policy inaugurated by the decree of 1890, opened no less than ten exchanges, its subscribers, after two years' working, only numbered 135! The State management appears to be on less liberal lines than that of the companies, since the statistics show that it possesses no public telephone stations, and that there is neither a telegram nor telephonogram service in connection with its exchange.

The Spanish system, although now modified on decidedly liberal lines—so liberal as to include the cheapest rate for telegrams in the world—is defective in one important particular. The concessions are for twenty years only, after which the whole system becomes the property of the State *without payment* to the concessionaries of any kind, unless the State is willing to take over the switch-boards and subscribers' instruments (a most unlikely contingency, seeing that most of this apparatus will be of old design and well worn), which will then be paid for at a rate to be settled by arbitration failing friendly agreement. This means that the concessionaries have not only to earn adequate interest on their capital, but are to get back the principal too, and that within twenty years. Such an arrangement must be bad for the subscribers during the latter half of the concessionary term, for it may be taken for granted that no improvements will be introduced and the service starved in every conceivable way. And eventually the State will come into possession of a system the upkeep of which has been so neglected that a thorough reconstruction will be the first thing it will have to set about. Technically, the future in Spain is not bright; for although metallic circuits prevail, we may be sure that Cheap Jack will rule the roast wherever possible. The concessionary system cannot produce the best, or even passably good, results with the bogle of confiscation growing bigger and more imminent every year.

SERVICES AND TARIFFS

1. **Rates for local exchange connections:**

CLASS 1. Connection to a private residence for the use of the subscriber, his family, and servants only.

CLASS 2. Connection to a place of business for the use of the subscriber, his partners, and employees only.

CLASS 3. Connection used by several occupants of the same building.

CLASS 4. Connection to a casino, club, place of amusement, café, theatre, or railway station, where it may be used by customers or visitors.

—	Per annum, quarterly in advance			
	Class 1	Class 2	Class 3	Class 4
	£ s.	£ s.	£ s.	£
In a town of less than 10,000 inhabitants	4 16	5 12	6 8	8
,, 10,001 to 20,000 ,,	5 12	6 8	7 4	12
,, 20,001 ,, 50,000 ,,	6 8	7 4	8 0	16
,, 50,001 ,, 100,000 ,,	7 4	8 0	9 12	20
,, 100,001 ,, 200,000 ,,	8 0	8 16	11 4	24
,, 200,001 and more ,,	10 0	12 0	14 0	32

These rates apply to subscribers located within three kilometers of the central exchange or of a branch switch-room. Beyond that distance an excess rate of 2*s.* 5*d.* per 100 meters must be paid. An extra set of instruments connected to the same line by a switch is supplied for 16*s.* per annum. The Government and provincial and municipal authorities enjoy a reduction of 40 per cent. on all local rates. A subscriber is entitled to deduct from his next payment in advance the proportion of his subscription proper to the number of days (if any) on which his line has been interrupted during the preceding quarter. On the other hand, he may be required to deposit 75 francs (3*l.*) to cover the value of his instrument.

2. **Rates for internal trunk lines.**—Time unit, three minutes.

	s.	*d*
Up to 50 kilometers	0	5·3
51 ,, 100 ,,	0	6·7
101 ,, 200 ,,	1	0
201 ,, 300 ,,	1	4·8
301 ,, 400 ,,	1	9·6
401 ,, 500 ,,	2	2·4
501 ,, 600 ,,	2	7·2
For each additional 100 kilometers	0	4·8

The State has reserved the right to use each trunk line for public purposes one hour every day free of all charge, and for a second hour at a reduction of 40 per cent. on the above rates. The concessionary has to pay a royalty of 10 per cent. on his net receipts—i.e. his profits—with a minimum payment of 16*s.* per kilometer of trunk line per annum. He has also to deposit with the Government a sum equal to 1*l.* 12*s.* per kilometer.

3. **Rates for public telephone stations.**—For local talks :

Subscribers	free
Non-subscribers, per three minutes or less	1·92*d.*

For trunk talks : Subscribers and non-subscribers, as per trunk tariff above.

4. **Telephoning of telegrams.**—There is no provision in the authorised tariffs for this service, but the statistics (see p. 331) show that it exists.

5. **Rates for the telephoning of messages for local delivery (telephonograms).**—The exchanges write down and deliver by messenger to non-subscribers located in the same town messages which may be dictated from a subscribers' office or from a public telephone station, or written and handed in at a public telephone station. This really constitutes a local telegram service. The rates are :—

For a message of 20 words or less	1·92*d.*
,, each additional 5 words	·48*d.*
When a message is addressed to more than one person : each extra copy	·96*d.*

The Spaniards may well be congratulated on having established a record in telegraphic rates. Twenty words for rather less than twopence is calculated to stir up feelings of envy in less fortunate people, such as those, for instance, who may not send written messages by telephone at all and have to pay 6*d.* for twelve words, however short the distance covered.

WORK

The author has not had an opportunity of personally inspecting the Spanish exchanges, which are, to a large extent, in the hands of French companies. The character of the work is, as is natural under such circumstances, decidedly French. In fact, the Société Générale des Téléphones, of Paris, supplied practically the whole of the material used up till 1891, when the customs war between the two countries interposed a barrier to the importation of French apparatus, which practically killed the trade. The business is now supplied, but principally on French models, from workshops established in Spain itself, although Belgian instruments are not unknown. The prevailing type of subscribers' apparatus comprises Ader transmitters, Ader receivers, push-buttons, trembling bells, and Leclanché cells. The usual class of switch-board is that designed by M. Berthon and used in Paris during the reign of the Société Générale des Téléphones, and which has been often described. An exception is the case of Madrid, which has recently been provided with a multiple board of the Western Electric Company's ordinary type. In regard to outside work, that at Madrid is remarkable as consisting chiefly of aerial cables, a form of construction necessitated by a municipal decree which forbids the employment of any open wire for a greater distance than 500 meters. The cables usually contain twelve wires, of a resistance of forty ohms per kilometer, insulated with rubber and wrapped in waterproofed tape. They are suspended from galvanised steel wires of 3 mm. diameter by steel hooks placed one meter apart. The Spanish system is, however, a model in one of the most important of all respects—it is metallic circuit throughout.

STATISTICS

At December 31, 1892, the date of the last published return, the position of Spanish telephones was as follows :—

	State	Companies
Number of exchanges	10	36
,, subscribers	135	10,849
Length of wire in use, in kilometers . .	390	22,432
Number of public stations	—	28
Number of local talks between subscribers, for year	73,258	1,237,235
Number of local talks from public stations, for year	—	26,538
Number of telegrams telephoned, for year .	—	13,088
Number of telephonograms from subscribers, for year	—	12,143
Number of telephonograms from public stations, for year	—	2,356

XXIII. SWEDEN

HISTORY AND PRESENT POSITION

In Sweden at the present day one may gain a glimpse of what telephony in the future will be everywhere, and an inkling of the kind of problem which awaits the coming telephone engineers. In population Stockholm is about 11,000 souls behind Edinburgh (Edinburgh, 1891, 263,646; Stockholm, 1892, 252,574). Both are capitals. In Stockholm at the end of 1894 there were 11,534 exchange instruments in operation; in Edinburgh about 1,000. In Stockholm each hundred inhabitants, including women, children, and babies, had 4·57 instruments between them—one and a fraction over to every twenty-five souls. In Edinburgh each hundred inhabitants had ·37—a little more than a third part of a telephone between them. Taking the population of London as 5,600,000, and imagining that London telephonically were on a par with Stockholm, what should we find? Why, that London would then possess

255,920

exchange instruments! What is the present number? About 8,000, or ·14 per hundred inhabitants.

The credit of the Swedish development is unquestionably due in a large measure to Mr. H. T. Cedergren, the managing director of the Allmänna Telefonaktiebolag (General Telephone Company) of Stockholm. He has truly been the Hotspur of telephonic warfare—ever in the front with extensions and improvements; ever devising new uses and applications for the telephone; ever appealing to the public for support, and, what is a great deal more to the purpose, ever deserving it. Mr. Cedergren was amongst the first to perceive the sufficiency of a low rate of subscription, and to appreciate its fostering power on the telephonic industry. At first a theory only, the keen competition which ensued in Stockholm when the original monopoly of the International Bell Telephone Company was attacked, provided the opportunity for its practical demonstration. The result of the low rates and Mr. Cedergren's unceasing energy has been to place Sweden in the foremost telephonic position in the world. 'And what,' the advocates of high rates will ask, 'and what about the poor unfortunate shareholders?' Well, as will be seen further on, those commiserated personages have received year after year better dividends than telephone shareholders in the United Kingdom ever did,[1] or are ever likely to. 'But,' say the advocates, 'Cedergren had everything his own way—no opposition—free way-leaves—low-priced labour—a benevolent corporation—a free-handed and complaisant Government.' Nothing of the kind—a mere collection of red herrings.

The pioneer in Sweden was the International Bell Telephone Company, which opened in Stockholm and Gothenburg in 1880 and soon afterwards in a few other towns. But the rates were high and development was slow until opposition appeared in Stockholm in 1883 in the guise of a local—Mr. Cedergren's—company, and in Gothenburg in the form of a co-operative telephone society, the idea of which was that each member should pay for the cost of his line, instrument, and proportion of switch-room apparatus, and contribute 3*l.* 6*s.* 8*d.* per annum towards the working and upkeep of the system, which contribution would be reduced, after the formation of an adequate reserve fund, whenever circumstances permitted. The idea was found to work out well in practice, and Sweden was soon dotted with co-operative telephone exchanges, even villages with names undiscoverable in the best gazetteers indulging in what was at first looked upon partly as a scientific curiosity and partly as a luxury, but which soon proved to be a useful adjunct of everyday life.

The extent of the mine waiting to be worked was soon demon-

[1] With the exception, perhaps, of those of the Dundee and District Telephonic Company, Limited (see page 7), which worked on a 5*l.* 10*s.* rate.

strated by Mr. Cedergren's methods. Instead of a yearly rental of 8*l.* 17*s.* 9*d.* (the Bell Company's rate) the new competitor asked 2*l.* 15*s.* 7*d.* down on connection, and thereafter an annual inclusive subscription of 5*l.* 11*s.* 1*d.* The Bell Company was, of course, convinced that Mr. Cedergren had simply discovered a royal road to ruin for himself and friends, and that all that was necessary to bring about his self-immolation was to allow him sufficient room to caper about in. So when at the end of 1883, after seven months' working, his exchange had 785 instruments connected, as many as the Bell had after three years, it was felt that he was advancing towards his inevitable goal with satisfactory rapidity. But when at the end of 1884 he had 2,288 against the Bell's 900 or so, and was moreover paying dividends, it was perceived that there was a certain--or rather uncertain, for it was not easily understood—method in his madness. Then the Bell Company began to wake up, but it was too late; and it never afterwards played but a secondary part in the telephonic game. Ultimately its Stockholm system, with the exception of the Östermalms district in the north-east of the town, was bought and incorporated by the General Company. The Östermalms exchange has preserved a separate organisation, but practically it forms part of the General system, since free intercommunication between the two prevails. As early as 1884 the General Company began to extend its operations to other towns in the neighbourhood of Stockholm and to erect trunk lines between them. This was found to be a remunerative undertaking, and in the next succeeding years was pushed to such an extent that the Government began to take alarm for its telegraph revenue, more especially after an application by the General Company for a concession to run trunk lines to Gothenburg, Malmö, and other of the principal towns. The question of the proposed concession became a burning topic in Parliament; special committees took it in hand; and deputations headed by Mr. Cedergren carried it even to the foot of the throne. Ultimately, it was decided to give the State post and telegraph department the exclusive right to erect inter-town wires except within a radius of seventy kilometers ($43\frac{1}{2}$ miles) around Stockholm, within which area the General Telephone Company was left free to do as it liked. Mr. Cedergren's long-distance ambition was thus baulked ; but the inhabitants of the 70-kilometer radius have no reason to lament the fact, for his energies, being concentrated within that circle, have led to its becoming, without any exception, the best-telephoned bit of country in the world.

But the jealousy of the telegraph department had now been thoroughly aroused. It was no longer content to erect trunks for the use of local companies and co-operative societies. It was felt that by doing so and nothing more it was taking most of the expense and risk and least of the profit, profit moreover gained (as was then imagined) by competing with, and murdering, its own telegraph revenue. So the State determined to go in for the better paying part—the local exchanges—also ; and began by purchasing the Gothenburg and other provincial exchanges of the International Bell Telephone Company. In Stockholm there was already existing at the central telegraph office a small telephone exchange for the use of the Government departments, and this was made the nucleus of a public system. The Swedish State telegraph department having definitely entered the lists, determined to do its work well. It made metallic circuits an inexorable rule, and underground work an end to be aimed at wherever possible. The experience of the General Company had demonstrated the feasibility and potency in developing custom of low rates, and the State started in Stockholm with a first payment of *2l.* 15*s.* 7*d.* on connection, and an annual subscription thereafter of 4*l.* 8*s.* 11*d.*, or 1*l.* 2*s.* 2*d.* below that of the General Company, which was to cover free communication not only in Stockholm, but within a radius of seventy kilometers around! It was a programme—metallic circuits against single wires, underground wires against overhead, direct connection with the long-distance trunks, all combined with an appreciably lower rate and a free 70-kilometer radius—that deserved success and was calculated to alarm the General Company. But Cedergren was used to competition. He had at this period over 5,000 subscribers working in Stockholm alone, and his service was as good as is compatible with single wires. But that was not enough ; and the State had scarcely got its exchange in operation before the General Company began to convert its system to metallic circuit, section after section of the multiple switch-board at the central station being altered to meet the new requirements, communication between the two sets being kept up by means of translators, until in 1894 there was not a single wire left in Stockholm. Probably the State had intended to intimidate the company into selling its system, and had there been a nervous man at the helm that result would probably have been brought about; but Cedergren picked up the proffered gauntlet and set about fighting the State as vigorously as he had done the Bell Company. He did not even reduce his subscription of 5*l.* 11*s.* 1*d.* to meet the State's 4*l.* 8*s.* 11*d.*, simply notifying that all subscribers' lines would be changed to metallic circuit without extra charge, and that the subscription would henceforth cover communication with all the company's subscribers within the 70-kilometer radius. The results are curious. The State opposition began to be pushed with energy in 1890, at the end of which year the General Company had 5,186 instruments connected. At the end of 1894, after four years of active rivalry, the General Company had 8,336 instruments and the State 2,400—that is to say, a respective increase of 3,150 and 2,000 since the end of 1890. Both systems have consequently found a field, just as the starting and rapid increase of the Mutual Telephone Company's exchange in Manchester took place without arresting the development of the National Telephone Company's system in the same town. The success of the General Company in its opposition is the more surprising since its subscribers are placed at a disadvantage (see *Tariffs*), as compared with those of the State, both in the use of the trunks and in telephoning telegrams. The result tends to confirm the often-expressed view that Government departments cannot successfully compete with properly directed private enterprise, a view which has also received practical illustration outside the precincts of Sweden.

In all the chief provincial towns the State now owns the telephone service, either by acquiring it from its original proprietors or in virtue of its own initiative. In some towns, Gothenburg for instance, there is opposition; but this is growing more and more feeble because the State declines (except in Stockholm) to allow its competitors to use the trunk lines, participate in the telegram service, or even, in some cases, to intercommunicate on any terms with its own subscribers in the same locality.

In many of the smaller towns and villages co-operative societies still afford the only means of telephonic communication, but they are gradually disappearing under the encroachments of the State. At the end of 1892, the latest available statistic, there

FIG. 113

were 158 co-operative exchanges, of which thirty were in towns and the rest in villages and rural communes. At the same date there were 466 telephone exchanges and 27,658 subscribers in Sweden. When it is recollected that the population is under five millions; that there are only eight towns of more than 20,000 inhabitants, and eleven more of between 10,000 and 20,000, this

Z

development is little short of marvellous. Compare the National Telephone Company's return for 1893—a year later—of 540 exchanges and 53,784 subscribers for the whole of the United Kingdom with its population of thirty-eight millions ! The constitutions of the Swedish co-operative societies are very similar. In the first place a member pays the whole cost of his connection to the exchange, and is annually assessed with his share of the working and maintenance expenses of the system, together with a contribution to the reserve fund. In the towns (as in Gothenburg) this assessment is sometimes as high as 3*l*. 6*s*. 8*d*., but in the villages it may be as low as 25*s*. or 30*s*.

In the Östermalms district of Stockholm, which is still worked by the Bell Company, the Swiss method of charging is in operation, the subscribers paying an annual subscription of 1*l*. 19*s*. 9*d*., which entitles them to a hundred free calls every three months each call over that number being charged 1·3*d*.

The success of the low rates in Stockholm, both State and company's, is rendered more surprising by the fact that the use of numerous submarine cables is rendered absolutely unavoidable by the geographical character of the locality. Not only is Stockholm itself built on several islands (fig. 113), but between the city and the Baltic, the islets, nearly all of which contain villages or at least summer residences, are several hundreds in number. A large proportion of them is in connection with either one or both telephonic systems, necessitating constant attention to submarine cable work. The General Company, in fact, keeps a small steamer specially for the purpose.

SERVICES RENDERED IN STOCKHOLM BY THE GENERAL TELEPHONE COMPANY AND THE STATE TELEGRAPH DEPARTMENT

1. **Local intercommunication between its own subscribers and public stations and those of the rival system.**

2. **Communication within a 70-kilometer radius around Stockholm.**

3. **Internal trunk service.**—Every Swedish town of note and many villages are in trunk connection.

4. **International trunk service.**—This exists to Norway and Denmark only. A line to Finland or Russia is not yet spoken of.

5. **Telephoning of telegrams.**—The State's own subscribers are switched through to the central telegraph office for the transmission of their telegrams, but this facility is denied to the General Company's supporters. But Mr. Cedergren has established an office adjacent to the central telegraph station where his subscribers' telegrams are written down by company's clerks and immediately handed in over the counter for transmission. Conversely, telegrams for his subscribers are delivered at the special office and telephoned to the addressees. This may, perhaps, be a little less rapid than direct connection with the telegraph department, but the difference is not great, and the subscribers are reconciled to it by enjoying the service free, while the State's subscribers have to pay ·66*d.* per message.

6. **Local message (telephonogram) service.**

7. **Messenger service.**

8. **Public telephone stations.**—In Stockholm a public telephone station belonging to the State or to the General Company is met with about every hundred yards in the principal streets, as nearly every hotel, restaurant, and tobacco-shop keeps one. These keepers pay the full tariff for their instruments and are allowed to retain all local receipts in the case of the company, and 25 per cent. of the receipts in the case of the State. Public stations are also numerous in the provincial towns. The company has tried and abandoned many forms of automatic slot machines; the State is now about to experiment with them. The General Telephone Company's services Nos. 3 and 4 have to be conducted through the intermediary of the rival exchange and paid for.

The State renders similar services in the other towns in which it is established, except that the international wires are not yet available from all points. The 70-kilometer radius is, of course, an arrangement peculiar to Stockholm, Gothenburg, Malmö, &c., having lacked local Cedergrens at the critical moment. The General Telephone Company is conducted on similar lines in Upsala and the other towns within the 70-kilometer radius in which it does business.

TARIFFS

1. **General Company's exchange.** *Rates for local exchange communication.*—Subscribers are divided into four classes.

	Admission fee £	s.	d.	Annual subscription £	s.	d.
CLASS I.—For a direct metallic circuit to any of the three principal switch-rooms	2	15	7	5	11	1
,, II.—Two subscribers on one metallic circuit or on separate metallic circuits joined to one jack and indicator at the exchange, each .	2	15	7	4	8	11
,, III.—Three subscribers on one metallic circuit or joined to one jack and indicator, each	2	15	7	3	6	8
,, IV.—For a direct metallic circuit to one of the branch switch-rooms with restriction to 100 free calls every three months, every extra call being paid for on the Swiss system at 1·3*d.* per call . . .	0	11	1½	1	19	9

In addition there is a ship tariff:—

	£	s.	d.	£	s.	d.
For one vessel on a direct metallic circuit .	2	15	7	4	8	11
For each additional vessel using the same metallic circuit	2	15	7	2	15	7

There is no extra charge if a line exceeds a kilometer in length. Contracts are generally for five years. The admission fee may, at the subscriber's option, be paid down on the connection being completed or spread over the five years of the contract. Classes I., II., and III. are allowed *unlimited* communication with all the General Company's subscribers in Stockholm and within the 70-kilometer radius, and with all subscribers to the Bell Telephone Company as well.

Communication with the State exchange subscribers in Stockholm or seventy kilometers around, 1·3*d.* per talk, no time limit, within the town ; 1·3*d.* per five minutes beyond. This charge is paid over to the State.

2. **General Company's exchange.** 70-*kilometer radius.*—The local rates, Classes I., II., and III., cover free and unrestricted com-

munication with any part of the 70-kilometer radius, in which, at the end of 1894, the company possessed 2,012 subscribers, besides the 8,334 within Stockholm city. Two subscribers located on opposite sides of the radius may consequently converse at will without extra charge over a distance of 140 kilometers (87 miles).

3. **General Company's exchange.** *Internal trunk rates.*— Same as those of the State, plus 1·3*d.* per connection, which also goes to the State. Accounts are collected every three months. The record of connections on which money is payable by the company to the State is kept by the State operator, and, as a rule, this record must be accepted as correct. The State pays the General Company 1·3*d.* for each conversation originated by a State with a company's subscriber.

4. **General Company's exchange.** *International trunk rates.*— The company's subscribers do not participate in this service.

5. **General Company's exchange.** *Rates for the telephoning of telegrams.*—This service is free, but subscribers using it must keep a deposit balance of not less than 5*s.* 6*d.* with the company.

6. **General Company's exchange.** *Local message service rates.* Same as the State's, which see.

7. **General Company's exchange.** *Messenger service rates.*— Same as the State's, which see.

8. **General Company's exchange.** *Public telephone station rates.*—For communication with any company's subscriber in Stockholm or seventy kilometers around, 1·3*d.* Time may be limited to five minutes if necessary. Connections to State subscribers, 2·6*d.* Trunk rates those of the State, plus 1·3*d.* per connection. Subscribers have no advantage over strangers in using the public stations.

1. **State exchange.** *Rates for local exchange communication.*—

	Admission fee £	s.	d.	Annual subscription £	s.	d.
For a business connection not over two kilometers from the nearest switch-room	2	15	7	4	8	11
For a private house connection, the State reserving the right to put two houses on the same line	—			3	6	8
Members of Parliament (four months in the year only)	—			1	13	2

Ship-owners who have their vessels fitted with telephones, so that they can connect with the exchange on coming alongside the quay at their usual berth, are charged *2l.* 15*s.* 7*d.* per annum in addition to the cost of the ship and shore connections. Contracts are for five years.

Communication with the General Company's subscribers, 1·3*d.* per talk, no time limit, within the town; 1·3*d.* per five minutes beyond.

2. **State exchange.** 70-*kilometer radius.*—The local charges named above cover communication with any State subscriber within seventy kilometers of Stockholm.

3. **State exchange.**—*Internal trunk rates.*—

	Per 3 minutes or fraction thereof
Up to 100 kilometers	2*d.*
100 ,, 250 ,,	4*d.*
250 ,, 600 ,,	6·6*d.*
600 ,, 900 ,,	9·9*d.*
Over 900	13·25*d.*

Talks may be extended indefinitely so long as the line is not wanted by others. There are no 'express' or 'urgent' connections, and the tariff is not reduced at night. Unless a caller's request can be met and satisfied, he is not charged anything, notwithstanding that the operators and wires are sometimes engaged a considerable time in vainly trying to arrange the connection; but a subscriber who engages a trunk for a certain time and fails from any reason to use it, is debited with the cost of a conversation.

4. **State exchange.** *International trunk rates.*—

	Per 3 minutes *s.*	*d.*
Stockholm to Christiania	1	8
,, ,, Drammen, Dröbak, Lilleströmmen, &c.	1	10
,, ,, Copenhagen	2	2½
Malmö ,, ,,	1	8

5. **State exchange.** *Rates for telephoning of telegrams.*—The charge for a telephoned telegram is ·66*d.*, irrespective of the number of words. Subscribers are not required to make a preliminary deposit, but the State charges 2 per cent. on the amount of accounts to cover the cost of keeping them.

6. **State exchange.** *Local message service rates.*—A subscriber may telephone a message of not more than forty words to a telegraph office, where it is written down and delivered by messenger ; or any person may hand in a written message of similar length at a telegraph or public telephone station and have it telephoned to a subscriber for 3·3*d*.

7. **State exchange.** *Messenger service rates.*—A non-subscriber may be called by messenger to a public station for 3·3*d*.

8. **State exchange.** *Public telephone stations rates.*—

Within Stockholm	1·3*d*.
Beyond Stockholm, but within 70-kilometer radius . .	1·99*d*.

Time unit, three minutes. Subscribers have to pay equally with non-subscribers. The trunk rates are the same as from the subscribers' offices. When a General Company's subscriber is called from a State public station the charge is doubled.

BELL TELEPHONE COMPANY

This company has but one rate, which is identical with the General Telephone Company's Class IV. It covers free communication with the latter company's subscribers.

WAY-LEAVES

The Government enjoys no special advantages except in respect to the State railways and the State lands, which, however, are very extensive. With private owners and with municipalities agreements have to be negotiated. In 1892 the Stockholm Town Council, owing to a difference of opinion about the laying of the State underground mains, withdrew a previously granted permission to open the streets, and the Government had to submit pending adjustment of the dispute. The Town Council has recently granted a corresponding way-leave for underground conduits to the General Company. The companies may not even cross the State railways and lands with their wires without permission ; in other respects they enjoy equal facilities. When the number of wires fixed is small, a nominal acknowledgment only is paid ; when large standards carrying one hundred wires or over are wanted, it is usual to give a free exchange connection in return for the accommodation.

The obtaining of way-leaves in Stockholm is much facilitated by the mode of roofing buildings. Slates or tiles are rarely employed, the buildings being covered with sheet iron, painted, which is not readily damaged by workmen. Complaints, so common in England, of leakage are consequently rare. Most buildings have also a common stairway from the street level to the roof, so that access can be had without passing through the interiors. Way-leaves are consequently not so difficult to obtain and retain as with us ; moreover, the mode of joining the squares of sheet iron results in a series of ridges which afford a hold to the linemen, and render the roof safer to work on.

SWITCHING ARRANGEMENTS

General Telephone Company's system.—Originally working with one central station, and after its fusion with the International Bell Company with two, the General Company has within the last two years entirely changed its plan, and simultaneously with its change to metallic circuits remodelled its switching arrangements. Fig. 113, which is a map of Stockholm city divided into eight switching districts, gives a clear idea of the existing arrangement, which, it will be seen, bears a strong resemblance to the plan—originally suggested by General Webber—advocated in the author's British Association paper of August 24, 1891, and which, had it not been for the death of the late Duke of Marlborough, would have come into operation in London on January 1, 1893. The adoption of some such plan is inevitable in the future, both on the score of expense, of accommodation for wires and of switching space. A central station may conceivably be arranged to take 30,000 or even 36,000 subscribers, if the wires could be got to it, but beyond that number the complications involved would be too costly to be faced. And even 36,000 is not enough, as it has already been shown that London, on the example of Stockholm, may reasonably be expected to require accommodation for 250,000 subscribers in the not distant future. The existing arrangements are ludicrously deficient as it is, and no extension of them could possibly meet the tenth part of such a demand; so the ultimate adoption of the British Association, or divisional, plan is inevitable. That such an authority as Mr. Cedergren has recognised the fact and adapted it to the needs of the most telephonically advanced city in the world, affords gratifying confirmation of the author's convictions.

The backbone of the Stockholm system is the line of what are called primary switch-rooms, known as Brunkeberg, Stortorget, and Maria. Subscribers of Classes I., II. and III. are only connected to these, so that they obtain amongst themselves a service which never brings into requisition any of the branch switch-rooms. Of these branch rooms there are four belonging to the General Company and one to the Bell Company; but as the working agreement between the two concerns is of the most intimate character, the Bell room practically forms part of the General Company's system. The only difference is that, whereas the General Company uses Ericsson's instruments for all its subscribers, the Bell Company supplies magnetos of the American type, Bell receivers, and Ericsson transmitters. To these five branch rooms only members of Class IV. are joined, it having been found by experience that it is only the smaller people who do not make frequent use of their instruments who choose this mode of subscribing; but this class is also joined to the three primary rooms when they happen to be the nearest. The three primary switch-rooms are connected together by a large number of junctions, and each branch or secondary room possesses junctions to every other room, both primary and secondary.

On December 31, 1894, the instruments connected to this extensive system numbered 9,136, divided as follows:—

General Company, Class I.	3,359
,, ,, ,, II.	1,847
,, ,, ,, III.	1,482
,, ,, ,, IV.	684
Extension lines	964
Bell Company (all like Class IV.)	800
	9,136

By adding the 2,400 instruments of the State exchange, with which all are also in connection, the telephonic circle of Stockholm city is found to possess a total membership of 11,536.

The largest switch-room is at the General Company's old central in the Brunkeberg division, where there are 5,547 subscribers actually connected. The board was originally a Western Electric single-wire, double-cord, series multiple of twenty tables and an ultimate capacity of 7,000. It was altered section by section in the General Company's workshop during the conversion to metallic circuit, and made to conform in pattern to eleven new tables, which, when added, raised the ultimate capacity to 12,000 lines. This great capacity is achieved by reducing the size of the jacks and by sloping some of them over the operators' heads in the manner shown at J in fig. 114, which is an end section of the board. The experiment is interesting, but a stretch above the floor of two meters (6 ft. $6\frac{1}{2}$ in.) will be required when the table is full. The length of each table, which takes 300 subscribers' lines and three operators, is 1·62 meters. The dimensions of the jacks are: each jack 11 × 11 mm., and each set of 100 jacks (five rows of twenty), with necessary space for screws, is 55 × 249 mm. The capacity of 12,000 is made up of 6 × 20 sets of 100 jacks.

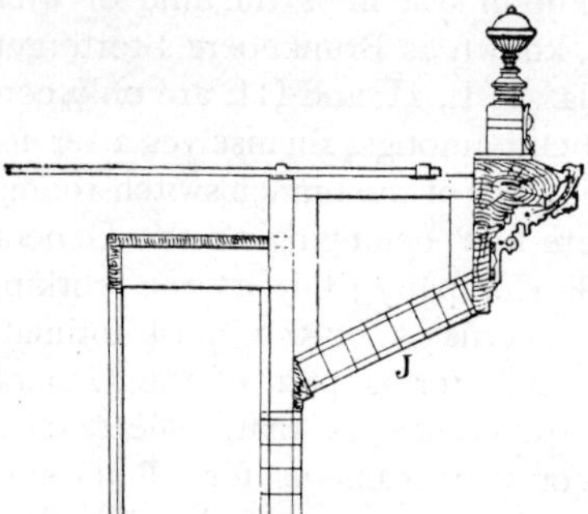

Fig. 114

During conversion to metallic circuits, a portion of this board was altered to single-cord, but, after some experience, changed again to double.

In the Southern, or Maria, exchange a switch-board, of six tables of 300-line capacity, possessing several novel features has recently been fitted. It is a metallic circuit, double-cord, parallel-jack multiple, with self-restoring drops of a new design. With the exception of these drops, which are manufactured by Ericsson & Co., the whole table was made in the workshops of the General Telephone Company. The self-restoring drop is shown in fig. 115. The signalling magnet M^1 is placed in front of the restoring one M^2 (see also fig. 117). The armature is a bent lever L^1 pivoted at *p*, which, when unattracted, engages with and holds up the shutter S working on the pivot *p*1. On dropping the shutter, its base B strikes against a pin which runs in a guide the whole length of the magnets and terminates at the back in a shoulder Y and a pointed head Z; forces the pin back, and closes the contacts C^1 C^2 of the night-bell and 'attention' indicator circuit. On operating the restoring magnet M^2, the armature L^2 is attracted, and its point, striking against the shoulder Y, forces back the pin,

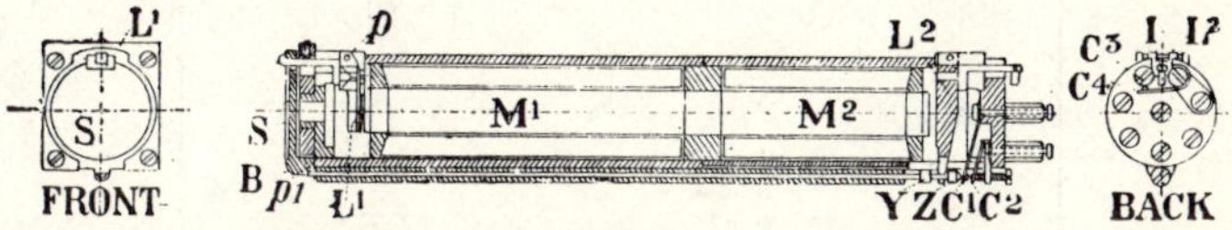

Fig. 115

which in its turn lifts the shutter S to its position of rest. At the back of the drop will be seen another pair of contacts C^3 C^4 and an ivory pin I attached to the armature L^2. While L^2 remains attracted under the influence of the restoring, which is also the test, battery (three Tudor accumulators of 175 amp hours), the pin I presses the contacts C^3 C^4 apart and breaks the circuit of the M^1 coils, thus cutting out the signalling indicators during connection, and leaving only the ring-off drop in derivation across the loop. The ring-off drops are also automatically restored, but mechanically. Fig. 116 shows the arrangement. L is a lever pivoted at *p*, which, when unrestrained by the weight of the plug P or pressure on the finger stud A, allows the plunger D to fall. The plunger presses against a spring C placed under the electro-magnet M. The shutter S is provided with a curved base piece B, which, on the shutter falling, depresses the spring and closes the circuit of the night-bell and 'attention' indicator. The restoration is effected by replacing the plugs or depressing the finger stud. The general arrangement of the jacks, test and restoring circuits is shown in fig. 117, which explains itself. The attention signal, included in the night-bell circuit, is intended to assist the lady superintendent. Each operating section has two—one with a white flag in connection with all the signalling indicators of that section, the other with a red flag in connection with all the ring-off drops of that section. Small white and red glow lamps

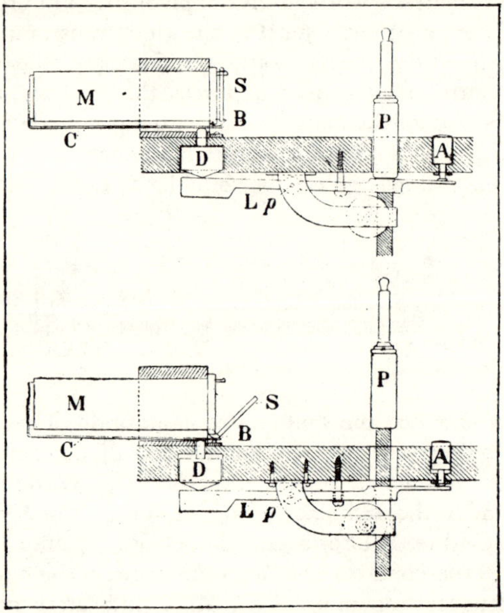

Fig. 116

have been tried instead of indicators with coloured shutters; they answer perfectly, and as self-restorers cannot be surpassed. By their position the superintendent can see whether any signalling drop has fallen and remains unanswered, or any ring-off has been given and left unnoticed. To enable operators to detect a disconnection on a subscriber's line, a polarised electro-magnet working a visual signal is included in the ringing circuit, the armature of which oscillates during ringing if the line is right.

Fig. 118 shows the operating connections, with details of the keys. It will be seen that the operator replies to a call by pressing the key and speaking on the right-hand cord, and that the desired subscriber is called by merely pressing the same key lower down while the operator is still speaking to the caller. None of the metallic parts of the keys can be touched. A connection counter, or at least a counter of the number of times the connection key is operated, is included in the arrangements. There are already over 1,000 subscribers working on this board, the ultimate capacity of which is 6 × 18 sets of 100 jacks = 10,800, and the designers are perfectly satisfied with the results obtained. The general outline of the board resembles that shown in fig. 114,

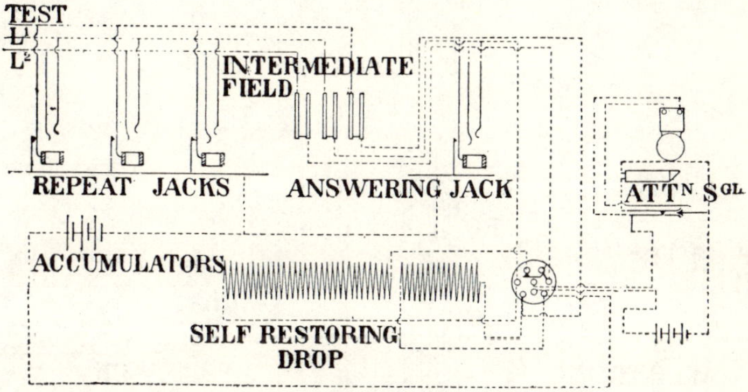

FIG. 117

without the overhanging projection. The measurements of tables and jacks are the same.

The Stortorget switch-board, by Ericsson & Co., consists of six tables of 300 lines, and is designed for an ultimate capacity of 7,800. It is on the single-cord principle, with jacks measuring 11 × 13 mm., the set of 100 occupying 70 × 249 mm.

The Bell Company's board, also by Ericsson & Co., has only two tables, each for 300 subscribers and three operators. The subscribers' lines are arranged for double-cord switching, but the inward junction lines terminate in separate cords. The jacks are of the same dimensions as those at Brunkeberg, and the board may be expanded to take 3,600 lines ultimately.

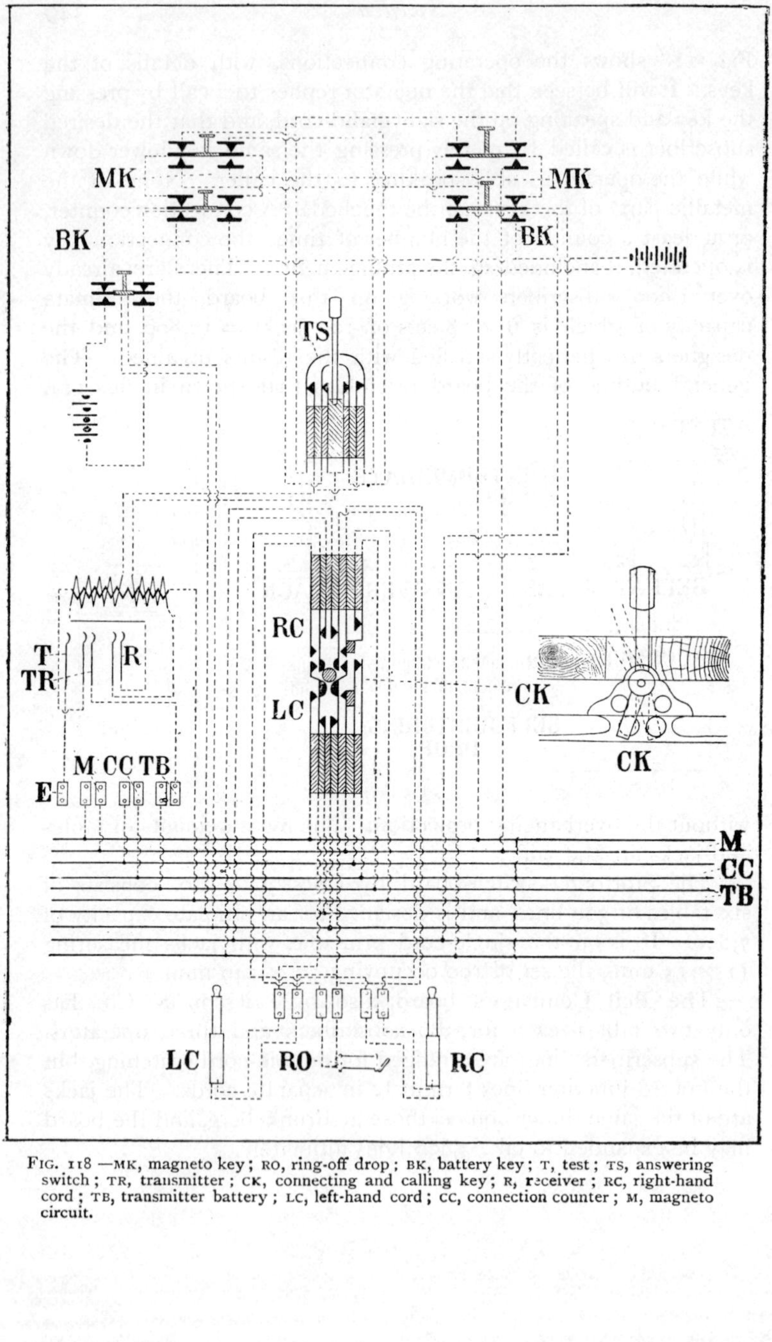

FIG. 118 —MK, magneto key; RO, ring-off drop; BK, battery key; T, test; TS, answering switch; TR, transmitter; CK, connecting and calling key; R, receiver; RC, right-hand cord; TB, transmitter battery; LC, left-hand cord; CC, connection counter; M, magneto circuit.

The traffic between the two Stockholm systems is large, both for local and trunk work, and the junction wires are consequently

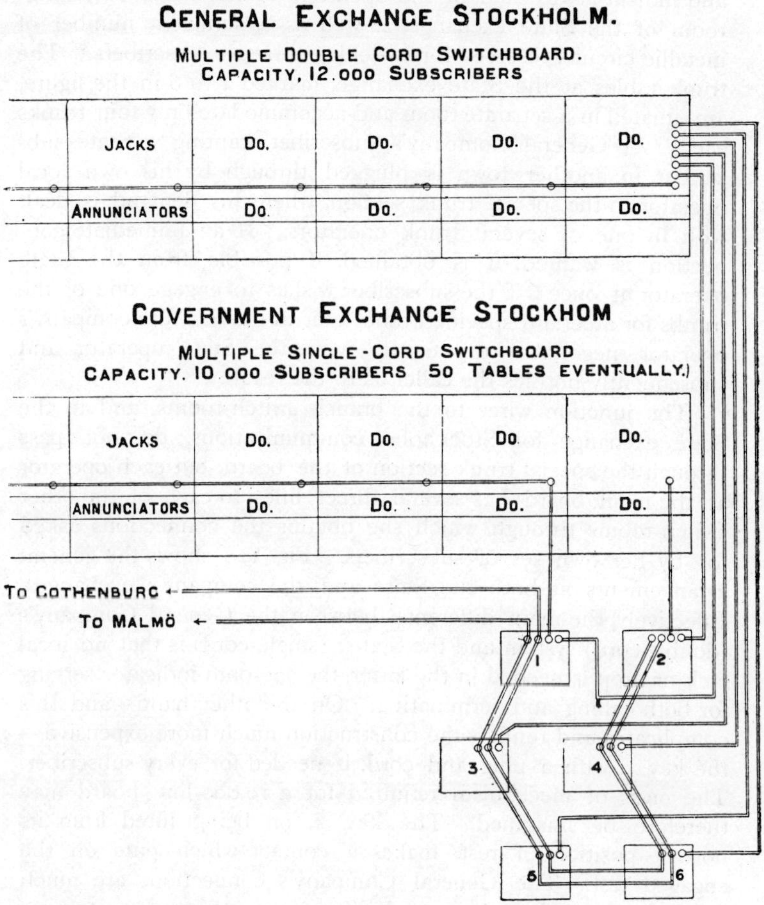

FIG. 119

very numerous. A general idea of the trunk arrangements between the two exchanges is given in fig. 119. The trunk connections are managed from a special section of the board on which all the local subscribers are represented by multiple jacks. From this special section proceed calling wires, operated by plugs and indicators, to each of the operators in the trunk switching-room of the State exchange, as well as a sufficient number of metallic circuits reserved for switching through subscribers. The trunk tables at the State exchange, marked 1 to 6 in the figure, are situated in a separate room and accommodate only four trunks each. A General Company's subscriber wanting a State subscriber in another town is plugged through by his own local operator to the special trunk section, where his demand is dealt with by one of several trunk operators. If an immediate connection is wanted, it is obtained, if possible, from the State operator at once; if the subscriber wishes to engage one of the trunks for a certain specified time later in the day, the company's operator negotiates the matter with the State operator and subsequently notifies the caller as to the result.

The junction wires to the branch switch-rooms, and to the State exchange for Stockholm communications, do not pass through the special trunk section of the board, but each operator at the main board has several direct lines to each of the other switch-rooms through which she obtains the connections asked for by her own set of subscribers. Fig. 120 shows the general arrangements at both the State and the company's exchanges. Effectively, the main difference between the General Company's (double-cord) system and the State's (single-cord) is that no local jack or drop is needed in the latter, the 900-ohm indicator serving for both calling and terminating. On the other hand—and this complicates and renders the construction much more expensive—the key *A*, with a plug and cord, is needed for every subscriber. The mass of mechanism required for a 10,000-line board may therefore be imagined. The key *A*, on being lifted from its normal position of rest, makes a contact which puts on the engaged test. The General Company's connections are much more numerous than those of the State, averaging at least ten per day. Each operator takes one hundred subscribers. As in the State system, the subscribers ring each other and drop the ring-off indicators at least once every connection. The time saved in shunting the ringing from the switch-room to the subscriber's

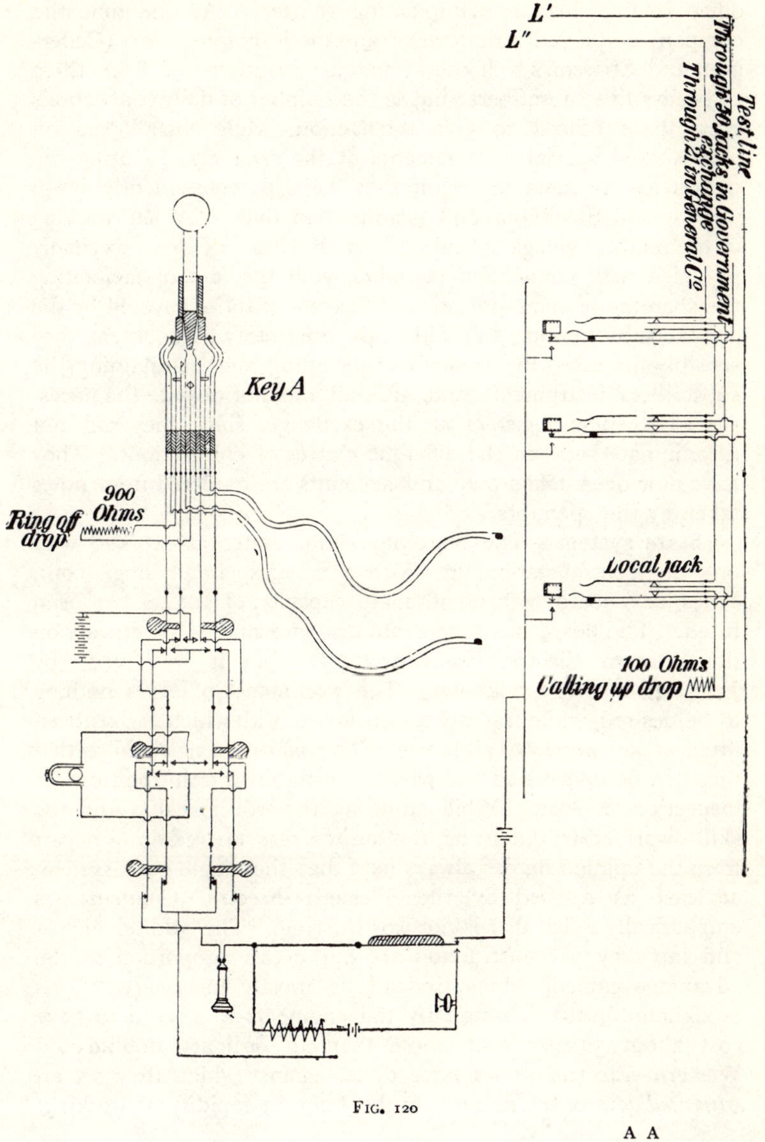
L′
L″
Test line
Through 30 Jacks in Government exchange
Through 21 in General Co
Key A
900 Ohms
Ring off drop
Local jack
100 Ohms
Calling up drop

FIG. 120

office is thus lost in manipulating shutters. At one time the company employed a number of automatic commutators (Cedergren and Ericsson's well-known patent) for groups of from three to twenty-five subscribers; but as the number of daily connections grew these ceased to give satisfaction, while their operation necessitated special arrangements at the exchange. During the conversion to metallic circuit they were all consequently swept away within Stockholm and vicinity, and only a few left working in the remoter villages. Subscribers of Class IV. were originally provided with connection counters, with the idea of facilitating the charging of communications in excess of those covered by the annual subscription; but, although satisfactory as counters, they sensibly increased the expense of installing and maintaining the subscribers' instruments, and, after all, did not obviate the necessity of keeping registers at the exchange, since they did not discriminate between the different classes of connections. They have now been taken out, and accounts are rendered from notes taken by the operators.

State system.—The State Stockholm system is worked with only one central station, in which a metallic circuit, single-cord, series jack board with an ultimate capacity of 10,000 has been fitted. The board has a separate test wire and is practically on the Western Electric Company's plan, but it was made by Ericsson & Co., Stockholm. The workmanship leaves nothing to be desired, while the care and neatness with which it has been fitted up are worthy of all praise. The jacks are arranged so that they can be unfastened and partially withdrawn from the front for inspection or repair. While admiring the workmanship and the skill displayed in the fitting, the author sees no reason to depart from the opinion he has always held that the single-cord system, at least as applied by the Western Electric Company, is emphatically a fish that is not worth frying. The additional cost and intricacy of construction are out of all proportion to the advantage gained, which, indeed, is mostly imaginary. This Stockholm board is stated by the engineers in charge to have cost about 50 per cent. more than an ordinary double-cord Western Electric would have done, against which they set an *estimated* gain of ten minutes in the hour in rapidity of working.

But on analysis it is difficult to understand where this gain comes in, the movements required from the operator for the double-cord being nine and for the single-cord eight, or a saving of one movement per connection. If it is true that the saving of one movement per connection equals ten minutes in the hour, what would be the saving accruing from the use of a board requiring only two movements per connection (and there are such)? The arrangements for trunk switching are of a familiar type. The trunk lines are brought to separate tables (which in Stockholm are in another room and out of sight of the local board), each table dealing with four trunks, and being under the charge of two operators, which means that each operator takes only two trunks (fig. 119). Actually, one girl operates four trunks, while the second keeps the very voluminous registers which are necessitated by the system of negotiating connections in advance. All the trunks are represented by jacks on each table. In addition to six separate trunk tables there is a special section of the local board through which all trunk connections must pass and on which all the subscribers are represented by multiple jacks, this special section also possessing ample communication with each of the trunk tables. A subscriber wishing trunk communication is turned on by his local operator to one or other of the trunk operators, who ascertains his wants and negotiates the necessary connections. A communication from a trunk to a local subscriber is obtained by the trunk operator concerned through the special section. The wires used by the operators for their communications are independent ones, special loops being reserved for the subscribers. Communications between operators are all conducted by dropping of shutters and plugging-in, no attempt being made to expedite matters by continuous listening, as to the practicability of which the Swedish engineers entertain serious doubts. The incessant dropping and replacing of shutters and movements of pegs must render this plan slower than a *vivâ voce* system of communication between operators. The fact that it necessitates an operator to every two trunks, besides those at the special section, must make it very costly.

There is no doubt that the practice of booking trunk talks in advance which prevails largely in Sweden adds greatly to the difficulties inherent in trunk and junction operating. A subscriber say in Stockholm, at 10 A.M. will call the exchange and book a talk to Gothenburg at 11.30 A.M. and another at 5.20 P.M., and perhaps other talks to Malmö and elsewhere at other stated times. The operator consults the list of booked talks already existing, and if the lines mentioned are not already engaged enters the orders. Then it is the business of the chief operators to have the lines ready for the caller at the times arranged. When the exchange wanted is intermediate with several others on one trunk line the difficulties multiply, and frequently the telegraph has to be used to transmit switching orders to stations that cannot be got at by telephone without interrupting talks in progress, and this in spite of the fact that when several stations exist on the same line each has fixed minutes in every hour for communicating with each of the others. The booking system has, however, become the rule, and the difficulties involved have to be fought and overcome. A noticeable feature of the State exchange is the arrangement of the lightning-protectors, and of the cross-connecting board, which, like the switch-board, is designed for 10,000 double lines. The protectors are made of carbon plates, kept from touching by thin strips of insulating material. The Swedish engineers were convinced that the carbons spark more freely than does any form of metal protector adapted for telephonic work, a conclusion the author has since confirmed by experiment. The cross-connecting board consists of two iron-tube frames arranged in concentric circles, the whole forming a neat and accessible arrangement.

The average number of daily connections dealt with is 5·5. Three operators are allotted to each 200 subscribers. The subscribers ring each other after being put through, a system which, owing to the absence of a discriminative ring-off indicator, increases the operators' work (restoring the shutters dropped by the ring through) and conduces to tapping.

HOURS OF SERVICE

Both the State and the General Company give a continuous service, night and day, in their principal towns. In the smaller places hours vary from 7 or 8 A.M. till 8, 9, or 10 P.M.

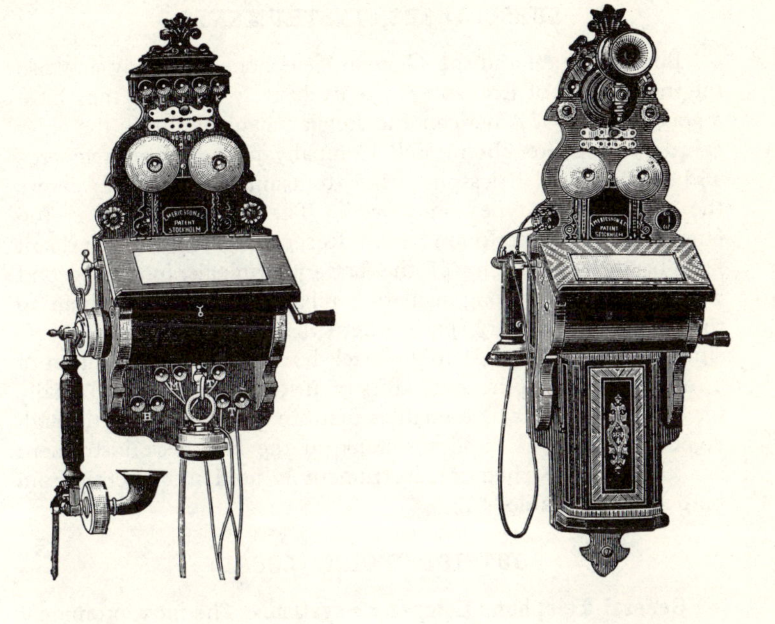

FIG. 121

SUBSCRIBERS' INSTRUMENTS

Both the State and the General Telephone Company now use the instruments of Ericsson & Co. exclusively, although they have a good number of American and Belgian manufacture in the older centres. The Stockholm Bell Company uses Belgian magnetos and receivers and Ericsson carbon transmitters. Fig. 121 shows the more usual types employed. Their construction is too familiar to need description here. Recent improvements in detail have been the mounting of the battery terminals, in the second wall instrument, on long insulating pillars so as to bring them to the front of the battery space, where they are much more accessible than when placed on the back-board; and the addition of a fourth magnet to the generators of instruments used habitually for long distances, which enables them to give a loud ring through 20,000 ohms. The duplicate crank on the first desk instrument is a convenience when the instrument is used indifferently from both sides of a table.

OUTSIDE WORK (LOCAL)

General Telephone Company's system.—The most prominent feature of this is undoubtedly the Bessemer steel tower (see frontispiece) at the Brunkeberg primary exchange, which is the largest telephone fixture in the world. It rests on special pillars built up from the ground, and rises 75 feet above the exchange roof. Its cost, 3,258*l.*, might well appal the telephonic financiers of Little Lilliput, but Mr. Cedergren is of opinion that no expenditure incurred by his company has ever yielded such a good return. Its building attracted attention to the telephone from one end of Sweden to the other; and when, in its finished state, it proved to be one of the landmarks of Stockholm and one of the best points from which to view the city, the identification of the General Company with the telephone became complete in the public mind, and is so yet, in spite of the State's opposition and lower rates. Figs. 122, 123, and 124, which show specimens of the company's smaller exchange fixtures, indicate a creditable fertility of design on the part of its engineers. Fig. 125 shows a

FIG. 122.—Telephone tower at Upsala.

Fig. 123.—Telephone tower in the Svartmangatan.

standard, with eight uprights, designed to carry 1,000 wires. The numerous other fixtures the presence of which a close examination of the picture reveals, and the manner in which the neighbouring buildings are dotted with insulators, afford some notion of the extent to which the upper air of Stockholm is netted with telephone wires. The system of roofing with iron plates which prevails in Stockholm is also clearly shown. Fig. 126 shows a type of standard employed at the junction of several routes, and fig. 127

FIG. 124.—Telephone turret at Södermalm.

one of the aerial cable rests that have become somewhat numerous since the reconstruction consequent on the change to metallic circuits and the re-grouping of the exchanges compelled a rather extensive resort to that mode of construction. The company's ground poles are not so noteworthy as its standards : indeed, there is nothing to pit against those of Belgium, Holland, or Switzerland, although solidity and strength are not wanting. Cross-arms on ground poles are often of angle-iron and not unfrequently of the

Fig, 125.—1000-wire fixture on No. 4 Myntgatan, Stockholm.

FIG. 126

Fig. 127

German double flat-bar type. In Stockholm the wire used is 1 mm. phosphor bronze of 30 per cent. conductivity and a breaking strain of 90 kilogrammes per square millimeter. Outside Stockholm, in Upsala and the other towns within the 70-kilometer radius, No. 11 B.W.G. galvanised iron wire is employed for the subscribers' lines. The insulators are small double-shed, fastened to their bolts with tow plugging. Joints in local wires are rarely soldered, the Macintyre dry joint (fig. 99) being found satisfactory enough for all purposes. Vibration in the houses is prevented or reduced by slipping a length of rubber tube on each wire and binding it tightly with leaden strip or wire (fig. 101). The aerial cables lately introduced to the extent of some twenty-five kilometers have been supplied by the Fowler-Waring Cables Company, Limited; W. T. Henley & Co., Limited; the Western Electric Company; Felten & Guilleaume, and Franz Clouth. The general specification of all these cables is 102 metallic circuits insulated with paper and enclosed in a leaden tube 2·25 mm. thick and 39 mm. exterior diameter. The conductors are copper of ·8 mm. diameter and 95 per cent. conductivity. The capacity of each single wire, all others being earthed, is ·05 mf. per kilometer. The company's underground system is intended to be of an extensive nature. The conduits are of the type invented by Mr. Axel Hultmann, formerly chief engineer of the State telephone system (see p. 369); the cables contain a hundred metallic circuits, with copper conductors of ·8 mm. enclosed in a leaden pipe 3 mm. thick and 50 mm. exterior diameter. They all have paper insulation and a capacity of ·05 mf. per kilometer. M. Aboilard, of Paris, has supplied some of the cable which has been so successful in the Parisian sewers for this underground work. On leaving the exchange, each route will consist of Hultmann concrete conduits containing eighty-six ducts of 75 mm. diameter, each duct capable of easily taking a 100-pair cable. The capacity of each route will be, consequently, 8,600 metallic circuits, which does not look as though Mr. Cedergren nourished any intention of hauling down his flag to the State, or had any misgiving of Stockholm's capacity and willingness to continue supplying him with subscribers *ad lib.* As they recede from the centres the conduits gradually decrease in carrying power, the successive sections having seventy-six, sixty-two, thirty-eight, and finally twelve ducts. Manholes occur about every 100 meters; they are cast in concrete of an elliptical shape and fitted with suitable cast-iron covers. This underground scheme, like all Mr. Cedergren's notions, is conceived on a grand scale, and will assuredly succeed. The submarine work necessary in and near Stockholm is usually done with armoured cable containing from four to fourteen pairs insulated with vulcanised rubber.

State system.—The State local work is very similar to the General Company's, except that the phosphor bronze is of 1·25 mm. diameter. The insulators are the same, and joints are not soldered. The standards, too, bear a family likeness, and fig. 128 will serve to illustrate those of both systems. The uprights are of double, the arms of single, channel steel. The fastening is done by sole-plates adapted to the slope of the roof and bolted through to the rafters. Sometimes, heavily laden standards are strutted in the Belgian fashion (figs. 22 and 23, Belgian section); if not so strutted they are carefully stayed. The Swedish and Norwegian mode of construction with channel iron or steel is unquestionably stronger, if heavier, than the tubular methods employed in Great Britain, Germany, and Holland. Tubes collapse when subjected to a sudden and heavy strain, such as is likely to result from the failure of a span of wires or of an adjacent standard, and crumple up beyond repair; the channel steel, being solid, may bend, but cannot collapse, and is consequently better adapted to withstand accidents, and, if injured, may be readily straightened again. On the other hand, it is more costly to make and transport, heavier to handle during erection, and permanently severer on the roofs. Fig. 129 shows a typical Swedish double ground pole, fitted with angle-iron arms, of solid and good construction. It is not the practice in Sweden to earth-wire either standards or ground poles. The State, like the General Company, has recently taken to aerial cables. Those erected are by Felten & Guilleaume, and contain thirty-eight pairs covered with jute and then with lead. The submarine type of cable is insulated with vulcanised rubber and armoured in the usual way.

A considerable proportion of the State local work is underground. The conduits are those originally designed for the purpose by Mr. Axel Hultmann, late engineer to the State telephone

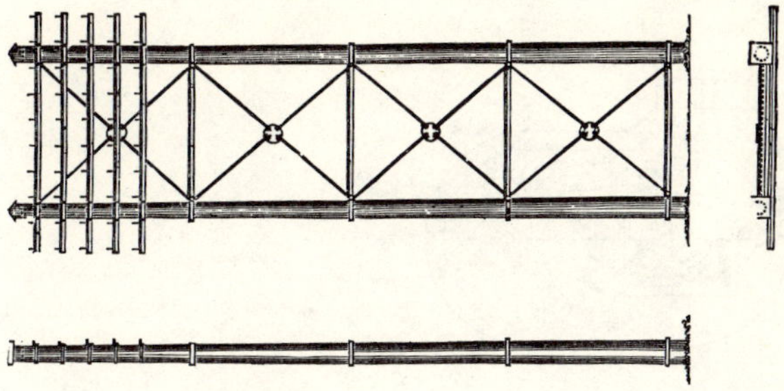

Fig. 129

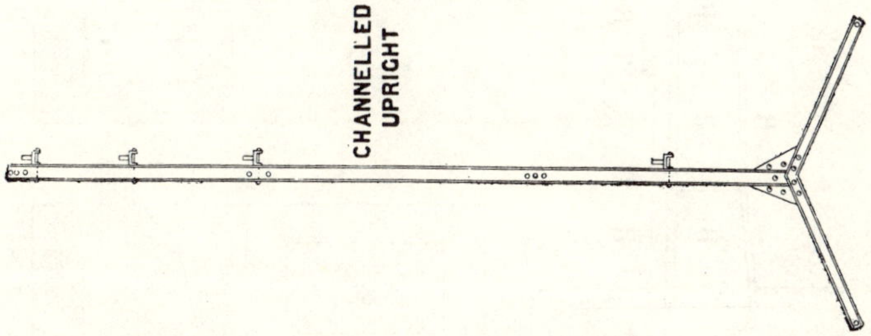

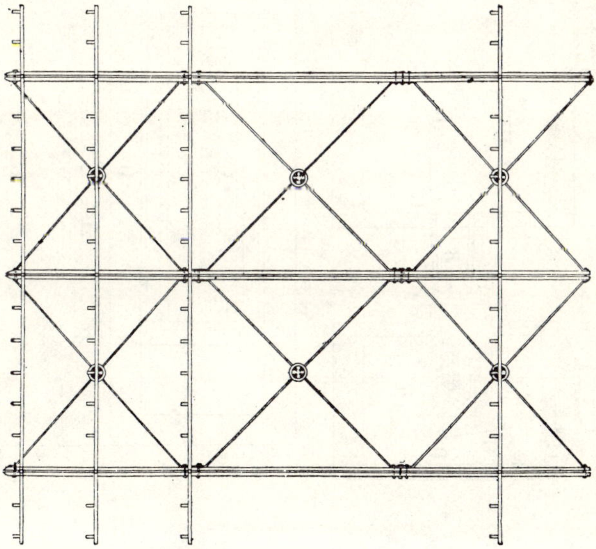

Fig. 128

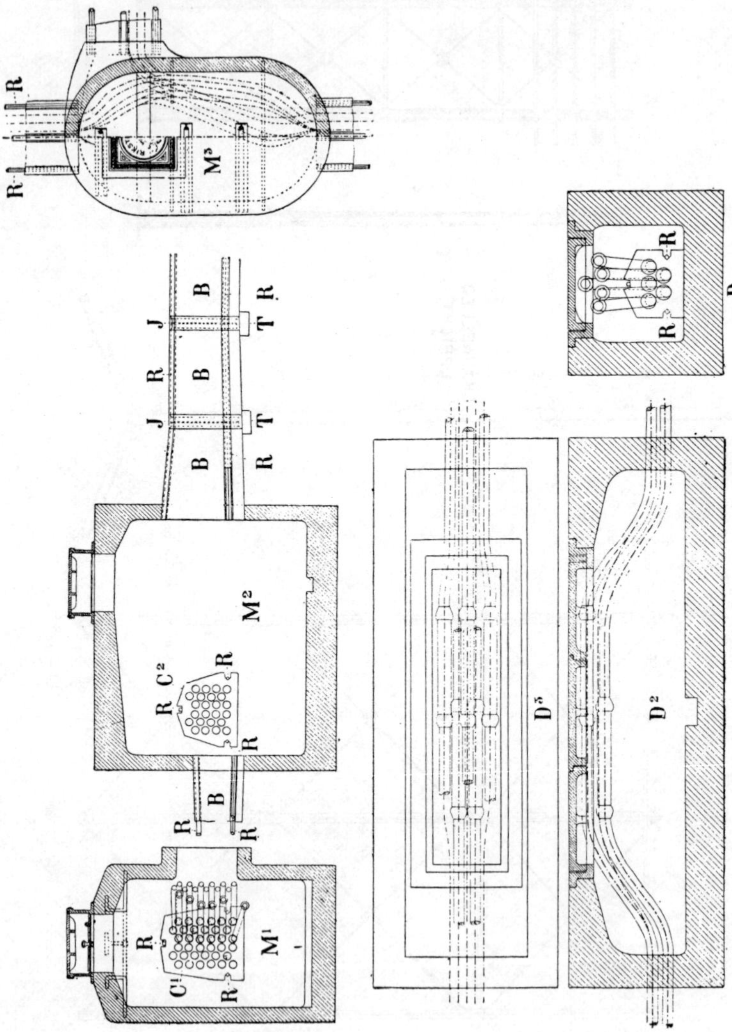

FIG. 130

department. They consist of cement blocks, pierced with a varying number of circular ducts, 75 millimeters in diameter. The blocks, which are from one to one and a half meters in length, are laid in the ground with the joints resting on cement base pieces of trough form, which keep them truly end to end. The blocks are made with three longitudinal depressions or furrows, into which, after the blocks are laid, strong iron bars are fitted. Thin plates of bitumen, having circular holes corresponding to the ducts, are placed between the blocks, several of which are forcibly clamped together, end to end, so as to compress the bitumen plates. The furrows containing the iron bars are then filled with cement, which, when set, binds the blocks rigidly together. Section after section is thus treated until a very solid conduit is produced, which, with the earth removed from beneath, is said to bear a direct weight of two tons without collapsing. The joints are made finally tight either with bitumen or cement. The ducts are made to correspond prior to clamping by inserting round rods made to fill them accurately through the blocks under treatment, the rods being withdrawn when the cement has set. No difficulty is stated to be experienced in obtaining correspondence between the ducts or in subsequently drawing in cable of a diameter of 52 millimeters to a length of 200 meters. The joints are said to be perfectly water- and gas-tight. The details of this system are made clear in fig. 130, in which M^1 M^2 M^3 are respectively cross, longitudinal, and horizontal sections of a concrete manhole with conduits and cables in position. The conduits are shown in cross-section at C^1 C^2, while B B B represent three blocks jointed together, as described, at J J, T T being the cement base pieces and R R the iron clamping rods. D D^2 D^3 are corresponding views of a draw-box adapted for a five-duct conduit. The General Company's conduit, while being essentially of the same construction, differs somewhat in form, the cross-section being as shown in fig. 131, with the iron rods passed through channels R R R R made in the interior instead of on the exterior of the blocks. Mr. Hultmann has unquestionably produced a strong and efficient conduit which has already stood the test of several years' service most successfully. The separate duct plan is almost essential to underground cable work, as it enables repairs and alterations to be carried out easily, which would be simply impossible when many heavy cables are superimposed in one large pipe. The General Company's 86-duct conduit measures 100 × 110 centimeters, or a little over three feet square, and contains accommodation for 86 × 100 = 8,600 metallic circuits; that is to say, all the telephone subscribers now existing in London, and more, could be provided with metallic circuits and concentrated within one such conduit. At the same time, the conduit is not so easily diverted for the purpose of avoiding obstacles as iron pipes are, and this would militate against its employment in London, at all events very near the surface. The cables used by the State were originally of the Pattison type; now all are insulated with paper. Fig. 132 shows a junction between an underground and an overhouse route, the test-box containing both terminals for testing and cross-connection and lightning-guards. The box illustrated is one of the General Company's, but the State's practice is essentially identical.

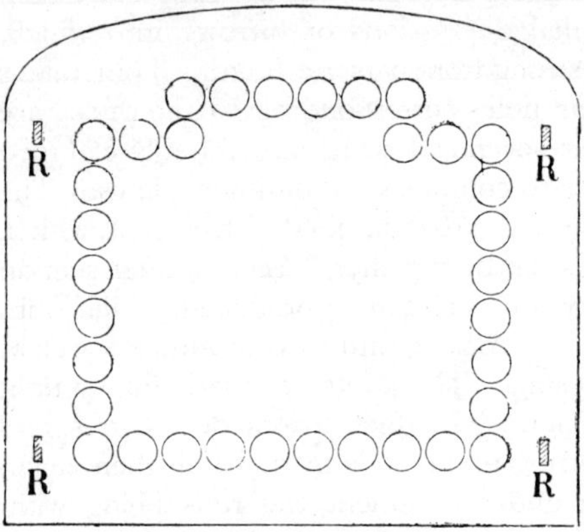

FIG. 131

OUTSIDE WORK (TRUNK)

General Company's system.—The company's trunks are of course restricted to the 70-kilometer radius, but are still very numerous. They are constructed of 2 mm. phosphor-bronze wire of 60 per cent. conductivity and a breaking strain of 80 kilogrammes per square millimeter. The wires are crossed at intervals to neutralise induction, with results that are completely satisfactory.

State system.—Much of the State trunk work was formerly run with the so-called bimetallic wire, steel coated with copper, of

1·9 mm. gauge ; but this has lately given way generally to high-conductivity bronze, although the Copenhagen trunk has been run in

FIG. 132

Sweden with 3 mm. hard copper. The shorter trunks are crossed and the longer revolved or twisted on the Moseley-Bottomley plan. Special fixtures are used to facilitate the twisting. They consist of iron frames, as in fig. 133, A and B, which, being made all exactly alike, secure the maintenance of perfect distance between the wires. When two loops run on the same poles the frames are modified as at C and D. The twisting system is reported to have given much trouble after breakdowns due to snow, the workmen, finding it impossible to restore the twist promptly, having had to

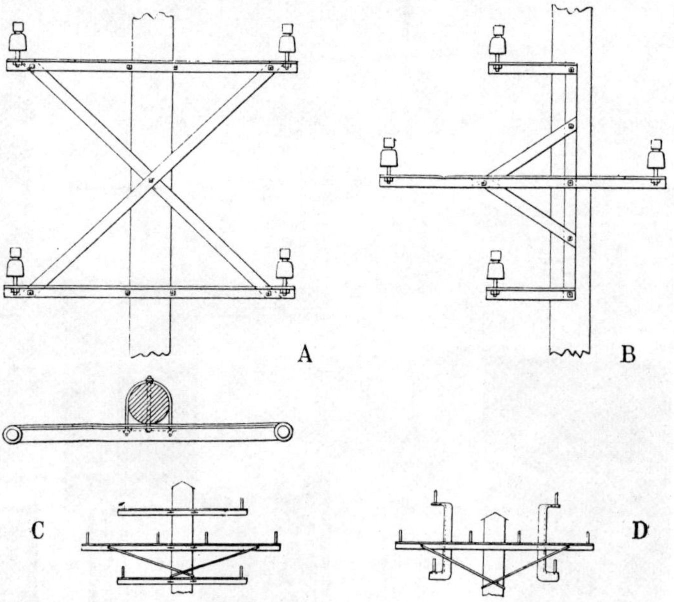

FIG. 133

run the wires straight through in order to re-establish communication, and to subsequently retwist them at leisure.

PAYMENT OF WORKMEN

General Company.—Foremen, 4*s*. 5*d*.; skilled wiremen, 3*s*. 4*d*.; labourers, from 2*s*. 2½*d*. to 2*s*. 8½*d*. per day. Country allowance, 2*s*. 2½*d*. per day.

State.—Foremen, 3*s*. 6*d*.; skilled wiremen, 2*s*. 9*d*.; labourers, 1*s*. 11*d*. to 2*s*. 2½*d*. per day.

Hours worked : in summer, 7 A.M. till 7 P.M., with one and a half hours for meals ; in winter, sunrise to sunset, with one hour for meals.

PAYMENT OF OPERATORS

General Company.—*2l.* 4*s.* 1*d.* to *2l.* 15*s.* 7*d.* per month, according to experience. Extra pay is given for night duty, which is performed by the girls in rotation. The hours worked are normally seven per day, divided into two watches with an interval of at least three hours between. Exceptionally the duty may be extended to eight hours, but never more. The girls, who are taken on at eighteen years of age, get fourteen days' holiday on full pay annually, and in case of sickness receive full pay for the first fortnight and half pay for a second. The lady superintendents receive from 3*l.* 17*s.* 8*d.* to 8*l.* 6*s.* 8*d.* per month, according to length of service and the importance of their charge.

State.—1*l.* 13*s.* 2*d.* to 2*l.* 4*s.* 1*d.* per month for ordinary, and 2*l.* 15*s.* 7*d.* for trunk operators. Extra pay is given for night duty. Girls are taken on at eighteen ; no examination is imposed.

STATISTICS

An enumeration of the exchanges in Sweden would be practically a list of the names of all the towns and chief villages in the country. At the end of 1894 the General Company, in addition to its Stockholm switch-rooms, possessed 113 exchanges within the 70-kilometer radius, having between them 2,012 subscribers. Of these, Upsala (population 21,000), with 363 subscribers, was the most important ; and Södertelge, with 145, the second. At both these towns the State is also established. The Upsala rate is 2*l.* 15*s.* 7*d.* per annum, without any admission fee. At Södertelge and the majority of the other places, the rate is the same, but with an admission fee, also of 2*l.* 15*s.* 7*d.* In a few instances this subscription covers only one hundred free connections per quarter, all over that number being charged 1·3*d.* each. In other cases, principally where submarine cable work is necessary, the admission fee is 2*l.* 15*s.* 7*d.* and the annual subscription 4*l.* 8*s.* 11*d.* A few of the smaller places are worked at an admission fee of 11*s.* 0½*d.*, an annual subscription of 11*s.* 0½*d*, and a charge of 1·3*d.* for *every* connection had.

Within the 70-kilometer radius the State also possesses seventy-five exchanges in addition to its Stockholm system, making a total for the radius of 188 exchanges, exclusive of the metropolitan, and entitling the area to the distinction of being by far the best telephoned piece of country in the world. The area represented by seventy kilometers round London is, on the other hand, probably the worst in the neighbourhood of an important city. The State's provincial tariff is the same as in the town.

For the rest of Sweden there are no statistics later than the end of 1892. At that time the State owned 288 exchanges and co-operative societies 158, but a number of these last have since been absorbed by the State. At the same date the State owned 15,416 kilometers of trunk lines. At the end of 1893 the General Company had 9,031 instruments working in connection with 95 switch-rooms, 15,259 kilometers of lines (not wires). The number of connections in Stockholm alone for the year was 25,060,715, or 9·05 per subscriber per day, dealt with by a total staff of 200 lady telephonists. The value of the company's Stockholm system at December 31, 1893, was 2,006,693 kronor, and of its country system 1,018,510 kronor, making a total of 3,019,203 kronor, or 165,890*l.* Adding the value of premises, workshop plant, stores, and raw materials in hand, the assets were brought up to 3,742,801 kronor, or 205,648*l.* All this had been brought into existence with a share capital of only 32,966*l.*, and the surplus of profits remaining after paying 8 per cent. per annum to the shareholders and creating a reserve fund, a renewal or deterioration fund, a fire insurance fund, an accident fund, an employees' benevolent fund, and a general purposes fund. At December 31, 1893, these several items stood as follows :—

	£
Reserve fund	9,890
Renewal fund	70,747
Fire insurance fund	1,257
Accident fund	1,362
Employees' fund	21,978
General purposes fund	3,177
	£108,411

The net profits each year since 1883 have been :—

	Kronor		Kronor
1883	6,528·68	1888	62,418·96
1884	36,059·22	1889	64,780·04
1885	49,559·82	1890	79,579·26
1886	56,005·94	1891	81,819·57
1887	58,843·86	1892	100,285·28

1893 ... ... 113,198·79 kronor (6,219*l.*)

These profits have sufficed to pay a steady dividend of 8 per cent. per annum (the maximum allowed) on the share capital, to extend the business to an extent unprecedented elsewhere, to convert the system from single to double wire, and to lay by money against deterioration and almost every possible contingency. And all on a maximum rate of 5*l.* 11*s.* 1*d.* operative over 140 kilometers !

XXIV. SWITZERLAND

HISTORY AND PRESENT POSITION

THE Swiss Government at an early date determined to control the telephones within its jurisdiction, and in 1885 took over the only exchange, that at Zürich, which it had permitted a company—the International Bell Telephone—to establish. For about nine years the administration has consequently been in the hands of the State, and the development attained is certainly most imposing, there being at the close of 1894, with a population of about three millions, nearly 20,000 subscribers.

The Swiss telephone system is remarkable in many ways. From the beginning of its management the Government has endeavoured to bring the telephone within the reach of all and to render the service as complete and satisfactory as possible. Originally, the annual subscription for an ordinary line and instrument within a radius of two kilometers was 150 francs (6*l.*), without restriction as to the number of communications; but Dr. T. Rothen, then director of the Swiss telegraphs, as early as 1883 pointed out in the 'Journal Télégraphique' that it was not more logical to accept an annual payment from a merchant to cover all his telephonic communications than to cover all his telegrams. The system, notwithstanding its convenience and almost universal application, is, in fact, inequitable—for a busy merchant, to whom telephonic communication is a necessity, obtains much greater value for his annual subscription than does a person whose business relations are neither so extensive nor so important. Dr. Rothen proposed, as the only just solution, to charge subscribers a fixed sum for every connection asked for and had, just as telegrams are charged for separately according to the tariff. The practicability of this plan was disputed on several grounds, and not without plausibility. Its probable effect on the revenue was feared, and the discontent of those busy subscribers whose payments would no longer be covered by 6*l.* per annum dreaded. However, by the 'Loi Fédérale' of June 27, 1889, the principle was definitely adopted in Switzerland with the modification that a foundation or first payment to cover 800 connections per annum was prescribed, all subsequent communications having to be paid for on Dr. Rothen's plan. The annual charge was fixed at 4*l.* 16*s.*, 4*l.*, and 3*l.* 4*s.*, for the first, second, and third and subsequent years respectively, while the connections had in excess of the 800 were rated at 4*s.* per hundred, or five centimes (·48*d.*) each. This law came into operation on January 1, 1890, and has led to an immense increase in the number of telephonic subscriptions.

Subscribers' wires are generally single with earth return, but all trunks and many of the junction lines to parochial stations are metallic circuits, translators being employed for the connections between the two. It is pleasant to know, however, that the Swiss are alive to the inadequacy of the single-wire system as a permanent institution, and have decided to gradually supersede it everywhere by metallic circuits. A very earnest and creditable beginning has already been made at Zürich, and similar changes are to follow immediately at Berne, Geneva, and Lausanne.

The cost of keeping the voluminous and complex accounts rendered necessary by recording the subscribers' calls and charging each individual every month for his local calls above a certain number; for his trunk calls; for his telegrams forwarded and delivered; and for his telephonograms, is unquestionably very considerable; and the question whether an automatic counter in each subscriber's office would not be a useful addition is being debated. Many such counters have been devised and tried, but a really trustworthy one has not yet been forthcoming, while the first cost of installing 20,000 such instruments would not be a negligible quantity. But the chief difficulty lies in the diversity of traffic which is liable to emanate from the same office. A counter that could not differentiate between telephonograms, telegrams forwarded and received, local calls, and trunk connections, would at best serve as a rough check, and the operators' records would still have to be relied on and the present laborious system of accounts preserved. On the other hand, the mind rather shrinks from the idea of fitting four or more counters in each office, and especially from the expectation that the subscribers would use them properly if fitted.

In Switzerland the adaptation of the telephone and telegraph to popular requirements has undoubtedly received its widest application. The consequence is that the country is covered with trunk wires altogether out of proportion—it must be remembered that there is not an ounce of native coal in Switzerland—to its industrial importance. One pauses in wonder at the idea of what might be done in the United Kingdom to facilitate intercourse under similar intelligent (not benevolent, because it pays) management.

But the Swiss public was not yet satisfied. It was held a grievance that a subscriber should be compelled to pay for 800 communications per annum whether he had them or not, and many found that they could manage with less. As a result of representations of this nature, a committee of the Federal Council was appointed, and after hearing evidence, reported in favour of a reduction of the first or foundation annual charge to 4*l.* for the first, 2*l.* 16*s.* for the second, and 1*l.* 12*s.* for the third year, the abolition of the free margin, and the rating of *all* connections at 4*s.* per hundred, or ·48*d.* each. According to this plan, a subscriber making two calls per day exclusive of Sundays, or 616 per annum, would pay after the expiration of his second year of membership only 1*l.* 12*s.* + 616 × ·48*d.* = 2*l.* 16*s.* 7*d.* per annum (a sum which has been proved remunerative in Holland) in lieu of the present minimum of 3*l.* 4*s.* If he can manage with one call per day his annual telephonic disbursement would be only 1*l.* 12*s.* + 308 × ·48*d.* = 2*l.* 4*s.* 4*d.* The committee's recommendations were adopted by the Federal Council and embodied in a law on June 13, 1894, which received the sanction of the Council of States on December 7, 1894. It is still liable to challenge until March 26, 1895, by a demand for a national vote on the subject, but no steps have been taken in this direction; and as the measure is a popular one it is considered certain to pass the critical date successfully, and to be added definitely to the statute book. In this case it will come into operation on January 1, 1896, at the latest. Switzerland will then enjoy the cheapest and, at the same time, the most rational telephonic tariff in the world; for after subscribing 1*l.* 12*s.* annually, a charge sufficient to maintain his line and instrument in good order, every man will pay exactly in accordance with the use he makes of his connection. By the same law the existing charges in connection with parochial telephone stations (see p. 386) are abolished, and the parish councils put on exactly the same terms—those just cited—as ordinary subscribers. The new charges, like the present, are to cover lines not exceeding two kilometers in length; excess rates for longer distances, both single wire and metallic circuit, are to remain unaltered. The present charges for telephoning telegrams, telephonograms, and public stations stand. It must be clearly understood that the new tariff, like the existing, covers erection, maintenance, and all expenses.

SERVICES RENDERED BY THE STATE TELEPHONE ADMINISTRATION

1. **Intercommunication locally between the subscribers and public telephone stations of a town or district.**

2. **Internal trunk line communication.**—There is scarcely a town or village of any size that does not participate in this service. The system is at present somewhat wanting in direct trunks between the more distant towns, intermediate switching—i.e. the joining of two or more short trunks to make up a temporary long-distance line—being requisite; but this defect is being gradually removed as traffic develops. The longest distances at present talked over are (as the wires go, the mountains, and lakes, which are too deep and uneven for cables, preventing direct routes in many cases) 166 miles, Geneva to Schaffhausen; 178 miles, Geneva to St. Gallen; and 239 miles, Geneva to St. Moritz. One of the regulations relating to trunks forbids the engagement of a line in advance for a conversation at a specified time, which is directly opposed to the Swedish practice of booking talks a long time beforehand.

3. **International trunk communication.**—The Swiss wires have already broken bounds in several directions by connecting with France, Baden, Würtemberg, Bavaria, and Austria. These international lines are not, however, with perhaps the exception of the French, of much importance as yet, communication on the German side being restricted to the Swiss towns—St. Gallen, Romanshorn, and one or two others—nearest the frontier. Consequently, when subscribers at Berne, Zürich, and of other exchanges west of St. Gallen wish to communicate beyond the frontier they must find somebody in one of the border towns to act as intermediary. These restrictions are understood to be due to objections raised by the Imperial Political Bureau at Berlin. Communication was also established *viâ* Basle with Alsace-Lorraine, but after a time had to be discontinued by orders from Berlin. The junction with the French lines is at Besançon; with the Baden, at Constance; with the Austrian, at Bregenz; and there is communication *viâ* Bregenz with Lindau in Bavaria, and Friedrichshafen in Würtemberg.

4. **Telephoning of telegrams.**—Subscribers are afforded every facility for forwarding and receiving their telegrams by telephone, as the State regards the telephone system as the natural feeder of the telegraphs, in the same manner as light railways are collectors for the heavier main lines, and accordingly cultivates an intimate connection between the two. All the exchanges have a connection with the nearest telegraph office, which is given to a subscriber who wishes to forward a telegram, and used by the telegraph office for obtaining communication with a subscriber for whom a telegram has arrived. The Swiss, however, are not so liberal in this particular matter as the Belgians and Bavarians, since the subscriber has to pay ·96*d*. for each telegram, in or out, transmitted by telephone. Copies of the telegrams telephoned to subscribers are afterwards delivered by messenger. This is not such a shrewd arrangement as that existing in Belgium, where copies are posted instead of delivered (unless the subscriber specially wishes otherwise). The Swiss plan saves nothing in messengers, and wins very little popularity, since in the vast majority of cases the receivers are quite content with the version telephoned. Telegrams for telephoning must be in the German

or French languages except in the Italian-speaking cantons, where Italian is also admitted.

5. **Telephonogram service.**—This facility, unknown to the National Telephone Company's subscribers in Great Britain, but largely patronised in many continental countries, is in Switzerland called officially the 'phonogram' service. It enables any subscriber using his own telephone, or any non-subscriber from a public one, to dictate a message to the operator addressed to any non-subscriber resident in the same town or district, which is written down like a telegram and delivered to the addressee by messenger. Telephonograms are subject to the same regulations respecting language as telegrams.

6. **Parochial telephone stations.**—An important feature of the Swiss telephone system is the parochial or communal office. It is no longer peculiar to Switzerland, having been adopted, with modifications, by France; but it originated there in the anxiety of the Government to make the people, as far as economically possible, participators in the public institutions, and in pursuance of the idea of utilising the telephone as a feeder of the telegraph. It enables a parish or commune without a telegraph or telephone station to provide itself with these conveniences in the following manner: The parish council undertakes to pay the State 120 francs (4*l.* 16*s.*) per annum for a wire to the nearest telegraph station or telephone exchange, the charge being increased by 2*s.* 5*d.* for each 100 meters in excess of two kilometers. The council provides a suitable room or office for its station, and pays the wages of the necessary operators and messengers, both office and servants being subject to approval by the State. The public may use the station as an ordinary telegraph or telephone office, paying 1·44*d.* on each telegram sent or conversation had, in addition to the ordinary tariff, which 1·44*d.* is the property of the parish council and goes towards covering its expenses. No charge is made on delivered telegrams within the ordinary free delivery radius. The facility is largely taken advantage of, there being nine parochial stations in the vicinity of Berne alone. When the traffic has grown sufficiently to justify such a course, the State takes over the station, and relieves the parish council of further responsibility.

7. **Connection of private groups of subscribers to an existing trunk or junction wire.**—This is another service which owes its initiation to the anxiety of the Government to bring the telephone to, or rather within, the doors of all. It provides for the wants of a community which has not yet attained to the dignity of a parish council. One or more persons resident on, or near to, a route of poles carrying trunk or junction telephone wires, excepting trunks intended for the direct service of important towns, may, if not numerous enough to justify the establishment for their benefit of a regular exchange, claim a connection with the system, either by means of an automatic commutator looped into, or tapped off, a wire going to the nearest ordinary exchange, or by means of a small switch-board placed in the house of one of them, or in that of a competent person, and attended to at the expense of the subscribers participating in the benefits secured. The State erects the wires, switch-board, instruments, &c., in return for the usual subscriptions, while the subscribers find house room, and do, or pay for, their own switching. They may talk amongst themselves without stint, but conversations over the connecting wire to the nearest regular exchange are subject to the 800 communications per annum rule. This service is widely patronised. It is not by any means a desirable one from the point of view of the telephone engineer, as it introduces complications and derivations inimical to the best talking and promptest switching; but when the convenience of the people living in out-of-the-way localities is considered, it is worthy of the highest commendation. The automatic commutators are not so numerous (there are only some fifteen in use) as ordinary switch-boards operated by hand, but they are the best of their kind (Cedergren and Ericsson's).

8. **Public telephone stations.**—These are very numerous, and may be divided into two classes: (1) those provided specially by the State at telegraph and railway stations, and the premises of non-subscribers; and (2) those at the offices of subscribers who, after having their premises approved as suitable, have contracted with the State to place their instruments at the disposal of all applicants in consideration of a commission on each sum collected. The public stations are available not only for speaking to subscribers in the same or other towns, but for the forwarding of

telegrams and telephonograms to all and sundry. What a boon it would be in Great Britain if it were possible to pop into a shop or office bearing the sign 'Public Telephone Station'—and several such should be found in every long street—and not only call up a telephone subscriber, but forward telegrams and telephonograms to anybody! And how the Post Office telegraphs would benefit, too, could the officials but see it.

TARIFFS

1. **Rates for local exchange communication.**—These are uniform throughout the country, and include every expense. Within two kilometers of an exchange a subscriber pays:

	£	s.	d.
First year	4	16	0[1]
Second year	4	0	0[1]
Third and subsequent years	3	4	0[1]

If the local connections he asks for do not exceed 800[1] in number per year, there is nothing more to pay. All in excess of 800 are charged 4*s*. per hundred, or ·48*d*. each. Trunk line talks, telegrams, and telephonograms are not reckoned in the 800 talks allowed. The chief Government office in each canton, and the chief office in each commune, is entitled to a simple connection to the nearest exchange as soon as it counts thirty paying members, for which nothing is paid unless the communications asked for exceed 800 per annum, in which case the usual fees are collected for talks in excess of that number. Institutions of public utility, not working for profit, pay 3*l*. 4*s*. per annum from the beginning, without restriction as to number of talks. Fire brigade stations pay 1*l*. 12*s*. per annum, and ·48*d*. per talk. Subscriptions are payable half-yearly in advance on January 1 and July 1. When a subscriber's distance from the exchange exceeds two kilometers he pays 2*s*. 5*d*. for each 100 meters in excess. When it is considered desirable, to prevent annoyances from overhearing, that a subscriber should have a metallic circuit, no extra charge is made up to two kilometers, but beyond that distance the sub-

[1] These charges are altered by the new law soon to come into force (see p. 373).

scriber has to pay 3*s*. 7*d*. instead of 2*s*. 5*d*. for each additional 100 meters.

A subscriber wishing to have a second instrument in connection with his exchange line pays 16*s*. per annum, with 2*s*. 5*d*. for each 100 meters of extra wire required. In such a case, the annual talks from the two instruments together must not exceed 800 without the extra charge being incurred. A drop indicator in connection with the subscriber's instrument, to show whether he has been called in his absence, costs 1*s*. 7*d*. per annum ; a two-indicator switch 8*s*., and a trembling bell 3*s*. 2½*d*. per annum.

The areas that may be spoken over without incurring trunk line charges are much more restricted than in Belgium. As a rule, communications outside the limits of a town and its suburbs, if obtained through a second exchange, are regarded as trunk messages. Should an interruption of a subscriber's wire continue for a longer period than five days, he is entitled to have his subscription refunded for every subsequent day that he is without communication. Government and police calls take precedence of all others. Subscribers may allow outsiders to use their instruments, but as all conversations go to extinguish the 800 free talks permitted, it obviously does not pay to admit much latitude in this respect. An outsider may arrange to use a subscriber's instrument and to have his name printed in the list on payment of 8*s*. annually to the State. He is left free to make his own arrangements, monetary or otherwise, with the subscriber, the latter being held responsible for all payments except the 8*s*.

There is no limit set to the duration of local talks. The shifting of subscribers' instruments is charged for. For a shift within the same building the actual expense incurred falls to be paid ; a removal to another house, whether within the same exchange area or another, is subject to a fixed charge of 16*s*., with excess mileage if the new line exceeds two kilometers in length. Each subscriber is furnished free with a list of members within his own district, but must pay 2·88*d*. for each copy of other district lists. Non-subscribers must buy all lists at 4·8*d*. per copy. When a subscriber wishes to figure in his list under more than one letter or denomination he can do so on payment of 1*s*. 7*d*. per additional entry.

2. **Rates for internal trunk lines.**—The time unit in Switzerland is three minutes. No person may retain a line longer than six minutes if it is otherwise wanted.

The trunk rates are :—

Up to 50 kilometers	. . . .	2·88*d*.
50 ,, 100 ,,	. . .	4·8*d*.
Over 100 ,,	. . . .	7·2*d*.

As previously mentioned, 239 miles may already be spoken over. Trunk charges, and all others involving the trusting of subscribers, must be covered by deposit on which no interest is allowed. Accounts are rendered monthly. Non-subscribers pay the same trunk rates as subscribers, but must of course make use of a public telephone station.

3. **Rates for international trunk lines.**—The rates between Switzerland and France were determined by the convention of July 31, 1892, and are regulated by the distance talked over. Within a radius of ten kilometers of the frontier the charge is 4·8*d*. per three minutes ; within a radius of 100 kilometers, 9·6*d*. ; within a radius of 200 kilometers, 16·8*d*. ; for each 100 kilometers of additional radius, 9·6*d*. extra. There is no restriction imposed as to the distances talked over, so that, electrical conditions permitting, all Swiss may converse with all French subscribers.

On the German side there is communication at 5·76*d*. per three minutes between Kreuzlingen and Constance (Baden).

On the Austrian side, St. Gallen, Romanshorn, and a few other Swiss towns near the frontier may speak with Bregenz, Dornbirn, and Feldkirch (Austria) at 1*s*. per three minutes.

The same towns may likewise speak, *viâ* Bregenz, to Lindau (Bavaria) and Friedrichshafen, Ravensburg and Langenargen (Würtemberg), at 14·4*d*. per three minutes.

Before the communication was discontinued by order of the Imperial German Government the rate between Basle and St. Ludwig and Mulhouse (Alsace) was 1*s*. per three minutes.

4. **Rates for telephoning of telegrams.**—Each telegram dictated to a telegraph office through a telephone exchange by a subscriber from his own office, or handed in by a non-subscriber

at a public or parochial office, is charged ·96*d.* in addition to the tariff cost of the telegram.

Each telegram dictated by a telegraph office to a subscriber is charged ·96*d.*, and a copy is forwarded to his address by messenger.

5. **Rates for telephonograms.**—Each telephonogram is charged 1·92*d.* plus ·096*d.* (·1 centime) per word, odd centimes being counted as five.

If the addressee is located within one kilometer of the nearest telegraph office or other available point of distribution, no charge is made for delivery; if beyond, the usual excess rate is collectable.

6. **Rates affecting parochial telephone stations.**—The parish council pays the State for installing the line and instrument 4*l.* 16*s.*[1] for the first, 4*l.*[1] for the second, and 3*l.* 4*s.*[1] for the third and subsequent years, increased by 2*s.* 5*d.* for each 100 meters over two kilometers. The parish council provides and furnishes a suitable house or room rent free, and pays the wages of the necessary operators and messengers.

As a set-off against these expenses the parish council is authorised to collect for its own behoof from persons using its station, in addition to the tariff charges:—

·96*d.* on each three-minute local talk had at its station up to 800 in number; if the talks in one year exceed 800, the balance must be charged only ·48*d.* each.
·96*d.* for each three-minute trunk talk.
·96*d.* for each telephonogram forwarded.
2·4*d.* for each telegram despatched forward.
·96*d.* ,, ,, received (collectable from the addressee).

If the delivery is effected beyond a distance of one kilometer, excess charges are made as follow:—

Up to 1½ kilometers	2·4*d.*
,, 2 ,,	4·8*d.*
For each additional kilometer	2·88*d.*

7. **Rates for private groups of subscribers looped into or tapped off an existing trunk or junction wire.**—Each subscriber pays the State the ordinary subscription of 4*l.* 16*s.*, 4*l.*, and 3*l.* 4*s.*, for the first, second, and third years respectively, for which he

[1] These charges are altered by the new law soon to come into force (see p. 378).

may talk to any extent amongst his own group, but is restricted in the usual way to 800 free conversations per annum through the ordinary exchange to which the group is connected. If the line by which the connection is effected exceeds two kilometers in length, each member pays an equal share of the extra annual charge of 2*s.* 5*d.* per 100 meters.

The State erects and maintains all wires and instruments ; the subscribers find a free location for the switch-board, and pay for all operating.

If the group is not composed of more than five subscribers the switch-board and operator may be replaced by an automatic commutator, which occupies little room and can be fixed in the house of one of them. When the automatic commutator can be placed centrally in respect to the group, so that none of the lines exceed two kilometers in length, the usual subscription is reduced by 16*s.*, and becomes 4*l.*, 3*l.* 4*s.*, and 2*l.* 8*s.*, for the first, second, third and subsequent years respectively. If one or more of the subscribers happen to be over two kilometers off, the extra distance is paid for on the usual scale, which also comes into operation if the commutator cannot be placed centrally. An extra annual charge, which is shared equally by the subscribers, of 3*l.* 4*s.* for a five-line and 1*l.* 12*s.* for a two-line commutator is made.

It is rather curious that the State makes a reduction in favour of automatic commutators, which are more liable to get out of order and require more attention than ordinary switch-boards. If the cost of operating these last fell on the State instead of on the subscribers, such a course might be justifiable ; but as it does not, the wisdom of the procedure is not very apparent. The first cost of the automatic instruments is much greater than that of ordinary switches, and they are not so quick or so effective in action, yet the State encourages their use by accepting lower subscriptions.

8. **Rates affecting public telephone stations** :

Local talks (per three minutes)	·96*d.*
Internal trunks .	·96*d.* in addition to the usual rates for these services.
Telegrams forwarded .	
Telephonograms .	

Subscribers who permit their instruments to be used as public stations are remunerated by being allowed to retain of these charges the whole of the amount for local talks up to 800, and half thereafter, together with the whole of the surcharges accruing on internal trunk talks, telegrams, and telephonograms. When the State arranges for a public station on the premises of a non-subscriber, that person keeps half the receipts for local talks and the surcharges on the others. Keepers of public stations may, if they make satisfactory arrangements for the purpose, also receive, write down, and deliver telegrams and telephonograms addressed to persons in their neighbourhood, in which case they get ·96*d.* for each message delivered. Public stations are never established in inns or restaurants. Automatic boxes for checking payments are not used. Subscribers enjoy no preferential treatment. Telegrams and telephonograms have to be handed in written out, and are telephoned forward by the attendant, not by the sender personally.

WAY-LEAVES

The position of the State in the matter of way-leaves is defined by the law of June 26, 1889, which provides :—

1. That the State has the right to use all public lands and places for the placing of telephone wires on paying for damage done, but must not do anything inconsistent with the purpose to which such public place is devoted.
2. That the State may pass wires without attachments over private property, provided the presence of such wires does not prejudicially affect the property.
3. No work must be done on public or private property without arriving at an understanding with the authorities or proprietors concerned. In the event of dispute the Federal Council will decide, if necessary on the advice of independent experts.
4. Proprietors of trees must cut any branches which interfere with State telephone lines. Notice that cutting is necessary to be given to proprietors through the local authority. If no notice is taken within eight days, the State may itself cut the branches.

5. Authorities or proprietors under Articles 1 and 2 may require removal of any wires calculated to interfere with projected building or other lawful operations. If the State removes wires to make room for such proposed operations, the proprietor will be debited with the cost if he does not begin to build within a year of such removal.
6. The State may build telephone lines along railways belonging to companies, provided such lines do not prejudice the railway in any way, nor interfere with the security of existing works. The company to be compensated for any damage done, but to be entitled to no payment in name of way-leave.
7. The State must carry out at its own expense such changes as may from time to time become necessary owing to alterations in the railways.

8 to 15 Deal with installations of electric light and transmission of power as affecting telegraphs and telephones, and the procedure to be followed in event of disputes.

The application of this law appears to have given rise to misunderstandings, for it was supplemented on December 7, 1889, by a rider which declares that Article 1 of the law is not to apply to buildings or to property not accessible to the public; on such buildings no supports may be placed without the consent of the authorities or proprietors; and that the right to pass over refers only to wires suspended in the air, and does not include the placing of supports. Proprietors of trees cut by the State to have a right to compensation, which must not exceed five francs per tree without the express approval of the Telegraph Administration.

It will be seen that the Swiss Government possesses no autocratic powers in respect to way-leaves. In effect, it can do nothing without the consent of the proprietors affected, and has to pay its way just like a telephone company in the United Kingdom. The way-leaves paid average one franc per wire per annum, and some standards cost as much as 400 francs (16*l.*) per annum. In one disputed case the Telephone Administration took advantage of the arbitration clause in the law, but was disgusted to find that the award was five francs per wire per annum in addition to the cost of the reference. Trouble was caused by the railway companies objecting to Article 6 of the law, and it was found advisable to pay them to watch the telephone lines and report faults. The State also pays full carriage and fares for all material and workmen, so that the railways do not suffer appreciably after all. The right to go along the railways is a most important one in connection with the extension of the trunk line system.

SWITCHING ARRANGEMENTS

The most recent switch-board in Switzerland is that lately installed at Zürich. It is an American-made (Western Electric Company) metallic-circuit, parallel-jack, multiple board with an ultimate capacity of 5,400 lines, but fitted at present for 3,400

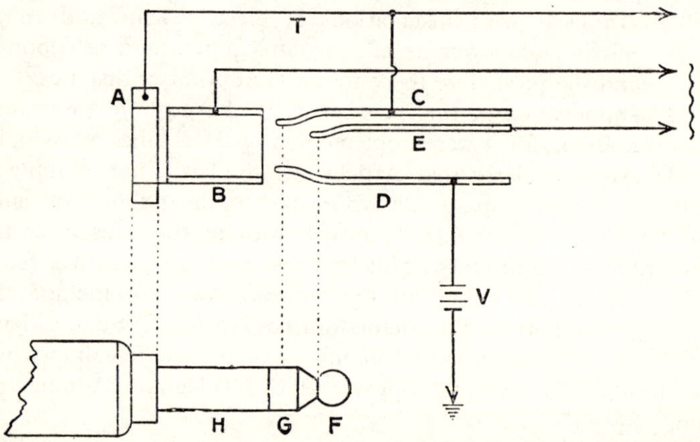

Fig. 134

only. Including the cross-connecting and lightning-guard boards it has cost 9,600*l.*, or 2*l.* 16*s.* 6*d.* per subscriber. The parallel connection of the jacks presents several advantages, such as the avoidance of multiple contacts, which are apt to become dirty, in the speaking circuit; the reduction in number of soldered joints; and the saving in length of the connecting wires. The scheme of the Zürich jacks is shown in fig. 134. A is a brass ring, in connection with the test wire T, which is touched in testing by the point of the plug. Behind this ring, and insulated from it, is a socket B, smaller than the ring in diameter, and in connection with one of the wires of the subscriber's loop. Behind the socket again are two springs C and D, C being in connection with A and with the test wire T, while D is permanently connected to one pole of the test battery V. Further back still is a third spring E, joined to the second wire of the subscriber's loop. The plug is divided into three conducting parts separated by insulating material—viz.,

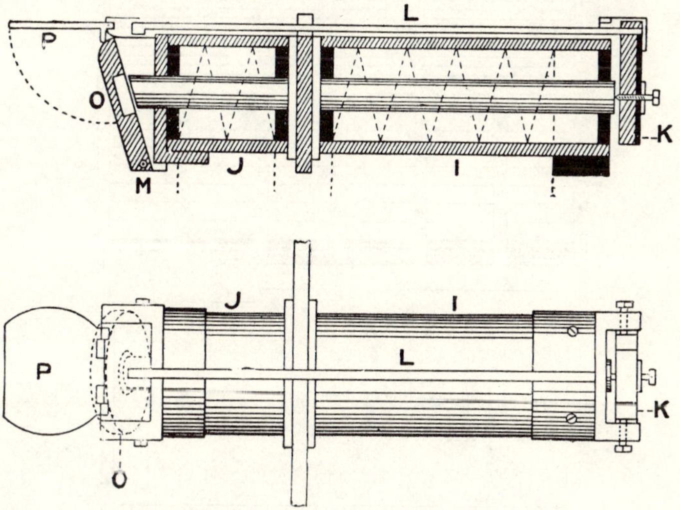

FIG. 135

F and H, which are in connection with the conductors of the cord; and G, which is a simple metallic ring. When inserted, the connections are effected as indicated in the figure, H and F making contact with the line through B and E, while G establishes connection between D and C, joining the battery V to the test wire T. The indicators are of the self-restoring kind, and are constructed as shown in fig. 135. There are two electro-magnets, I and J, mounted one behind the other: I, which is linked into the subscriber's loop, being wound to 600, and J, which is in circuit with the test wire T (fig. 134), to forty ohms. When a ringing current arrives from line and traverses the coil I, the armature K is attracted and the lever L attached to it lifted, releasing in the ordinary way the heavy iron shutter O turning on the pivot M. The shutter falls, however, only a short distance, about five millimeters, just far enough to strike against a small projection on the aluminium

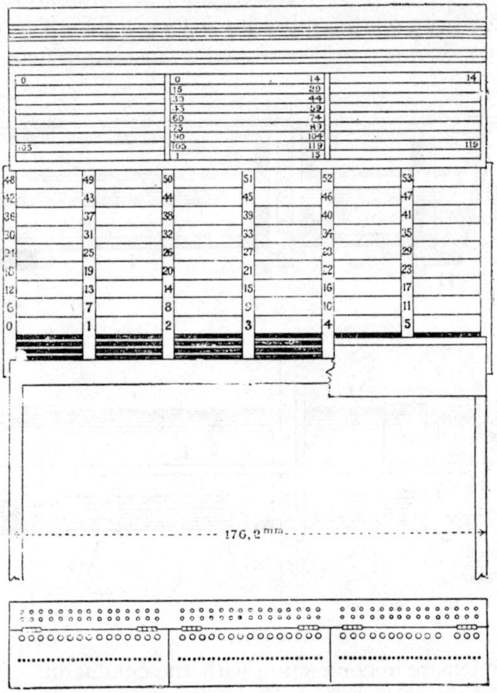

FIG. 13

plate P, which is cocked up to a horizontal position by the shock and discloses the number on the shutter O which it had previously covered. In the back of the shutter O is a hole into which the projecting and sloping end of the core of the electro-magnet J fits when O is upright. It does not fall far enough to remove it from the attractive influence of J, so that when a plug is inserted and the test line and battery joined (fig. 134), J is excited and draws O back to its upright position, the aluminium plate P then falling and covering the number. This plan relieves the operator of the work of restoring shutters after use ; it also enables the shutters to be removed out of reach, thus affording more space for the jacks. Once adjusted, the drops act well, probably better than ordinary ones, which are subjected to careless and sometimes rough handling by the operators. Figs. 136 and 136A represent front and end plans of the table. The indicators are mounted above in sections of 120 lines, having below them a strip of fifteen ring-off indicators for each operator. Then come the repeat jacks in sets of 100, each operator having 1,800 before her ; and below, the local jacks. Fig. 137 gives a good idea of the general appearance of the table. Owing to the length of cord necessary to reach over so many jacks, the shelf supporting the keys and plugs is one meter above the floor, and the operators' seats have consequently to be very high—80 centimeters. Fig. 138 shows the connections of an operator's apparatus. The left and right keys are for ringing in either direction, the middle one for cutting in the operator's phone. The ring-off indicators, like the calling ones, are in parallel between the cord conductors, so that when a connection is on there are three indicators in derivation across the circuit. But when a ring-off comes, only the proper drop falls, as the two others are held up by the test current circulating through their restoring coils. The ring-off drops are

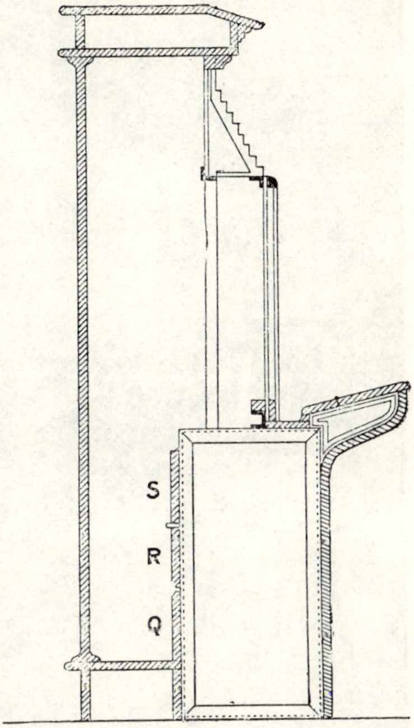

FIG. 136A

restored by depressing the operator's speaking key, which, by means of a special contact, sends a momentary current through

FIG. 137

the restoring coils. The ringing and cutting-in keys are of ordinary make ; they are mounted on hinged flaps which are ordinarily locked down, but can be turned up for the purpose of inspection or repair. The telephone and its induction coil are each wound in two equal parts, the middle point being earthed. When the telephone is cut in, the connections are as indicated in the figure. The condenser stops the test current from going to the plug used for answering calls. For answering, the left plug must always be used, while the right is employed for testing and completing the connection.

The wiring is effected with flat cables, 3 mm. thick and 60

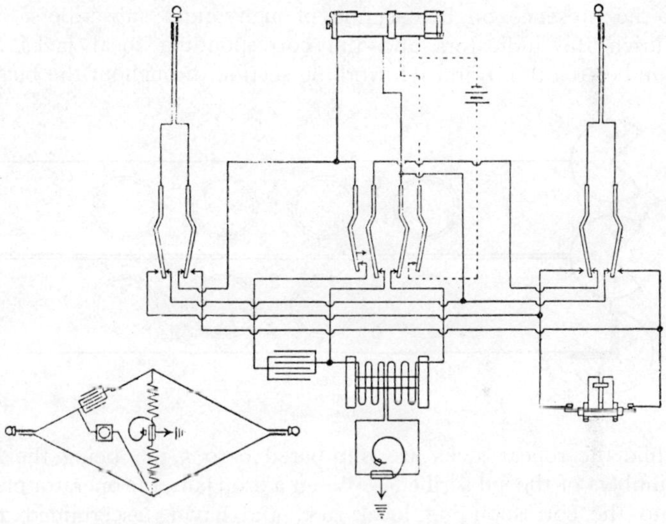

FIG. 138

deep, each containing sixty wires. As the ranges of spring-jacks have a height of 13 mm., three superimposed cables are not so thick as a row of jacks. As each series of jacks occupies six vertical divisions, two beds, placed one behind the other, each of three flat cables, suffice. This arrangement allows of rows of jacks being withdrawn from behind for cleaning or repair. With this view, the rows of jacks are kept in place by circular nuts having a rectangular notch cut in each. By turning the nuts until the notches coincide with the square end of the strip, it is freed and may be withdrawn. Fig. 139 shows this arrangement. When it is desired to withdraw one of the lower strips it is necessary to lift the superincumbent layers of cables on a steel stirrup or frame.

All modern boards are provided with means for distributing the work with some approach to equality amongst the operators, for when this cannot be done it frequently happens that several very busy subscribers are grouped together on the board and provide more work for the operator of that section than she can properly attend to, while her neighbour may be almost idle owing to the presence on her section of many quiet subscribers. At Zürich the indicators and the corresponding local jacks are numbered 1 to 119 in each working section throughout the board,

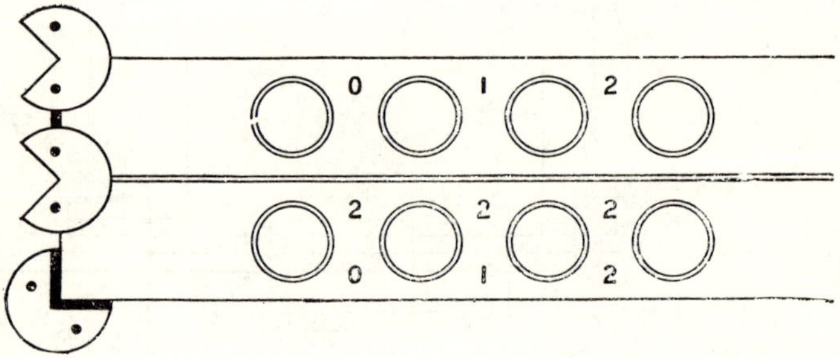

FIG. 139

while the repeat jacks are numbered 0 to 5,399, being the list numbers of the subscribers. When a drop falls, the operator plugs into the corresponding local jack, and having ascertained the number wanted, completes the connection through the repeat jack which bears it. She has no occasion to know the list number of the calling subscriber unless the connection demanded cannot be given at once, when it must be asked for in order that he may be rung up later. In a busy exchange this may, however, become an important point, and it would be an improvement to add a second number (which might be movable) to the indicator showing the true list number of the caller. The equalisation of work is effected by an intermediate field in the following manner. The repeat jacks of each group of subscribers are connected in parallel in the ordinary way; then, from one or the other end of the table is brought what is called a return cable to the section occupied by the indicators and local jacks of the group. Behind the table, below the level of the jacks, are groups of terminals, Q and S (fig. 136A), divided by a horizontal box or channel R. The wires in the return cable are soldered to the terminals of Q, while those of S are in connection with the indicators and local jacks. If no distribution is necessary to equalise the work, the two groups of terminals Q and S are simply joined across with short pieces of wire; if otherwise, any desired adjustment can be effected by long wires laid in the box R. There are cross-connecting and lightning-guard boards of familiar types. Notwithstanding the self-restoring drops, the number of movements required to make and undo a connection is only one less than that necessary on the old Western Electric double-cord board. They are as follow:

1. On receiving call, operator plugs into caller's local jack.
2. Turns down key and speaks.
3. Tests line called for.
4. Plugs into called subscriber's jack.
5. Rings called subscriber.
6. Turns up key (connection completed).
7. Removes both plugs.
8. Depresses speaking key to restore ring-off drop.

It will be seen that, good as this type of board is in several respects, the chief advantage generally claimed for it—that it reduces the number of movements necessary on the part of the operator is chimerical. Another grave drawback is that subscribers cannot ring through to each other without dropping the ring-off indicator and running the risk of getting disconnected. The switch-board that finally comes to stay will have to meet this difficulty, for there is no privilege more appreciated by subscribers than the power to hold one another within call until their conversation is finished.

The switch-room is lighted with incandescent lamps, the current for which, together with that required for the operators' transmitters, ringing keys, test, and replacement of indicator drops, is furnished by two batteries of accumulators, one of sixty-one cells for the lighting and ringing, and one of two cells for the transmitters, test, and drops. The accumulators, which have a capacity of 127 ampère-hours, are charged by a 12-h.p. gas-engine driving a 140-volt dynamo. The smaller battery is used in parallel for the transmitters, and in series for its other work. The necessary alternating current for ringing the subscribers' bells is provided by means of an electro-motor driven by the accumulators. Two opposing segments of the commutator are connected to two insulated metal rings on the other end of the motor spindle, on which rings collectors in connection with the ringing keys are always pressing. As the opposing segments come alternately under the + and – brushes, the current in their rings is reversed and the necessary alternations produced. This arrangement is shown in fig. 140. The voltage required for ringing

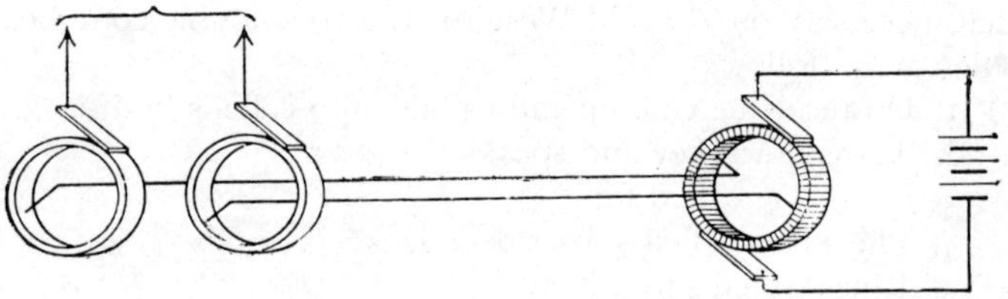

FIG. 140

being only 60, resistance has to be interposed between the motor and the keys.

Zürich is the most important telephonic centre in Switzerland, although it is run closely by Geneva and Basle. At the end of October 1894 there were 2,769 subscribers, together with thirty-five trunk lines, operated by the switch-board. The population being about 130,000, there are thus 2·13 telephones for each hundred inhabitants. The operators are thirty-two in number, or one to every eighty-six lines, besides which there are three girls occupied in registering calls of various kinds that are subject to special charges. The number of local connections from January 1 to June 30, 1894, was only 809,807, while the trunk communications mounted up to 233,213—more than a fourth. The number of telegrams telephoned to the telegraph office was 8,842. It thus seems that the effect of the Swiss local tariff is to reduce the traffic, since during the period named the local talks amounted to only 630 per subscriber per annum, just over two per subscriber per working day. Many of them were, of course, far busier than that ; but the majority were evidently trying not to exceed their 800 free talks per annum. On the other hand, the trunk connection average was very good.

The switch-board is placed in a large and well-ventilated room, and everything is arranged in readiness for the ultimate and inevitable advent of metallic circuits.

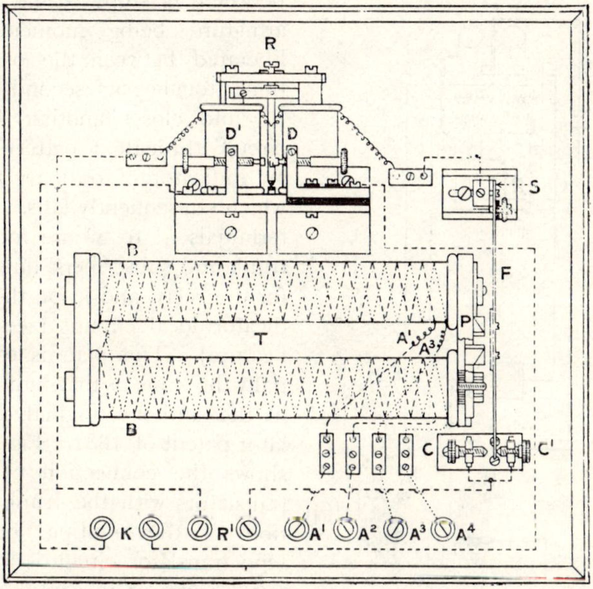

FIG. 141

The arrangements for the trunk line service have severa features of interest. Translators are interposed between the trunks and subscribers' lines, even when these last are metallic circuits. A peculiar arrangement (figs. 141 and 142) of translator is adopted with the view of excluding from the circuit all other coils and electro-magnets. The translator itself is of the Landrath pattern, and consists of two bobbins, B B, with iron wire cores, placed side by side, the cores being joined by a yoke at one end, and at the other furnished with pole pieces P. The primary and secondary circuits are of the same resistance, 170 ohms, and are equally divided between the two bobbins. An armature, F, hangs from the support S, and is adjusted to make contact normally with one of two stops, C C[1], which closes the circuit of the local battery and the relay R, the armature of which is kept attracted against the dead stop D until a magneto current traversing the coils of the translator sets the armature F oscillating between its stops. The relay armature being momentarily liberated between the oscillations, touches its second stop D[1], and closes another local circuit through a battery and an indicator of ordinary type, which consequently falls. The terminals A[1] to A[4] are for the trunk and subscribers' lines, R[1] for the relay, and K for the indicator local circuits (see also fig. 142). This plan is essentially the same as that indicated in the author's original translator patent of 1881. Fig. 142 shows the connection of the translators with the trunk section of the multiple board. One translator circuit is joined to the wires of the trunk; the other, on the single-cord plan,

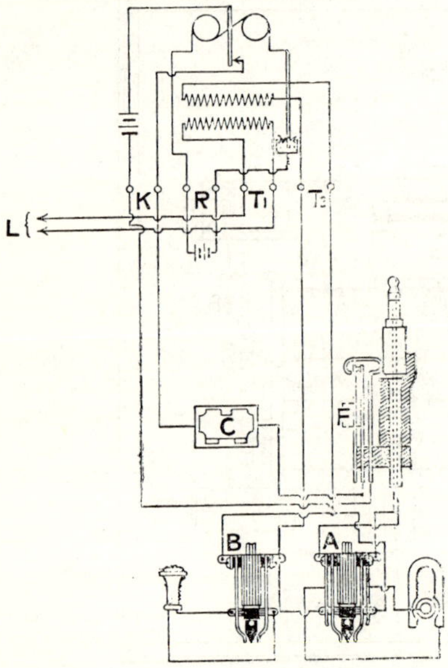

FIG. 142

to a double-conductor cord and plug. On inserting this latter in the spring-jack of the local wire, translation between the two lines is effected. The double-conductor cord is provided with two switches, A and B, for ringing and speaking respectively. The ring-off drop C is worked by the translator as described, but as it is only wanted after a connection has been put through, two springs, F, are provided which keep apart so long as the plug is in its idle position, but which touch and loop

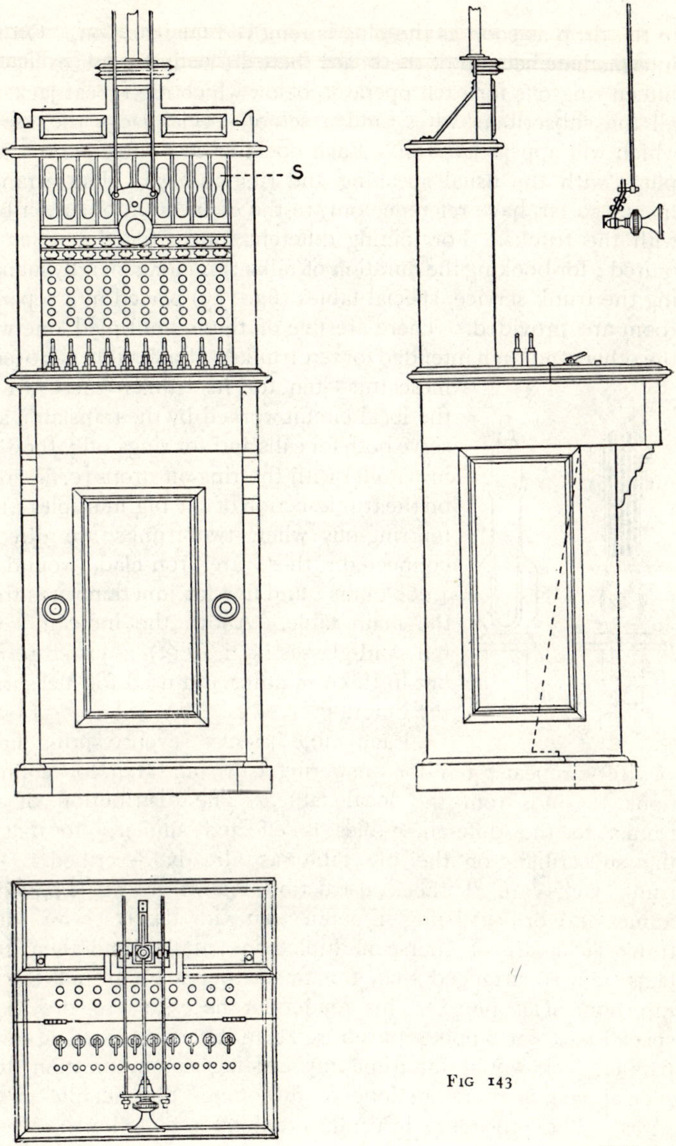

FIG 143

in the drop as soon as the plug is removed for insertion. On the intermediate section of the board there are no indicators excepting fifteen ring-offs for each operator, below which are repeat jacks for all the subscribers' wires, and a set of special jacks the use of which will appear later on. Each operator has fifteen cords and plugs, with the usual speaking and ringing keys. The arrangements, so far, have reference only to the connection of subscribers with the trunks. For joining different trunks together when required ; for booking the duration of talks ; and for generally managing the trunk service, special tables (fig. 143) placed in a separate room are provided. There are five of these, multipled one with the other, and each intended for ten trunks. Each table has twenty indicators—ten for its trunks (these are in the local circuit worked by the translator, and serve both for calls and for rings off ; they are in parallel with the ring-off drops [c, fig. 142] on the trunk section of the big multiple) ; five for ring-offs when two trunks are directly connected : these are iron-clad, wound to 1,000 ohms ; and five for junction wires from the main table. Above the indicators are ten sand-glasses (s, fig. 143), adjusted to run out in three minutes and used for measuring the duration of talks.

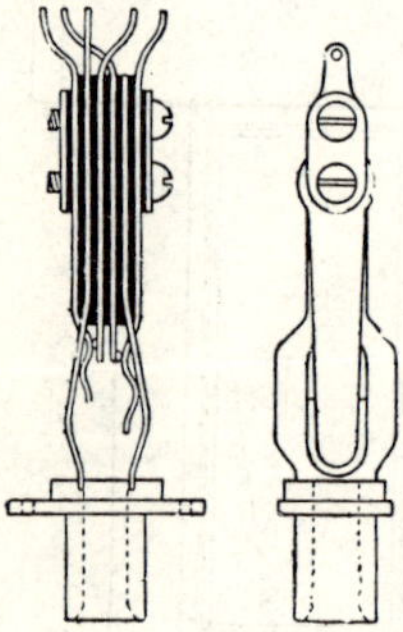

FIG. 144

Each table has also seventy spring-jacks, viz. fifty repeats, ten for answering calls, and ten for connections to and from the local table. The distribution of the trunks to the different tables is effected similarly to that of the subscribers on the big table as already described. The trunk jacks are of the construction shown in fig. 144, their frames and orifices being in connection with the test wire. The trunk jacks are of course multipled in parallel, and their contacts are so arranged that the translators are cut out by the insertion of a plug. This renders it necessary to provide a special jack for tapping purposes, as an operator plugging into a parallel jack would interrupt any existing communication ; this special jack is therefore looped into one of the metallic circuit wires. The procedure in trunk switching is as follows : A local subscriber, A, wanting a trunk, rings and says 'long distance,' whereupon he is joined through to an operator in the trunk room, who makes a note of the name or number and town of the person wanted and sends it to the operator controlling the trunk affected. When A's turn arrives, this operator rings the intermediate section of the multiple and asks for him. While A is being rung, the person he wants is demanded of the operator at the distant town. As soon as the two subscribers reply, the lines are joined, a sand-glass reversed, and the operators turn up their keys. When a request for a local subscriber comes from a trunk line, the trunk operator rings the trunk section of the local multiple, asks for the person wanted, and joins the trunk to the junction wire. As all the local subscribers have jacks on this section, the operator there has only to ring, and when a reply comes to go off the line, the duration of the talk in this case being noted at the distant end. The system appears to work well and smoothly, but the communication between the operators at the local multiple and those in the long-distance room, and consequently the service, would certainly be accelerated if it were conducted on the listening plan instead of by the constant dropping of indicators. Much work would be saved, too, if the subscribers, or, at all events, those among them who habitually use the trunks, had repeat jacks on the trunk tables. At present, when a connection is ready the caller has to be notified through the intermediate operator, which means a certain loss of time repeated hundreds of times a day. When three trunks exist between two towns, A and B, it is found advantageous to use one for the calls from A to B, a second for those from B to A, and the third for communications between other towns whose traffic passes by that route. When the third line is otherwise free, it is used as a service wire between the operators at A and B, who are, by its aid, able to get through more connections on the other two than would be otherwise possible.

The installation at Zürich, both as regards the switch-room and the outside work to be described later on, undoubtedly reflects the greatest credit on Dr. Wietlisbach, director and chief technician to the Swiss Telephone Administration, and Mr. Homburger, the local manager and engineer.

In obtaining connections, the subscribers ring the exchange

and put the telephone to the ear without waiting for a ring back. On hearing the operator's voice, the number and name, or (in the small centres) the name only of the person wanted is given. He is rung by the operator, and, taking down his telephone, replies without ringing, so that the caller, who is still listening, hears his voice. As soon as she finds them in touch, the operator retires from the line. When finished, the caller rings off in the ordinary way. This is no doubt the best form of procedure when ringing through is liable to give rise to mistakes. In trunk switching the caller is put through to the town wanted and asks the operator there for his client. Sometimes a caller must speak to three exchanges, as in getting through from Zürich to Lausanne : Zürich gives him the Berne operator, who gives him the Lausanne operator, who gives him the Lausanne subscriber.

HOURS OF SERVICE

All exchanges possessing two hundred or more subscribers are open all night and on Sundays. The smaller ones close at 9 or 10 P.M., but where a caretaker resides on the premises he is not prohibited from answering calls and giving connections after hours. Such calls are charged extra at the rate of 2·4*d.* each if made within one hour of closing time, and 4·8*d.* each afterwards. These surcharges apply to all the different kinds of connections.

SUBSCRIBERS' INSTRUMENTS

These consist, in all the larger centres, of magneto, base-board and battery-box of Swiss manufacture ; a granular transmitter, usually of Western Electric Company's type ; and a double-pole receiver. In a few of the smaller towns, battery calls are still employed. The instruments are solidly constructed and well fixed. Sand-glasses are attached to some of the subscribers' instruments for the purpose of measuring the duration of trunk talks. Lightning-guards are also supplied to the subscribers' offices. The leading-in wire is of 1·3 mm. copper, insulated with vulcanised india-rubber and protected by a braided covering steeped in preservative compound. From the lightning-guard to the instrument the wire has a skin of india-rubber covered with braided paraffined cotton. The earth-wire is covered with paraffined cotton only. In connection with the lightning-guard there is a fusible wire calculated to go at one ampère; this is to protect against the consequences of possible contact with an electric light or power system.

OUTSIDE WORK (LOCAL)

In Zürich there is much to remark, for the double problem of metallic circuits and underground wires has been boldy and ably tackled. The desire to keep to one central station led to a great convergence of overhead wires at one spot, and it became more and more difficult to find room on the houses for the rapidly augmenting number. Besides this, electric lighting on the high-tension alternating system is rife in Zürich, and with an overhead electric tramway, a duplicate of that at Leeds, tended to make things more lively than agreeable for the single-wire earth-return subscribers. The disturbance from the tramway was greatly reduced by laying a 7 mm. copper wire between the rails to help the return, and by removing (at the cost of the tramway company) all wires running parallel to the tramway route. But extensions of the latter are promised, and parallelism cannot be avoided indefinitely; so it was determined to place all telephone wires in the centre of the town underground in cables containing twisted pairs, and to distribute overhead to the subscribers from suitably placed towers or columns made as sightly as possible. A considerable portion of this work has already been completed with most satisfactory results. The town council objecting to cement conduits, cast-iron pipes of thirty, forty, fifty, and sixty centimeters diameter are used to contain the cables, the joints being made tight with lead caulking. The pipes are laid at depths varying from ·8 to 1·5 meters, sometimes under the street and sometimes under the footpaths; they are kept straight and horizontal, manholes being provided at each change of direction or of level. On the straight, manholes are placed every 100 meters. These manholes are of concrete, are generally one and a half meters square and two meters deep, arched at the top, and closed by a disc of cast iron roughened at the top. Figs. 145 and 145A show the construction of the holes for the road and footpath respectively. The cables are drawn in by a capstan and iron wire having a breaking strain of 3,500 kilogrammes, rollers being temporarily fixed in the intermediate manholes to lessen friction. The length drawn in at one time is 600 meters as a maximum. The cables used for the main routes

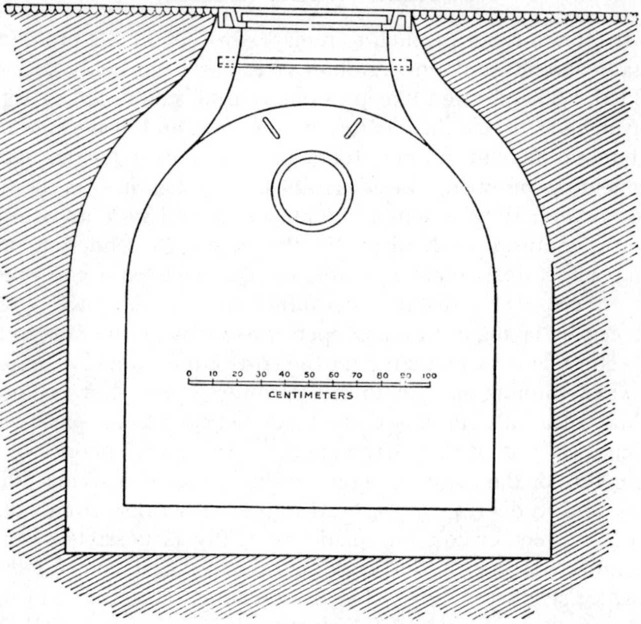

FIG. 145

contain twenty-seven and fifty-two twisted pairs, the wires being ·8 mm. gauge, loosely insulated with paper, so as to leave plenty of air space. The protection consists of cotton yarn dried at a high temperature; then a leaden tube about 2 mm. thick; then a serving of jute tape impregnated with preservative compound; and finally an armour of flattened steel wire laid on spirally. Each flat wire has an external width of 4·7 and an internal width of 4·3 mm., and is 1·7 mm. thick. The outside diameters of the finished cables are 40 and 50 mm. respectively. The copper resistance is 34·4 ohms, the insulation 5,000 megohms, and the capacity ·055 microfarad per kilometer, and the cable stands a pull of eight tons with an elongation of only 1 per cent. The maximum strain sustained in drawing in has been ascertained not to exceed two tons, and the elongation to be only ·3 per cent. Such a cable as this, if perfect to start with, once

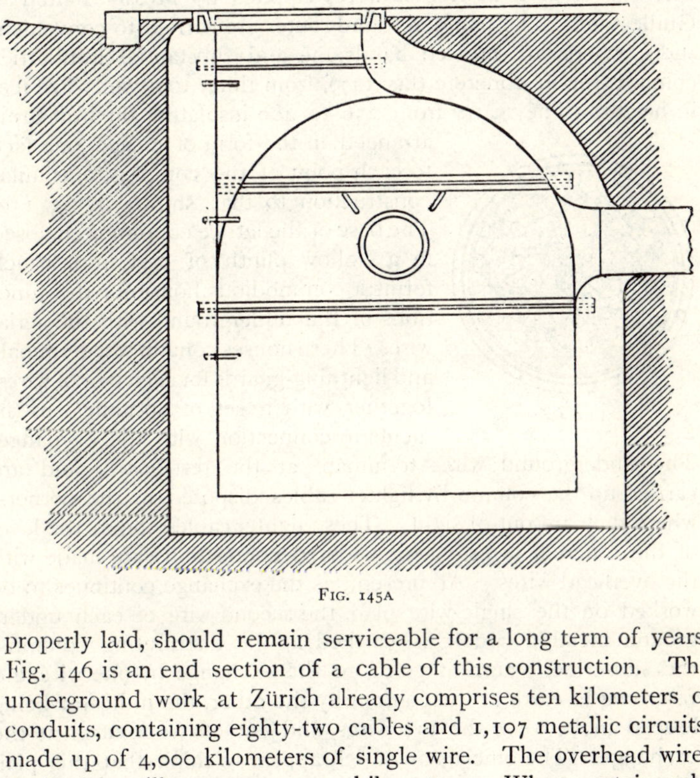

Fig. 145A

properly laid, should remain serviceable for a long term of years. Fig. 146 is an end section of a cable of this construction. The underground work at Zürich already comprises ten kilometers of conduits, containing eighty-two cables and 1,107 metallic circuits, made up of 4,000 kilometers of single wire. The overhead wires in Zürich still measure 5,200 kilometers. When certain subscribers are connected, as much as six kilometers of underground line is spoken through, the transmission being indistinguishable from that over a corresponding length of overhead metallic circuit. The only criticism that need be offered in respect to this underground work is that, when a mass of cables has been laid in an iron pipe, the weight of the upper ones will render it impossible to safely withdraw any of the lower ones that need replacement. The engineers expect that as many as 3,000 metallic circuits, say fifty-seven 52-pair cables, can be placed in the 60-centimeter pipes. Perhaps so; but once there they are fixtures. The cables used at Zürich were supplied by Messrs. Felten & Guilleaume. These underground routes are carried to convenient spots, where are erected handsome and substantial iron-lattice columns set in concrete (fig. 147), from thirty to seventy-five feet in height. They carry from 256 to 400 insulators on iron arms arranged in the form of a cage, one face to each point of the compass, of similar construction to that shown in fig. 150. The base of the lattice column is enclosed in a hollow plinth of cast iron, which forms a commodious house for the junctions of the underground with the aerial wires. These houses contain test-terminals and lightning-guards for each pair of wires, together with a set of speaking instruments in connection with the exchange.

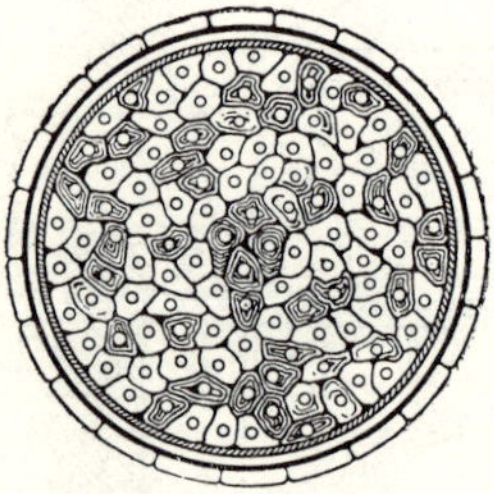

Fig. 146

The underground wires terminate at the test-board, and are carried up the column by lighter cables disposed in the corners, where they are out of sight. These lighter cables end at the level of the different arms, where soldered connections are made with the overhead wires. At present, as the exchange continues to be worked on the single-wire plan, the second wire of each underground metallic circuit is earthed at the distributing columns, the subscriber's current going to the exchange by one wire and returning to earth at the column by the other, the indicator being looped in between the two wires, and cut off from earth at the exchange. The officials at Zürich appear to think that this plan of doubling back to earth helps to reduce disturbance materially. It no doubt assists in reducing disagreeable inductive effects, but, except to subscribers doubling back to the same column and

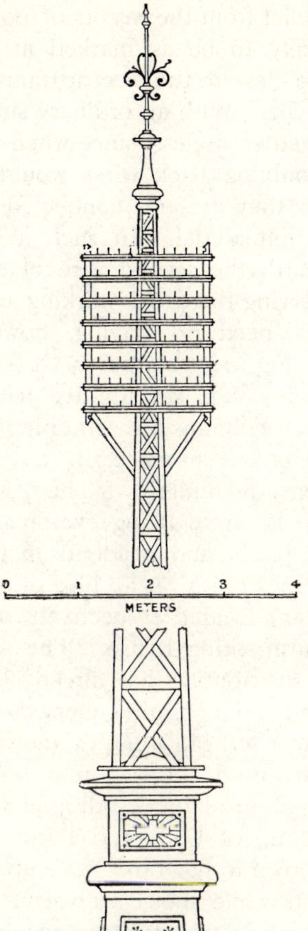

FIG. 147

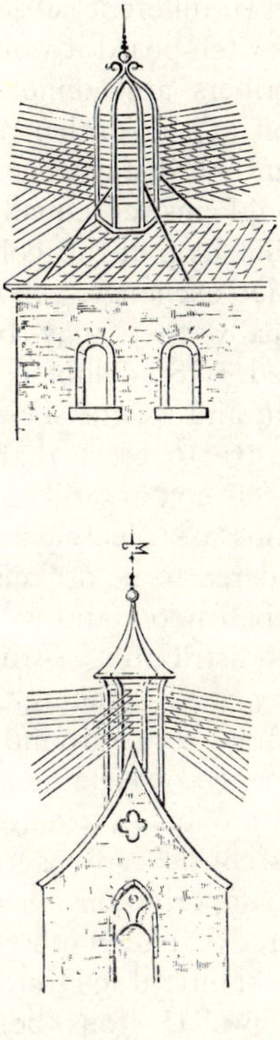

FIG. 148

earth-plate, it can scarcely afford relief from the results of polarisation of earth-plates, which is likely to be as marked at two different columns, some of which are close to the electric tramway, as at two different subscribers' stations. With an ordinary single-wire switch-board it would be altogether useless, since when two subscribers are connected their doubling back wires would be cut off at the exchange, and when they are not connected the amount of disturbance present is immaterial. In such a case it would answer equally well to earth the second wire of each metallic loop at the columns, preserving the usual working earth at the exchange. The columns need no staying, however unequally they may be loaded. Fig. 147 shows the column erected at Stadelhofen Platz, Zürich; it is seventy-five feet in height, and weighs seven tons. Such columns are certainly more expensive to erect than creosoted poles, but once up, a yearly coat of colour will preserve them indefinitely. The Zürich columns are nicely painted, and, so far from being eyesores, are considered to be ornamental by the public and residents in their neighbourhood, and with reason. Fig. 148 gives an idea of three other distributing fixtures in Zürich, located respectively on a railway shed, a church, and a warm-spring house. The same system of underground work and distribution has already been commenced in Berne and Lucerne; and Lausanne, Geneva, and Basle are being arranged for. The overhead work in the Swiss towns consists—no aerial cables are used—of 1·25 mm. bronze wire, supported on small double-shed insulators. All joints are soldered. The standards are built up of U, L, and T iron. A single standard for thirty wires is shown in fig. 149. The upright is of two U irons bolted together, while the T iron arms are stiffened by two vertical pieces of smaller U section, which are likewise connected to the main upright by L iron brackets. Fig. 150 shows one face and plan of a four-faced junction standard employed at the meeting of several routes. There are also double and triple standards, amplifications in all essential details of Fig. 149. The Swiss standards are always taken through, and rigidly fastened to, the roofs; they are of strong construction, well stayed, and of neat appearance. They are usually connected to earth as a precaution against lightning. None of the exchange

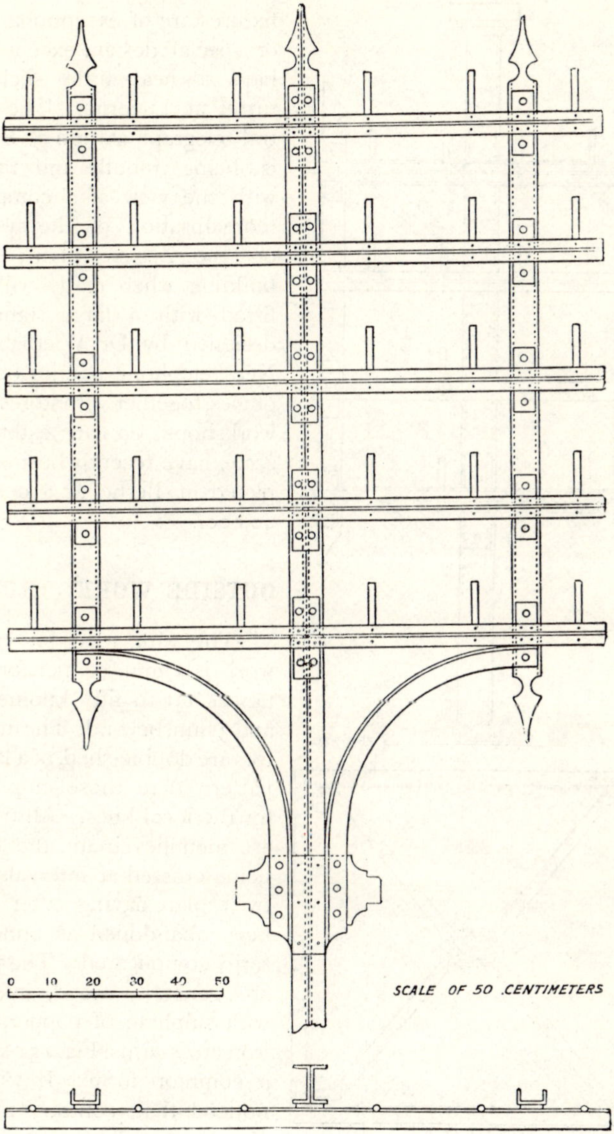

FIG 149

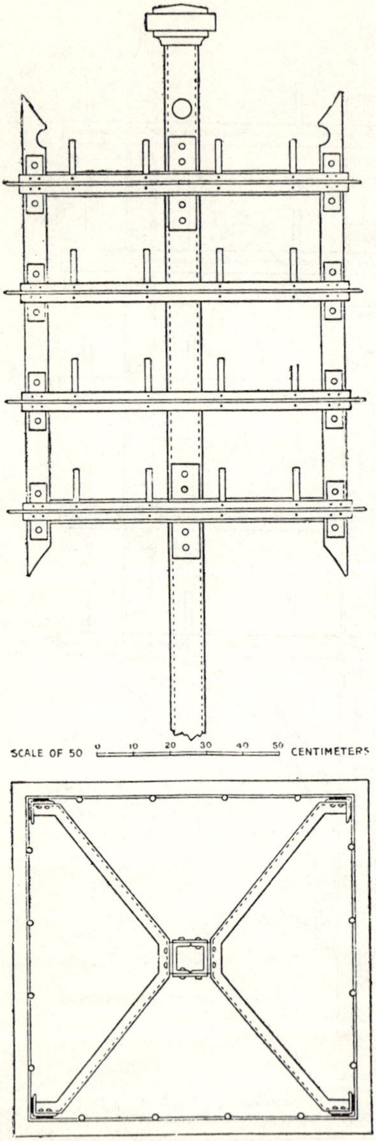

FIG. 150

fixtures are of exceptional size or special design, except perhaps a neat little skeleton turret at Lucerne. The central telegraph station at Berne is being rebuilt and raised with the view of a complete reorganisation of the system on the Zürich plan; this building when ready will be fitted with a large standard designed by Dr. Wietlisbach. New telephone administrative offices, together with stores and workshops on an extensive scale, have recently been completed at Berne at a cost of 40,000*l*.

OUTSIDE WORK (TRUNK)

The wire used for trunk work is 2 mm. copper for distances up to fifty kilometers, and 3 mm. beyond. The insulators are double-shed, of a larger pattern than those employed for the local lines. All trunks are metallic circuit, the wires being crossed at intervals, the twist plan having, after trial, been abandoned as unnecessarily complicated. The poles are generally wood, injected with sulphate of copper, with iron cross-arms. Fig. 151 shows a common form. It will be noticed that, contrary to the

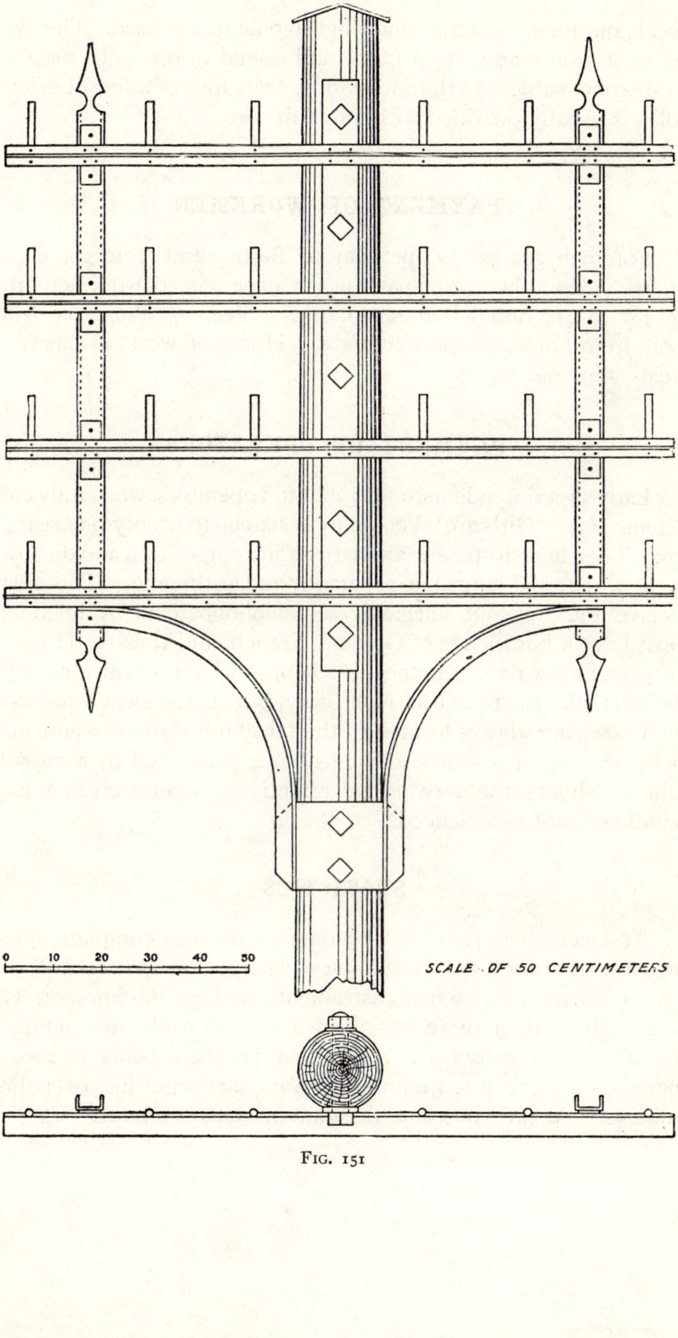

FIG. 151

usual continental practice, the English pole-roof is used. The arms are of T iron made into a frame and bolted to the pole together. In districts subject to thunderstorms, every fifth of a line of ground poles is usually provided with an earth wire.

PAYMENT OF WORKMEN

Foremen get 5*s.* 7*d.* per day in Berne, and from 4*s.* 6*d.* to 5*s.* 2*d.* in the other principal towns; experienced workmen from 3*s.* 7*d.* to 4*s.*, and labourers 2*s.* 9½*d.* Sleeping allowance when away from home, 1*s.* 7*d.* per night. Hours of work, exclusive of meals, nine per day.

PAYMENT OF OPERATORS

Lady superintendents, 6*l.* per month; operators, when fully competent, 3*l.* 4*s.* Girls are taken on from sixteen to twenty-one years of age. They have to pass examinations in composition and dictation in their maternal language, geography and arithmetic. Those who receive and transmit telegrams or telephonograms by telephone must have a knowledge of German, French, and Italian. Hours of duty, eight per day. At those exchanges which are open all night the girls take their turn at night duty, but as the switch-rooms in such cases are always located in the telegraph stations where male clerks are on duty—the two rooms being connected by a message tube or shoot—the nervousness attendant on isolation in a large building is not experienced.

STATISTICS

At December 31, 1893, the date of the last complete official report, there were 155 telephone exchanges in Switzerland, with 14,675 subscribers, 16,929 instruments, and 33,266 kilometers of wire. Since then there has been a very considerable increase, the number of subscribers at October 31, 1894, being 19,300, an increase of 2,371 in ten months on a population of just over three millions. At the same date the nine principal exchanges were:

—	Town	Number of subscribers	Population	Number of telephones per 100 inhabitants
1	Zürich . . .	2,769	130,000	2·13
2	Geneva . . .	2,648	78,777	3·36
3	Basle . . .	2,075	73,958	2·8
4	Berne . . .	1,190	47,270	2·5
5	Lausanne . .	1,070	33,340	3·2
6	St. Gall . .	825	28,000	2·9
7	Lucerne . .	649	22,000	2·9
8	Chaux-de-Fonds .	601	26,000	2·3
9	Neuchâtel . .	439	17,000	2·58

In 1889, the last year of the 6*l.* inclusive tariff, the total receipts of the telephone system amounted to 1,275,906 francs (51,036*l.*) Under the new tariff, which involved a very serious reduction, they had risen in 1891 to 65,340*l.*, having recovered lost ground and gained 14,304*l.* into the bargain. For the last two years the receipts have been—

1892 ... ... 74,091*l.* | 1893 ... ... 111,740*l.*

'Very good,' a Post Office protectionist will doubtless cry; 'but how about the poor telegraphs? They were built with the Swiss people's money, and the Swiss people have a right to be guaranteed against the ruin of their property.' Well, here are the telegraph receipts:

1891 ... 110,171*l.* | 1892 ... 111,034*l.* | 1893 ... 116,623*l.*

Notwithstanding the enormous increase of telephone accommodation, the telegraphs have continued to gain ground. In 1893 the telephone receipts had increased 37,649*l.* over the previous year, and for the first time equalled and surpassed the telegraph, yet the telegraph receipts increased also! The reason was that everywhere the telephone fed the telegraph, and the telegraphs of the world were brought to the firesides of nearly 20,000 Switzers.

The following statistics from the last available official returns are of interest as showing the comparative extent of the different classes of traffic and the rate of growth :—

Traffic	1892	1893	Increase	Decrease
Local talks:—				
Free (i.e. included within the 800 covered by the annual subscription) .	5,588,556	6,480,488	891,932	—
Charged at ·48*d.* each .	1,535,188	1,902,277	367,089	—
	7,123,744	8,382,765	1,259,021	—
Trunk talks:—				
Up to 50 kilometers .	655,647	954,628	298,981	—
51 ,, 100 ,, .	156,878	231,718	74,840	—
Beyond 100 ,, .	21,149	38,307	17,158	—
	833,674	1,224,653	390,979	—
International talks (those originating in Switzerland only) . . .	2,594	2,801	207	—
Telephonograms . .	7,377	6,526	—	851
Telephoned telegrams .	170,771	181,758	10,987	—
Total of communications of all classes . .	8,138,160	9,798,503	1,660,343	—

The increase under all headings for 1894 is understood to be far in excess of that in 1893, but the exact figures cannot be learned until the middle of 1895.

XXV. TURKEY

No telephone exchange work has yet been undertaken in Turkey, nor is likely to be, as a prejudice against it for political reasons is said to exist in high quarters. Many efforts have been made by French and other continental financiers to obtain a concession for Constantinople, but, so far, absolutely without success, the terms proposed by the Government being, most probably intentionaliy, altogether prohibitive.

XXVI. WÜRTEMBERG

HISTORY AND PRESENT POSITION

THE ubiquitous International Bell Telephone Company tried hard to win a concession for the telephone system of Würtemberg, but the policy of all the German States was to preserve the new means of communication to the Governments, and the company's efforts made no more impression here than in Berlin or Munich. But the Government, notwithstanding, had no idea of burking the telephone, and soon set about the business themselves, with results that cannot in any sense be deemed unsatisfactory. The rates have been reasonable and the service fair, while the linking up of the various towns to the capital, with one another, and with neighbouring States, was commenced early and carried out systematically. The consequence has been a very extensive exchange in Stuttgart and a satisfactory development throughout the country. It may be regretted that the single wire has heretofore been considered good enough for the subscribers' lines, but the necessity of a change is now recognised, and in future every development will be effected with the inevitable triumph of the metallic circuit in view. The extension of the trunks and the growing necessity, in Stuttgart at all events, for underground work, leaves no alternative possible to thinking men.

SERVICES RENDERED TO THE PUBLIC

1. **Local exchange communication.**—The local rate is 5*l.* per annum, including all charges, for any distance not exceeding three kilometers from the central station. In the case of Stuttgart,

seeing the extent of the exchange, this is remarkably liberal. In many countries the attempt to confine the use of subscribers' iustruments to those who pay for them has been abandoned either openly or tacitly as impracticable, but in Würtemberg the strictest regulations still exist on the subject. Subscribers are not allowed to use their instruments except for their own affairs, nor to permit strangers to use them—on pain of disconnection without return of money paid in advance—unless in the case of sudden illness in a lonely locality, or of accident. Even then the circumstances have to be explained to the operator, who may give or withhold permission. If the talk is allowed to take place, the subscriber whose instrument is used has to pay the amount that would have been collectable at a public telephone station. A subscriber becomes entitled to the refund of a proportionate part of his subscription when his line has been interrupted longer than four weeks from the date of notice. Subscriptions will also be refunded should the State at any time exercise its right to permanently or temporarily close the whole or any part of the telephone system. When subscribers change offices or houses, their new premises are connected to the exchange without charge if situated within the three-kilometer radius.

2. **Intercommunication between the town and its suburbs.**—In the case of Stuttgart this means Cannstatt, Feuerbach, Untertürkheim, Zuffenhausen, Waiblingen, Degerloch, Backnang, Vaihingen, and Böblingen. The town subscribers may ring up any suburban subscriber without additional charge, but, conversely, the suburban man has to pay 1*l.* 5*s.* per annum extra for the privilege of initiating conversations with the town. The excess charge is small, but it seems rather unjust to saddle the suburban subscriber with it. He necessarily cannot use his connection locally to the same extent (the largest suburban exchange is Cannstatt, with 190 subscribers; the others are much smaller) as can a subscriber in Stuttgart; consequently it is of less monetary value to him, and it would be more equitable to put him on the same footing exactly, especially in view of the desirability of encouraging the connection of suburban residences. The same arrangements apply between Heilbronn and Sontheim, Reutlingen and Pfullingen, and Ravensburg and Weingarten.

3. **Intercommunication between town and suburbs and more distant exchanges within the district or vicinity.**—No hard and fast radius is imposed in determining the limits of such a district, as trade and other local requirements are taken into consideration. The group round Stuttgart comprises Esslingen, Ludswigsburg, Sindelfingen, Hohenheim, and Castle Solitude. Other 'vicinity' groups are Reutlingen with Pfullingen and Tübingen; Ulm and Waiblingen; Friedrichshafen and Langenargen. The connecting lines are all metallic circuits, and are really extra-suburban or short-distance trunks. The charge for utilising them is generally 3*d.* per five minutes, but for some there are also annual subscriptions. (See *Tariffs.*)

4. **Long-distance trunk communication within the limits of the kingdom.**—Every town and many villages are in telephonic communication. The time unit is five minutes, and the charge is uniformly 5*d.* As the distances talked over are considerable (as Trossingen to Langenargen, 166 miles; Heilbronn to Friedrichshafen, 129 miles), this is one of the most liberal trunk rates in Europe. Talks are limited to five minutes if the line is wanted by another. There is a system of express talks by which a subscriber can take precedence of all others by paying triple the ordinary rate. A subscriber in one town may likewise demand simultaneous connection with two or more in another town in order that he may give them the same message or that all may consult together. Twopence per five minutes per extra subscriber connected in compliance with such a demand is the not extravagant charge levied. The records of the telephone operator must be taken as decisive as to the duration of talks, but complaints are inquired into, and any reasonable grievance that may be proved, rectified. Within Würtemberg itself, talks which are not, for any reason, actually held are not usually charged for, even if the wires are in order and the telephone officials have done everything that it was necessary to do to effect the connection. In the interest of good discipline amongst the subscribers this rule is more liberal than politic, since it permits a man who has asked for a trunk connection and caused the line to be occupied with the necessary communications between the operators, to change his mind or to leave his instrument and neglect the connection signal. In con-

nection with the Würtemberg trunk service some subscribers have sand-glasses timed to run out in five minutes attached to their instruments. This assists them to regulate their talk and to check the accounts rendered.

5. **International trunk communication.**—This already exists with Baden, Bavaria, Austria, and Switzerland, but the intercourse is not unrestricted, and is subject to seemingly strange limitations and variations, especially with Austria and Switzerland. All subscribers in Würtemberg may be connected with those in Pforzheim and Mannheim (Baden), and in Augsburg, Munich, and Lindau (Bavaria). The subscribers in Heilbronn may also talk to Heidelberg. Stuttgart and Ulm may alone speak with Stamberg, Tutzing, and Feldafing (suburbs of Munich). Again, only the subscribers in Ravensburg, Friedrichshafen, and Langenargen may converse over the Swiss frontier to St. Gallen, Romanshorn, &c. These restrictions are understood to be due to the Imperial Political Bureau at Berlin, and no doubt are justified by excellent, if inscrutable, reasons. The time unit with Baden and Bavaria is five minutes, except with Heidelberg and Mannheim, where it is only three. Three minutes is also the unit with Austria and Switzerland. The rates are uniform, being 1*s.* per unit to Baden, Bavaria, and Austria, and 1*s.* 2*d.* to Switzerland. With Baden all talks that are asked for are charged, whether had or not, unless the line or apparatus is at fault. Thus a subscriber at Stuttgart asking for one at Pforzheim who does not answer when called has to pay the fee all the same. He is also mulcted if, after asking, he leaves his instrument and the connection is made in his absence. The first rule is calculated to discourage the use of the trunks, since it fines the caller, who is not to blame ; it would be better for the State to take the risk of the occasional absence of a called subscriber. But the second is quite justifiable, and its enforcement tends to foster that spirit of attention and intelligence amongst the subscribers which is so helpful towards a satisfactory service. A stupid or careless person who either cannot or will not (and there are plenty such) learn the rules for using his telephone is an abomination, and more to be dreaded than half a dozen busier men who know exactly what they are about. The author has known several directors of telephone companies who did not

know, after years of experience, how to use their instruments, and who, with all seriousness, persisted in blaming the operators for the consequences of their own shortcomings. Being directors, they perhaps considered it superfluous to read the rules.

6. **Public telephone stations.**—These are not so numerous as in some other countries, and are invariably located at the State post, telegraph, and railway offices, no subscribers being licensed to keep stations. There are five in Stuttgart, two in Ulm, Heilbronn, and Ludwigsburg respectively, and one in each of the smaller places. These stations are sometimes, for the convenience of residents in the locality, converted into branch switch-rooms, a small switch-board being fitted up and operated by the attendant. This plan enables persons located not more than one kilometer from an outlying public telephone station to escape the excess mileage rate to the central; on the other hand, they have to pay the public telephone station fees for all talks they originate in addition to the usual annual rental, the public station line to the central being utilised as a junction wire. No automatic check-payment boxes are used, an attendant being always provided.

7. **Telephoning of telegrams.**—Subscribers may telephone telegrams to the telegraph office for despatch to all parts, and receive by telephone telegrams arriving for them.

8. **Telephoning of mail matter.**—Subscribers may telephone messages to the central station, which are written down and posted as post-cards or letters, as may be directed.

9. **Telephoning of messages for local delivery.**—Such written messages, instead of being posted, may be sent out at once by special messenger if the subscriber so instructs. A local telegram or telephonogram service is thus created.

10. **Fire service.**—The exchanges in Würtemberg being closed at night, special means have to be adopted to bring the fire-brigade within call when wanted after hours. Rather unwisely, it may be thought, this important service, so fraught with weal or woe to the community at large, is confined to those subscribers who pay an extra annual fee of ten shillings. The telephone system still being on the single-wire and earth-return system, it would not do to simply plug all the subscribers entitled to the service through to the fire-station at night, since the number of

derived circuits so created would render the action of the fire indicator uncertain; so each subscriber is provided with an earthing peg, with which he grounds his instrument by day, keeping it in a non-contact hole at night. So, normally, the fire-station is connected to a number of lines insulated at their further ends. When an alarm has to be given, the subscriber shifts his peg from its dummy hole to the earthing contact, and is enabled to ring the fire-station without loss of current through other subscribers' lines and instruments. In the morning all pegs have to be shifted to the earth contacts before communication with the exchange can be had; in the evening, at closing time, all pegs must be shifted to the dummy holes. Before joining to the fire-station the operator tests each line, and any found still to earth are left unconnected unless the subscriber can be got to answer his bell and remedy his mistake, or unless the subscriber has instructed the central office beforehand to advise him by special messenger at his expense of the occurrence of such an omission. There is little to be said in favour of such a system as this. It is too complicated, requiring apt attention at many hands and at stated hours. A clerk's forgetfulness overnight may deprive his employers of the prompt assistance of the fire-brigade, and in the morning (through leaving the line insulated) of important messages. A far more satisfactory plan, and one to which Würtemberg will no doubt come before long, is to arrange for an all-night service at the exchanges. It costs little, and enhances the usefulness and popularity of the telephone immensely. It is not surprising to find that the State disclaims all responsibility for the failure of the system to act.

TARIFFS

1. **Rates for local exchange communication.**—If within three kilometers (1·7 miles about) the rate is 5*l.*, payable annually in advance, although the State may, if thought fit, demand payment every six months. For this the State finds, instals, and maintains the line and instrument. Beyond the limit an excess rate of 1*l.* 5*s.* per kilometer or fraction thereof is levied.

Extra instruments on the same line :—

	Per annum *s.*	*d.*
If in one building or in the same locality . . .	1 0	0
,, different buildings widely separated . . .	2 10	0

When a building let off in flats is in connection with the exchange, extra instruments may be placed in each flat at an annual charge of 1*l.* 5*s.*, but with a minimum of 2*l.* 10*s.*

The owner of such a building may have instruments fixed in all the flats, offices, workshops, &c., and put in communication with the exchange through a switch-board suitably placed and operated at his expense. On paying to the State the whole of the tariff charges, he is permitted to let such instruments out to his tenants. Table instruments are charged 1*l.*, and extra bells 5*s.* per annum.

The State specially reserves the right to debit the subscribers with any special way-leave charge that may be incurred in reaching their premises. It is understood, however, that this is rarely, if ever, done.

New subscribers have to sign for two years if within the three-kilometer radius, and for four years if without ; subsequently the contracts run year by year, subject to three months' notice.

2. **Rates for suburban exchange connections.**—Within a radius of three kilometers :—

	Per annum £	*s.*	*d.*
To cover communication in suburb only	5	0	0
,, ,, with town and other suburbs .	6	5	0

3. **Rates for district or 'vicinity' exchange connections** :

Between town or suburban subscribers and district subscribers, per five minutes . . . . 3*d.*

Between Stuttgart, with suburbs, and Esslingen, free intercourse can be had for an annual payment of 2*l.* 10*s.* This payment entitles a Stuttgart or Esslingen subscriber to ring up and be rung up by any person in the opposite town. Similar arrangements are in force between Stuttgart and Ludwigsburg, and Stuttgart and Sindelfingen.

4. **Rates for internal trunk communication.**—Between any two exchanges in Würtemberg outside the suburban and district limits, a uniform charge of 5*d.* per indivisible unit of five minutes is levied. Longer talks are allowed if no one else wants the line. When several are waiting their turns, a subscriber may gain precedence of them all by demanding an 'express' or 'urgent' talk, for which he is charged triple the usual rate. A subscriber may be connected simultaneously to two or more in another town on paying 2*d.* per five minutes extra for each additional connection. Trunk charges, and all others involving the giving of credit by the State, must be covered by deposit. Accounts are rendered monthly, but may be required to be settled sooner if the amount reaches 2*l.* 10*s.*

5. **Rates for international trunk communication.**—The time unit with Baden (except Heidelberg and Mannheim) and Bavaria is five minutes; with Austria, Switzerland, Heidelberg, and Mannheim, three minutes.

The rates between such places as are permitted to talk (see p. 421) are 1*s.* to Baden, Bavaria, and Austria, and 1*s.* 2*d.* to Switzerland.

Express talks are not allowed with Austria and Switzerland. When an intermediate country is traversed, as is Bavaria when Würtemberg talks to Austria, and as are Bavaria and Austria when Würtemberg talks to Switzerland, a proportion of the through rate is paid to those countries for the use of their lines, apparatus, and operators.

6. **Public telephone station rates.**—Payments may be by talk, or by monthly or annual subscription.

Five-minute local talk, subscriber	1*d.*
,, ,, non-subscriber	2*d.*

Subscription entitling to use of all public telephone stations in Stuttgart and its suburbs:

	£	*s.*	*d.*
Per month	0	4	0
,, annum	2	0	0
Five minutes' district talk, no distinction between subscribers and non-subscribers	0	0	3

Trunk line talks per unit time (see p. 425), no reduction to subscribers :

	s.	d.
Within Würtemberg	0	5
Out of Würtemberg, excepting Switzerland . . .	1	0
To Switzerland	1	2

When an outlying public station is fitted with a switch-board for the use of one or more subscribers in the locality who want telephones in their own premises, the station line to the central is used as a junction wire, and the above fees are payable by such persons in addition to 5*l.* per annum for the use of an instrument and one kilometer of wire, longer distances being charged 1*l.* 5*s.* per kilometer or fraction thereof extra.

7. **Rates affecting the telephoning of telegrams.**—Each telegram dictated to a telegraph office by a subscriber is charged one pfennig per word, with a minimum charge of ten pfennige (1*d.*), in addition to the tariff cost of the telegram. Odd pfennige are counted as five.

Arriving telegrams dictated to subscribers through their telephones are taxed 1*d.* each, irrespective of the number of words.

8 and 9. **Rates affecting the dictating of mail matter, and of messages to be delivered by special messenger.**—Messages dictated to the central to be written down and posted as postcards or letters, or delivered by special messenger, are also charged one pfennig per word, with a minimum of ten pfennige, the total number of pfennige being divisible by five in all cases. The postage or charge for messenger is of course added.

10. **Rates in connection with the fire service** :

	s.	d.
For connection with the fire-station after the telephone exchange is closed, per annum	10	0
Advising a subscriber by special messenger when his earth peg has been left in	0	3

WAY-LEAVES

Contrary to what has often been alleged and believed, the Government of Würtemberg possesses no compulsory powers to place poles, standards, and wires on private property. It simply does what the National Telephone Company practises in this country—that is to say, inserts a clause in its agreements by which the subscriber binds himself to allow the erection of fixtures and wires, not only for his own accommodation, but for the general use of the exchange. If a would-be subscriber refuses to sign the agreement he does not get his telephone. The difference between Würtemberg and England, if there is any, consists in the fact that the Würtemberg Government adheres rigidly to the rule 'no way-leave, no telephone,' while the company only enforces it when it thinks itself strong enough to do so. In respect to lands and buildings beyond the control of its subscribers, the Government has to ask, and frequently to pay, for permission in the usual way.

SWITCHING ARRANGEMENTS

The only multiple switch-board is an ordinary Western Electric series double-cord at Stuttgart. It has been quite full for some time, and is temporarily supplemented by some ordinary boards. A new switch-room for 7,200 subscribers is in contemplation, but the plans have not yet been got out. Everything will be arranged for metallic circuits, however. There is only one switch-room in each town (excepting a few subscribers connected here and there to outlying public telephone stations), and it is intended to adhere to that plan as far as possible. The trunk-line switching is effected at a separate table, but the arrangements are in no wise remarkable. Two types of translators are used, those of Siemens & Halske and Zippernowsky, the latter being wound with two equal circuits of sixty ohms resistance. During thunderstorms the operators leave the tables and the service ceases, although every wire is provided with a lightning-guard. Subscribers are also recommended to leave their instruments alone until the storm has passed. The distribution and lightning-guard boards are of ordinary type. The connections asked for at Stuttgart average 20,000 per day, or eight per subscriber.

Subscribers are asked for by number and name, and are called by the operator, the caller meanwhile standing with his telephone to his ear. The called man replies to the ring by taking his telephone off its hook and speaking. This plan minimises the ringing, the tapping, and the risk of a premature disconnection that is such a grave defect when the subscribers ring through to each other in the absence of a proper ring-off system. Subscribers who have to leave their instruments for a few minutes to consult books, &c., are warned against touching their bells when ready to recommence; and after having rung-off are counselled not to ring for a new connection before the lapse of half a minute. To help the operators tapping to ascertain the stage which a conversation has reached, subscribers are requested to terminate every question or sentence that is not the final one with the words 'Please answer,' and at the end of the talk to say 'Finished!' In asking for suburban and short-trunk talks the caller first mentions the switch-room to which his client is connected, and keeps his telephone to his ear until he finds himself in communication with that switch-room; he then gives the number and name of the person wanted, and again waits with his telephone to his ear until he hears his friend's voice. In long-trunk talks the subscriber mentions the town, number and name of the person he wants, and hangs up his telephone till his bell sounds. The plan of waiting with telephone to ear is no doubt tiring and trying to the patience, but it is probably the quicker, and more satisfactory in the long run than such a perpetual sounding of bells, mostly without any ascertainable significance, as prevails, for example, in London. At all events it saves the generators and bells from needless wear and tear. But there will be no approach to perfection in telephone switching in Würtemberg or anywhere else without a disconnection signal that cannot be confounded with a call or a ring-through.

HOURS OF SERVICE

In this particular Würtemberg lags behind many other countries conspicuously. Stuttgart exchange is open from 7 A.M. till 10 P.M. all the year round; the other exchanges, from 7 A.M. in summer, or 8 A.M. in winter, till 6 P.M. This limitation of the service is regrettable, seeing the many uses to which the telephone is put at night.

SUBSCRIBERS' INSTRUMENTS

These comprise magnetos, Berliner transmitters, and spoon-shaped double-pole receivers. Some are fitted with sand-glasses to enable subscribers to time their trunk conversations. The magnetos are made in the State telegraph workshops at Stuttgart—which are extensive and well appointed—and are strong, well-made, and handsome instruments. The generator coils are cut into circuit when required for use, not automatically as in most other countries, but by means of a button contact in the front of the instrument which the subscriber has to press while he rings. Subscribers are responsible for any damage that may happen to their instruments, but are not called upon to insure them against fire.

OUTSIDE WORK (LOCAL)

The wire used for local work is galvanised steel, 2·2 mm. in diameter. The reason assigned for adhering, or rather for reverting, to steel is the bad behaviour of bronze during a severe snowstorm in Stuttgart some winters back, on which occasion it was found that the steel spans stood much better than the bronze. This was not, of course, a unique experience, although the difference in behaviour between the two metals under such circumstances is not generally held sufficient to disqualify bronze from an employment for which its other good qualities specially recommend it. But in the clear atmosphere of Stuttgart steel lasts for many years ; so one of the strongest original reasons for introducing bronze—the rapid decay of iron and steel in the atmosphere of our manufacturing towns—does not apply there. The local insulators are small double-shed. There are some twenty aerial cables in Stuttgart, each containing twenty-seven wires. One of these, erected in 1884, manufactured by Felten & Guilleaume, has still every wire working ; another of the same date, by Siemens & Halske, is still serviceable, although several of its wires are useless. The great feature of the overhead work in Stuttgart is the handsome dome of iron ribs erected at the central post office (fig. 152). It is capable of carrying 14,000 wires and is of graceful design, harmonising well with the building on which it is erected. Its designer, Herr Ockert, the State architect, may well be congratulated on having produced a telephone-wire support which is not only strong and suitable, but ornamental into the bargain. There is, however, no intention to attach 14,000 wires to the dome, since, of the 2,500 subscribers which Stuttgart boasts, no less than 1,000 are already wired underground by means of cables containing twenty-five or twenty-eight twisted pairs each, placed in cement conduits.

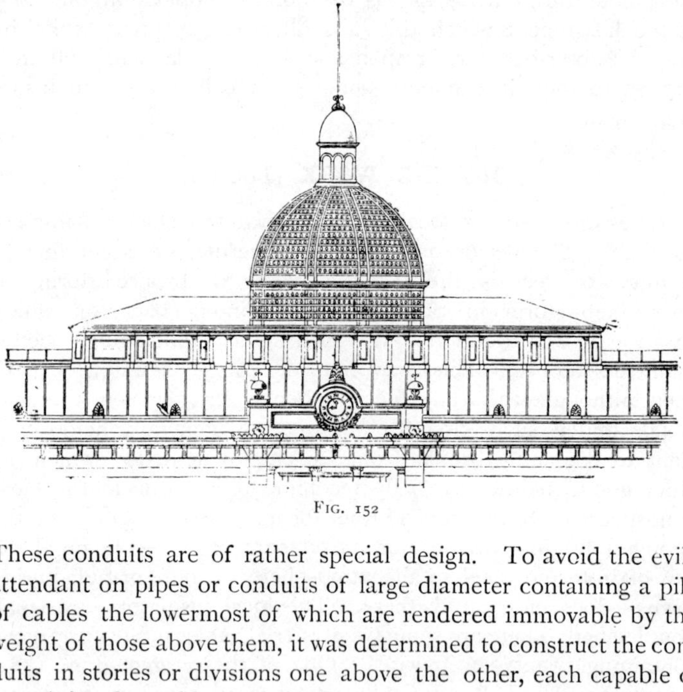

FIG. 152

These conduits are of rather special design. To avoid the evils attendant on pipes or conduits of large diameter containing a pile of cables the lowermost of which are rendered immovable by the weight of those above them, it was determined to construct the conduits in stories or divisions one above the other, each capable of containing five cables laid side by side. The removal and replacement of any particular cable becomes therefore a matter of easy accomplishment. The details of the conduits, which have proved satisfactory in every way, are shown in figs. 153 to 155. They are built up of inverted cement troughs 270 mm. wide, 75 mm. deep, and 1 meter long, piled one above the other. These dimensions are varied somewhat on different routes. Fig. 153 shows the end section of such a trough, with the method of joining two lengths. Fig. 154 shows a complete conduit for thirty cables, composed of six such troughs superimposed. A trench is dug, and lined at the bottom with concrete which is slightly raised along the middle so as to afford a hold to the sides of the first inverted trough. Subsequent troughs are added till the desired capacity is attained. When laid under the footpath, the trench is then filled

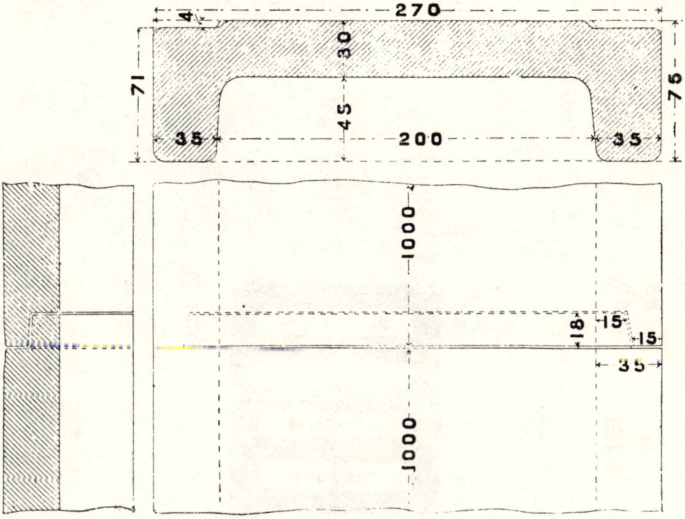

FIG. 153

in with soil, and a layer of concrete added immediately below the flags. This construction is shown in the left-hand half of fig. 154. When laid under the roadway, as in the right-hand half of the figure, injury from the weight of the traffic has to be provided against, and the trench itself is filled up with concrete, between which and the paving stones a layer of sand or gravel intervenes. When a conduit of very large capacity is required, two tiers of troughs are laid side by side. Fig. 155 shows the manholes and draw-boxes, in plan and section, as arranged for a conduit of two tiers. When passing a manhole the cables are diverted round the oval walls, on which they are supported by brackets. About the

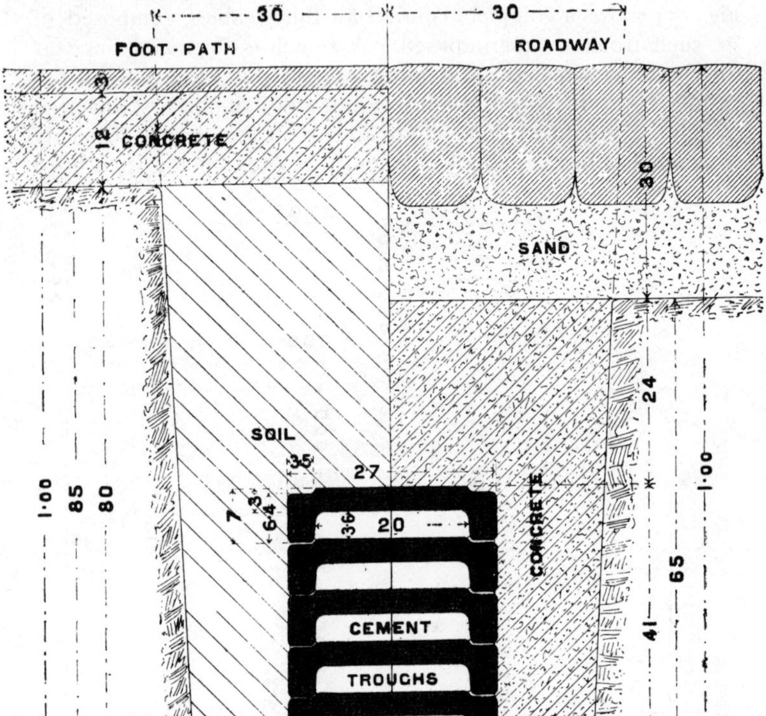

FIG. 154

superiority of such a system as this, when room can be found for it, over pipes or simple channels, there can be no doubt, as the facilities afforded for handling the cables are perfect. The work at Stuttgart is very well done, and reflects great credit on its designers and constructors. The cables laid in these conduits are some of twenty-five, others of twenty-eight pairs. Each trough can contain therefore 28 × 5 = 140 pairs, and a six-trough conduit 840 pairs. The cables themselves are of various types.

Lightning-guards, contained in weather-tight iron boxes and provided both with fine fusible wires and toothed dischargers, are always placed at the junction of overhead with cable lines. Fig. 156 is a cross-section of such a box, showing the connections of one

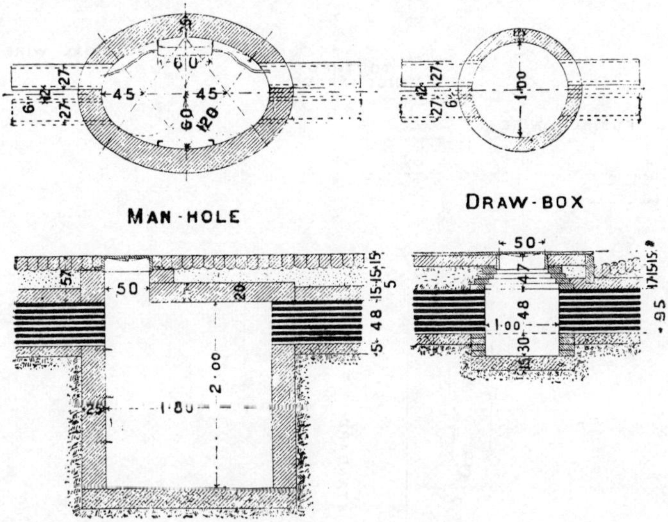

FIG. 155

wire. The cable end is sealed with insulating material, the wires spreading out, each to its lightning-guard, the other side of which (the box being fixed to the standard) is joined by a rubber-covered wire to the open wire beyond the insulator. All joints are soldered. Instead of making the junction between the copper and steel in the running wire as shown, where voltaic action to the detriment of the galvanised steel is bound to take place, it would be better to leave a long tag of steel wire after making off the turn round the insulator, thread it through a vulcanised india-rubber tube, and take it straight to the binding screw in the box, where no weather could reach it. The standards and cross-arms are strongly constructed of angle-iron. Ground poles are generally of wood and present no unusual features. Subscribers' wires are usually led down the front of the houses by means of open wires and insulators. As is commonly practised in Germany, the joint between the bare and covered wire is made inside

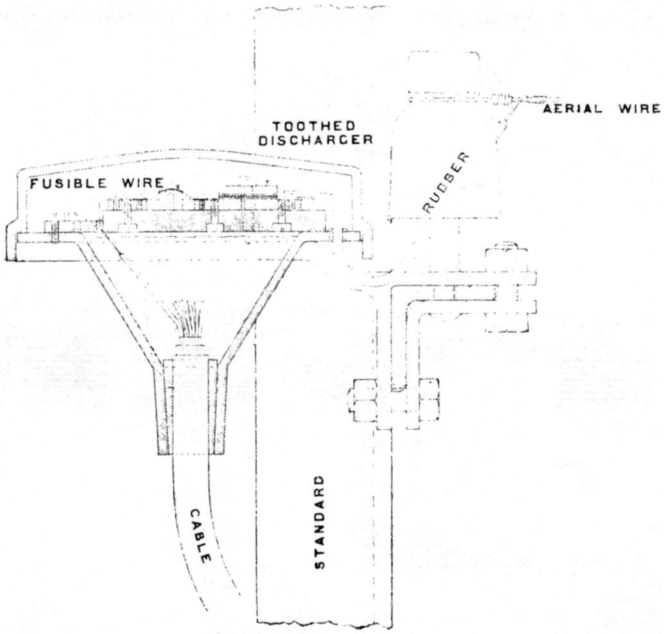

Fig. 156

an ebonite cup, which protects it from the weather and prevents surface leakage over the exterior of the insulated wire. The cup is light and hangs on a tag of the line wire. The covered wire is usually led into the building by means of an ebonite or china tube let into a hole made through the wall. The drop wires and insulators, which are specially shaped to receive them, are very neatly arranged and are by no means unsightly.

OUTSIDE WORK (TRUNK)

As the Würtemberg railways belong to the State, the telephone trunk lines naturally follow them for the most part, and, except for the crossings, are indistinguishable from the telegraph wires. The wire used is 2·5 mm. high-conductivity bronze for the short, and 3 mm. for the long distances, strung on large double-shed insulators. All trunks are metallic circuits crossed at intervals; the twist has never been employed. There are three circuits between Stuttgart and Ulm, and one between Stuttgart and Munich. The trunk traffic is considerable and continues to increase, but without prejudicially affecting the telegraph revenue, which likewise continues to grow, although not so rapidly as it did before the advent of the telephone. This satisfactory result is no doubt owing to the fact that the telephone is utilised, as in most other continental countries, as a feeder to the telegraph, and not treated as a pernicious rival to be discouraged and, wherever possible, excluded or suppressed. The telegraph tariff in Würtemberg is 50 pfennige (5*d.*) for ten words, each additional word being charged 5 pfennige (½*d.*), the minimum being 5*d.* This is the same charge as for a five minutes' long-distance telephonic conversation; but in the latter case the payer obtains a great number of words and also a reply for his money, and probably, in the majority of instances, greater speed. The speaking over the trunks is good, and undisturbed by external noises. The steel local wires do not appear to influence the service deleteriously, but, of course, the distances, even to Bavaria and Munich, are not great. The trunks are exclusively telephonic, no attempt being made to utilise them simultaneously for telegraphy.

PAYMENT OF WORKMEN

Foremen are paid 4*s.* per day; the men from 2*s.* 6*d.* to 3*s.* 6*d.*, according to length of experience. Sleeping allowance 1*s.*, and day allowance when working away from home, 6*d.* Hours of work, 6 A.M. till noon, with half an hour's interval for breakfast, and noon till 6 P.M. This gives a long working day of eleven and a half hours.

PAYMENT OF OPERATORS

Girls are taken on at seventeen years of age. If they have passed the usual school course no examination is enforced. They work eight hours per day, and are paid from 2*s.* 6*d.* to 3*s.* according to length of service. The lady superintendents get from 3*s.* 6*d.* to 3*s.* 10*d.* per day.

STATISTICS

The Würtemberg exchanges and their subscribers, at the end of 1894, were:—

Stuttgart (population 140,000)		2,500
Backnang	Suburbs of Stuttgart	3
Böblingen	Suburbs of Stuttgart	6
Cannstatt	Suburbs of Stuttgart	190
Degerloch	Suburbs of Stuttgart	20
Feuerbach	Suburbs of Stuttgart	40
Untertürkheim	Suburbs of Stuttgart	20
Vaihingen	Suburbs of Stuttgart	10
Waiblingen	Suburbs of Stuttgart	10
Zuffenhausen	Suburbs of Stuttgart	12
Dürrmenz-Mühlacker		15
Ebingen		30
Esslingen		110
Friedrichshafen		15
Gmünd		170
Göppingen		77
Hall		37
Heilbronn		320
Sontheim		6
Ludwigsburg		78
Ravensburg		65
Reutlingen		140
Pfullingen		18
Rottweil		25
Oberndorf		9
Schramberg		20
Schwenningen		12
Trossingen		12
Schorndorf		20
Sindelfingen		15
Tübingen		75
Ulm		350

32 switch-rooms and 4,430 subscribers

Spottiswoode & Co. Printers, New-street Square, London.

The Development of the Telephone in Europe . .

. . By . .

HERBERT LAWS WEBB, M.I.E.E.

Author of "A Practical Guide to the Testing of Insulated Wires and Cables," "The Telephone Handbook," "The Telephone Service Its Past, Present and Future," etc.

WITH AN INTRODUCTION
BY
HAROLD COX,
Formerly Member for Preston.

LONDON:
ELECTRICAL PRESS LIMITED,
37 38, STRAND, AND 1, BUCKINGHAM STREET, W.C.

ARTICLES REPRINTED

FROM

"ELECTRICAL INDUSTRIES."

1910.

INTRODUCTION.

OF all modern inventions there is none that has so greatly facilitated rapid inter-communication as the telephone. The writing of a letter takes time, and the conveyance of the letter to its destination takes much longer time. The telegraph is instantaneous so far as the conveyance of the electric signal between point and point is concerned, but it can only be used by expert operators, and the original message must be written at the dispatching end and transcribed afresh at the receiving end. The telephone, on the contrary, is instantaneous between individual and individual. The expert operator is only required to effect the communication between the two individuals wishing to speak to one another. The moment that has been done they can talk together almost as easily as if they were in the same room. There is no other means of communication which confers advantages even approaching those given by the telephone.

One might, therefore, imagine that the Government of a presumably intelligent country like Great Britain would do all in its power to facilitate the use of this important invention. In the following pages, Mr. Laws Webb, who has devoted special attention to the study of telephonic problems, lucidly tells the story of the dealings of the Government with the telephone, and shows how the public has been deprived of the full benefit which it might have derived from this wonderful invention. The trouble began with the purchase of the telegraphs. In 1869, the Government decided to buy up the telegraphic systems of the United Kingdom, and an Act was passed which established a State monopoly in electric signalling. This measure was advocated and passed in the belief that the State would give a better service than the private companies were giving, and would, in addition, make a handsome profit for the benefit of the taxpayer. It is impossible to test the former assumption, because there is no means of knowing what improvements the telegraph companies would have effected if they had been allowed to continue the work which they had already advanced to such a high state of efficiency. The second argument in favour of State purchase can be more easily brought to a test. It was anticipated by the advocates of State purchase that the whole of the capital required to buy out the companies would be paid off out of profits in some fifteen years, and that the State would then be in possession of a magnificent uncharged property. As a matter of fact, after the first two years of State working all profits ceased, and in most of the subsequent years the Government has not even been able to pay interest upon the original capital. In addition, the taxpayer has constantly been called upon to spend large sums of money on capital development, for which not a penny of interest has ever been allowed. Finally, Parliament has latterly had to vote, year after year, a considerable sum of money to make up the deficiency in working expenses. If allowance be made for unpaid interest, the net annual loss on the working of the State telegraphs now exceeds a million sterling. Including the original purchase money of about £10,800,000, the aggregate loss in the forty years that have elapsed since the telegraphs were taken over by the State exceeds £35,000,000. As a commercial venture the State purchase of telegraphs has been a colossal failure.

That, however, is only one side of the loss inflicted

upon the British public by the mistaken policy of 1869. As soon as the Post Office had acquired the telegraphs it became necessary for it to keep at arm's length any possible invaders of its monopoly. The telephone was not invented till four or five years later, and on its appearance was treated by the officials of St. Martin's le Grand as a scientific toy. The moment it became evident that this toy could be used for commercial purposes the Post Office applied for an injunction against its use, and succeeded in the application. The result was that the development of the telephone became subject to the control of the Postmaster General, who, for many years looked at this new invention solely as a disagreeable rival to his telegraphic monopoly. The companies which were licensed to work the telephone were compelled to pay an exorbitant royalty of 10% of their gross receipts, and it is interesting to observe that down to the year before last the Postmaster General, in his published accounts, treated his receipts from this royalty as part of the revenue earned by him on working the telegraphs. In spite of this heavy handicap, and in spite of the unreasonable opposition of many local authorities to the granting of underground wayleaves, the telephone has made enormous progress, and the National Telephone Company is able to pay regular dividends on its invested capital. Next year that company will cease to exist, and the whole telephone system of the country will pass into the hands of the Government. Unless the experience of the telegraphs is totally reversed, we may expect in a few years' time that the telephones, instead of yielding a substantial profit to the investor, in addition to a heavy royalty paid to the State, will be a cause of heavy loss to the British taxpayer.

Even if this loss were accompanied by some compensating advantage to the users of the telephone it would not be justified. The persons who pay taxes and the persons who use telephones are not always the same people, and it is unjust that the millions of taxpayers who do not make continuous use of the telephone service should be compelled to pay for the convenience of other persons—necessarily a minority—who do. The same consideration applies, of course, to the telegraphs. Only a minority of persons use the telegraph to any appreciable extent, yet every telegram dispatched is subsidised by the general body of taxpayers. There is, moreover, no reason to believe that even with the aid of the taxpayers' money the State will give a better service than private companies are capable of giving without State aid. In the United States, where both telegraphs and telephones are entirely in the hands of private enterprise, the service has reached the highest state of efficiency, and under similar conditions of freedom British commercial enterprise is certainly capable of attaining equally satisfactory results.

The root objection to State control of a telephonic or telegraphic system, as Mr. Laws Webb so admirably shows in the following pages, is that the State by its organisation is ill-adapted for commercial enterprise. Ministers and Parliaments necessarily look at all problems from the political point of view, and no commercial enterprise can be successfully managed if its control is swayed by political influences. This consideration applies not only to matters determined by the higher authorities, such as questions of rates and facilities, but also to the detailed conduct of the business. For in a State undertaking every employee knows that his position is controlled by political rather than by commercial forces, and he therefore has not the necessary motives to induce him to assist in making the business a commercial success.

These general considerations are illustrated by Mr. Laws Webb in detail. He shows, point by point, how political control prevents commercial success, and he shows how strikingly the extension of the telephone system in the United States contrasts with its relative stagnation in this country and in Europe generally. Such a work is peculiarly valuable at a moment when the Government of Great Britain, in continuation of the gigantic blunder made in 1869, proposes to make an even more serious blunder, and to subject to the benumbing influence of State management one of the most marvellous inventions of the human mind.

HAROLD COX.

LONDON, *January*, 1911.

CONTENTS.

CHAPTER VII.

CHAPTER VIII.

CHAPTER IX.

CHAPTER X.

CHAPTER XI.

CHAPTER XII.

CHAPTER I.

THE RISE OF THE TELEPHONE.

The rapid expansion of the Telephone in America, and its slow growth in Europe. The universal need for Telephone Service—the quickest means of communication. The early history of the Telephone. The causes of low telephone development in Europe.

IT is now approaching thirty years since the telephone service was established as a means of intercommunication, as the first telephone exchanges were built in the early eighties in almost all the principal cities of the civilised world. A statement of telephone statistics recently compiled from official returns shows that at the beginning of 1910 the whole of Europe contained in round figures 2,584,000 telephones, equivalent, for the aggregate population of Europe of 400,000,000, to 6·4 telephones per thousand of population. In the United States there are about 24,000 telephone companies or systems, having in round numbers a total of 7,000,000 telephones, equivalent, for a population of 80,000,000, to over 87 telephones per thousand inhabitants.

It is well-known that there is no State monopoly in the telegraph or telephone in the United States. Thus there is practically "free trade" in telephony in America, and consequently there is a large number of telephone companies operating all over the country, in rural as well as in city districts; and in some of the large cities there are two companies working independent services. But what is known as the "Bell" system, comprising the original telephone companies connected with the American Telephone and Telegraph Company, the parent telephone concern of the United States, contains a total of 5,142,692 telephones, scattered over the whole continent and interconnected by toll lines and long distance lines so as to form a universal system of telephonic intercommunication. Taking this general system alone and eliminating the independent companies, which are not connected with the general telephone system of the country, we find in America 64 telephones (of the type having universal service) per thousand inhabitants, as against only 6·4 telephones per thousand inhabitants in the whole of Europe.

This comparison is so striking, the development of the telephone in America is so greatly superior to that reached in Europe, that it is a matter of serious interest to set out the causes which have produced such widely different results.

The Essential Characteristics of the Telephone Service.

That telephone communication fills a universal need will not be contested by any thinking man. The high development of means of communication is an index of high civilisation, and the figures of railway, post office, and telegraph traffic are customarily accepted as evidence of the commercial and social activity of a country or a community. The telephone service is the most rapid means of communication there is, and it is difficult to estimate how much the efficiency of the average business man is raised by the enormous facility for communication placed at his disposal by a highly efficient and highly developed telephonic service. In social as well as in business affairs the telephone service saves time and abridges distance in a manner approached by no other means of communication. How much this is so can be realised when it is considered that the work done by a telegram and reply is done more effectively by telephone in about a thirtieth part of the time. While many people do not realise these simple facts until they are thoroughly explained and demonstrated, it is fair to assume that had the general public of the principal European countries had the same opportunities of availing themselves of telephonic communication as have been offered to the public of the United States, and had the telephone in Europe been made as efficient as it has been made in America, there would have been something approaching the same development of the telephone in Europe as in America.

The actual telephone development in Europe is, as the above figures show, only a tenth of that of the American Bell system, in proportion to population, and only a fifteenth of the whole telephone development of the United States. Taking the European countries individually one finds that there are only six which have a development of over ten telephones per thousand inhabitants. These are Great Britain, Germany, Sweden, Switzerland, Denmark, and Norway. France, Holland, and Belgium come a long way behind the above-named countries; and in the rest of Europe, including such important countries as Russia and Austria-Hungary, the development of the telephone is insignificant.

The Early History of the Telephone.

A brief sketch of the history of the telephone in Europe will show the fundamental reasons why this most useful means of communication has hitherto had such a stunted development in the greater part of the European Continent. When the telephone was first brought over from America there were few, except the American enthusiasts who came over as pioneers and a sprinkling of European business men, who could see any commercial future in the wonderful invention of Graham Bell. The scientific beauty of the invention was everywhere enthusiastically acknowledged, but the practical value of it was openly derided; and the opinion that the telephone was merely a scientific toy—a phrase given currency by telegraph officials—was very generally adopted.

Here we have an illustration, at the very beginning of the history of the telephone, of one of the greatest difficulties which has obstructed its development in Europe. It was most naturally assumed at the beginning that the telephone was an offshoot of the telegraph, and as the telegraph is throughout Europe a State monopoly, it was to Government telegraph officials that the telephone was first brought. Government officials are not famed for being the most far-sighted people, and in this case there was a pardonable touch of human nature in their attitude of distrust and dislike towards a competitor which might have in the future a serious effect on the business in which they were engaged.

Although European telegraph officials at first derided the possibility of the telephone being developed into a practical means of general intercommunication, they were, in most European countries, strongly opposed to any opportunity being given for the commercial development of telephonic communication. Broadly speaking, though there may have been some differences of detail in different countries, the story of the beginning of the telephone in Europe was this : The telephone was taken to the Government Telegraph Departments and offered for sale, but the Telegraph Departments declined to take the risk of developing a totally new business. At the same time, however, they assumed control over the telephone and issued licences to companies formed to exploit the new invention, these licences being generally for restricted periods and surrounded by the most onerous conditions.

A little later, when practical experience had shown that telephone communication was not only thoroughly practical but filled a distinct public want, the jealousy of the State Telegraph Departments asserted itself, and, with few exceptions, the telephone companies were absorbed by the State, and the telephone became, like the telegraph, a rigid Government monopoly. Thus, to-day the telephone is a State monopoly in Germany, France, Austria-Hungary, Switzerland, Belgium, Italy, Greece, Bulgaria, Servia, and Roumania ; and it may be added that, with the exception of Germany and Switzerland, the development of the telephone in all these countries is extremely low. The exceptions to exclusive State monopoly up to the present are Great Britain, Russia, Spain, Portugal, Sweden, Norway, Denmark, and Holland, although in these countries the State has assumed a very large measure of control over the conduct of the telephone business, and will no doubt eventually convert it into a State monopoly.

In the countries where the telephone has so far been largely under commercial management the development of the telephone is considerably higher than in those where for a long time past it has been a State monopoly. Thus, in Great Britain there are 2½ times as many telephones in proportion to population as in France, where the telephone has been a Government monopoly since 1889 ; and in the Scandinavian countries, where private enterprise has had a large measure of scope, and commercial methods have to a certain extent impressed themselves even on Government Departments, the relative development of the telephone is very largely superior to that of Belgium, France, Germany, Austria-Hungary, and the remaining countries of Europe which are afflicted with a State telephone monopoly.

CHAPTER II.

THE EVOLUTION OF THE TELEPHONE SERVICE.

Great development and high efficiency of service attained in America, under private enterprise, by efficient organisation and advanced technical and commercial practice. The restriction of telephonic development and efficiency in Europe by political control; the repressive influence of the policy of State monopoly.

TO explain the reasons why the exercise of a Government monopoly in the telephone and of excessive Government regulation of the telephone, such as we have had in this country, have so largely restricted the development of the telephone service, it is necessary to describe briefly the evolution of the telephone business. The evolution of the telephone industry may be roughly divided into three periods. First, there was an experimental period when the whole business was entirely new, and when both apparatus and commercial practice suitable to the conduct of the telephone service had to be developed. Second, there came a period when most of the principal difficulties involved in constructing and maintaining an efficient telephone system had been conquered, and third, there came a period in which the correct commercial methods of dealing with the telephone service on a large scale and of giving it the greatest possible development were arrived at. This is briefly the history of the telephone in America, which is the only country where it has been developed, or is in process of being developed, to its utmost capacity.

For about ten years the service was extremely crude, and the commercial practice was equally crude. Everything in a totally new business had to be learned. Apparatus which would work accurately and consistently had to be developed, systems of line and cable construction which would satisfactorily transmit telephonic currents had to be designed, and innumerable electrical and mechanical problems had to be solved, largely by trial and error. During this period the performance of the telephone very often failed to satisfy the public ; and while the facilities for communication were so great—when things worked well—as to cause a considerable demand for telephone service, things so frequently worked badly that the service was often abused in unqualified terms.

At the same time the commercial practice was almost as crude as the technical part of the work. At the beginning of the telephone exchange business it was considered that a telephone was merely an instrument to be hired out for so much a year, regardless of whether much or little use was made of it, and the telephone and its connecting line were the principal units on which the charge for service was based. As in most towns only a single exchange was established, and as the initial rate for service within a small radius of the exchange was fixed at a high enough point to give what the managers considered a safe average return, the price demanded for the service was considered high by a large proportion of the public and, owing to the mileage charges on long lines, became actually very high in all parts of the towns far distant from the central exchange, so that the telephone was considered a luxury, and the development in its use was proportionately small.

This experimental period of inefficient service, relatively high charges and low development, lasted in America for about twelve or fifteen years after the establishment of telephone exchanges, but by the end of that period sufficient progress had been made with telephonic apparatus to insure a much more rapid and reliable service, and sufficient experience of the commercial side of the telephone business had been gained to enable a more practical tariff to be devised. In the 'nineties very rapid advance was made both on the technical and commercial side, and with the modifications of the tariffs which brought the telephone service within reach of much larger classes of the community than had hitherto availed themselves of it, the ideas of telephone men as to the future possibilities of the telephone became greatly enlarged and an era of rapid development set in—to such an extent that within the past ten years the Bell system in America has increased from 400,000 telephones to over 5,000,000.

The high pitch of efficiency and the high pitch of development to which the telephone service has been brought in America are very largely due to the spirit of organisation and unity of purpose which has inspired the management of the American Bell Companies. These companies, working under a centralised management, have in reality been brigaded together as one great organisation with a single policy—the technical and commercial advancement of the telephone service. The impulse given to the industry by a dominating policy, by uniformity of technical and commercial practice, is perhaps the most effective element in producing the sound and efficient development of a complicated and highly specialised service. Without this centralised organisation and persistent policy of improvement the advance of the telephone in America could not have been so rapid or so effective as it has been.

The Status of the Telephone in Europe.

Turning now to Europe, we find that to a large extent the telephone in Europe, at least in many countries, has not sensibly progressed beyond the first or experimental period; at any rate, it has not approached yet the period of high efficiency of service, scientific tariffs and enlightened commercial practice which began in American telephony from twelve to fifteen years ago.

In most countries where the telephone is a rigid State monopoly the telephone plant has not been brought up to date, but still largely consists of apparatus of a type in use fifteen or twenty years ago, and, of course, relatively inefficient. The single exchange district, with relatively high rates for distant parts of the towns, is still the practice, the old flat rate scheme of tariffs is still the practice, and the active commercial policy of advertising the telephone service and catering to the wants of all sections of the public, which has been so largely the cause of the rapid development of the past ten years in America, is entirely neglected. In the countries where private enterprise is still largely responsible for the conduct of the telephone service, the principles which have had such successful results in America have been adopted, and this accounts to a large extent for the relatively higher development of the telephone in these countries than in those where a rigid State monopoly has existed since the early days of the telephone. Even in these more enlightened countries, however, the expansion of the telephone has been much restricted by State regulation, and it has not been possible to attain the same rate of increase as is shown by America.

Even when the technical work is well done and the service is efficient and well maintained, and a tariff and other features of commercial practice are adopted which make for sound progress, telephone development on a large scale must still always be principally a matter of capital investment, and a free flow of capital cannot be obtained for a business which is hedged round by all sorts of regulations and restrictions, and is conducted under a limited tenure. In the United Kingdom, for example, a large proportion of the exchange systems are fully as well equipped as those of the United States, are in fact of identical type, while the tariff and general commercial practice are fully as progressive as those of the American telephone companies ; but the fact that the telephone company here has been working under a limited licence, which terminates shortly, and is to be terminated under conditions of uncertainty as regards repayment of capital, has made it impossible to develop the telephone system of the whole country on the same broad scale as has been the policy of the American telephone companies for years past. To develop efficiently a city telephone system requires the constant expenditure of capital years in advance of the actual requirements of the moment, and it stands to reason that certainty as to the future is one of the essential requirements to encourage this constant flow of new capital.

Wherever the telephone business be closely investigated it will be found that even when the plant is relatively efficient and the organisation good, a low rate of development is principally due to the lack of capital expenditure to provide facilities for the constantly increasing public demand. This is, of course, one of the most important causes of the low development of State telephone monopolies. State telephone monopolies are invariably operated as part of the Post Office and telegraph departments, and as these departments throughout Europe do not generally pay their way, notably the telegraph and telephone branch, capital for further development of the telephone system is not easily forthcoming. Therefore, in the case of a Government monopoly we have not only the restricting effect of antiquated plant and technical methods (and consequently inefficient service), the restricting effect of a total absence of scientific tariff and of modern commercial methods, but also the absolute brake on progress of lack of capital for development.

It is not to be questioned that many of the engineers and administrators in charge of European telephone systems are enlightened telephone men, fully alive to the possibilities of wide development of the telephone service when operated on commercial lines ; but their hands are tied by the routine of a great Government administration and by the insuperable difficulties of obtaining capital in sufficient quantity to reconstruct antiquated plants and to build telephone systems on the comprehensive scale required by modern ideas of the telephone business. The curse which lies over the telephone business throughout Europe is political control, which absorbs an industry through mere jealousy of its possibilities as a competitor of another State monopoly, only to throttle it and to stunt its growth.

There are a few rare exceptions where the Government officials who have had charge of the telephone have to a certain extent realised its possibilities and developed it even to a greater extent than their first love, the telegraph. This is notably the case in Switzerland, which provides a brilliant exception to the ordinary run of State telephone monopolies. In Switzerland the telephone has been largely developed as a substitute for the telegraph, and the telephone has become so much more popular than the telegraph that in the past year the receipts of the Swiss Administration from the telephone were more than two and a half times as great as the receipts from the telegraph. In Switzerland distances are relatively short, and the low rates imposed on the telephone department by the political authorities encourage a large short-line interurban traffic, the telephone rate for short distances being actually cheaper than the telegraph rate, though the service rendered is greater and the plant used more expensive.

General experience is that the telephone does not compete so actively with the telegraph as the telegraph officials of most countries at first feared it would. The telephone has created new ways of doing business, and has created its own traffic; and telegraph and telephone traffic have grown side by side. The direct competition occurs where short distances are concerned, and is naturally favoured by artificial rates, fixed, as generally occurs under Government control, without regard to the real cost of conducting the business.

In Sweden, Denmark, and Norway, the political authorities have had more enlightened views on the utility of telephone development than have obtained in the larger European countries, so that greater scope has been given to private enterprise; while also it is fair to say that the Government officials who have had direct charge of telephone work have been more progressive in the Scandinavian countries than in some other parts of Europe, so that in Scandinavia both the company systems and the Government systems are better developed than in the central and southern parts of Europe, and the general efficiency and utility of the telephone service are vastly superior.

CHAPTER III.

AMERICAN AND EUROPEAN TELEPHONE DEVELOPMENT COMPARED.

How European communities are deprived of the most rapid means of communication. Wasted opportunities. Early telephone work in Europe. The restrictions imposed by short-sighted political authorities.

PERHAPS the most effective illustration of the much more restricted progress made in Europe with the telephone as compared with America is to show for the different European countries the number of telephones which would be in service if the rate of development in proportion to population had been the same as in America. In this comparison the figures relating to the Bell system will be used, and not the figures of the total development of the telephone in America, which include the statistics of the various independent companies. In many cases the independent telephones are duplicates of the Bell telephones, and as the Bell system is a general system, giving universal service, the fairest comparison is between that system and that of any European country; a comparison of the total telephone development of America with the total development of every European country would naturally give much more striking results.

In Great Britain there were at the beginning of 1910 602,209 telephones in service. If the telephone system of Great Britain were developed on the same scale as the Bell system of America, there would now be about 2,670,000 telephones in service in Great Britain.

In the German Empire, including Bavaria and Wurtemberg, there were 940,966 telephones in service at the beginning of 1910, but if the development were on the same scale as that of the American Bell system there would now be 3,700,000 telephones in service in the German Empire.

In France at the beginning of 1910 there were 211,728 telephones, whereas if the French telephone system were developed on the same scale as the American Bell system there would be over 2,500,000 telephones in service in France.

In Sweden at the beginning of 1910 there were 173,784 telephones in service. If the Swedish telephone system were on the same scale as the American Bell system, there would be 346,000 telephones in service in Sweden. Here it will be noted that the difference is not so great, as, for the reasons given above, the development of the telephone in Sweden is exceptionally good among European countries; the Swedish results, which are relatively high for the country as a whole, are somewhat aided by the very aggressive competition in Stockholm between the State telephone department and the Stockholm Telephone Company. These two have between them 66,900 telephones in Stockholm alone, or more than a third of the telephones in all Sweden.

Russia had, at the beginning of 1910, approximately 127,000 telephones in service. If the telephone development of Russia were on the same scale as that of the American Bell system, there would be more than 6,000,000 telephones in Russia, instead of a little over 100,000.

Austria-Hungary had, at the beginning of 1910, 146,156 telephones. Telephoned to the same extent as the United States, Austria-Hungary would have 2,950,000 telephones.

Switzerland, as already noted, one of the comparatively well telephoned countries of Europe, had, at the beginning of 1910, 73,758 telephones. If, however, the telephone development were on the same scale as that of the American Bell system, Switzerland would have 225,000 telephones.

Denmark, another comparatively well telephoned country according to European standards, had, at the beginning of 1910, 73,843 telephones. If the development were on the same scale as that of the American Bell system there would now be in Denmark 164,000 telephones

Norway, also high up in the scale for Europe, had, at the beginning of 1910, 58,026 telephones, but telephoned on the same scale as America, it would have 154,000 telephones.

Now we come to a group of far more backward countries, and the figures will show what vast scope there is for telephone development in several European countries having numerous large towns and cities and a large industrial population.

Italy, at the beginning of 1910, had 62,266 telephones. If the telephone development in Italy were on the same scale in proportion to population as that of the United States, Italy would have no fewer than 2,150,000 telephones.

Holland, at the beginning of 1910, had 51,421 telephones, but if telephoned on the same scale as the United States, would have 352,000 telephones.

Belgium, at the beginning of 1910, had 42,540 telephones. Telephoned on the same scale as the United States, Belgium would have 450,000 telephones.

Spain, at the beginning of 1910, had approximately 19,000 telephones. Telephoned on the same scale as the United States, Spain would have about 1,800,000 telephones.

The comparison for the remaining small European countries, such as Portugal, Greece, Servia, Bulgaria, and Roumania, would work out in about the same proportions as for Spain. Each of these countries has in proportion to population about one telephone for over sixty in America.

If the whole of Europe were as well telephoned as America, there would be about 25,000,000 telephones in Europe in place of the 2,584,000 which actually exist.

It is quite true that in several countries of Europe the nature of the population and the habits and requirements of the people do not favour the rapid development of such a means of communication as the telephone service, but this qualification does not apply to the greater proportion of European countries, and certainly it does not apply to the larger and more highly civilised and industrial countries, such as England, Germany, France, Russia, Austria-Hungary, Belgium, Holland, and Italy. The inhabitants of most of these countries would probably resent the imputation that they were less highly civilised than those of America, that their commerce and industry were less well organised, the life of their cities and towns less active, and their general capacity for taking advantage of improved means of communication lower than that of the inhabitants of any other country. And, indeed, they would be quite right, for it is an axiom with all means of communication and public facilities of the kind, that the existence of the facilities creates the demand. If an efficient and well-developed telephone service is supplied, the inhabitants of any town or city on the face of the earth will gladly use it and quickly come to rely upon it in their daily life. There is little doubt that the inhabitants of Rome, Madrid, Paris, or St. Petersburg, or even Constantinople would use the telephone just as freely as those of Boston, Philadelphia, or San Francisco do, if an efficient service were supplied, and if the system were developed in the progressive fashion typical of American telephone enterprise.

Indeed, although the peasant farmer of European countries is usually supposed to be about the most conservative individual on earth, it is highly probable that if the opportunity were put before him he would learn to use the telephone almost as rapidly as the American farmer has learnt to use it during recent years—so great is the value of telephonic communication in rural and sparsely settled districts. It is not to the lack of enterprise or of adaptability on the part of the general public of European countries that the low development of the telephone in Europe must be attributed, but rather to the lack of enterprise and energy on the part of those who have assumed control of the telephone in most European countries. The field is there in every case, but in most instances the tillers have stood still and let the plough rust.

Early Telephone Work in Europe

If, then, we study the history of the telephone in the various European countries, we find that with one exception the work of establishing and developing telephonic communication was at first taken up by commercial companies working under a licence or concession from the Government. The exception is Germany, where from the beginning the Imperial Post Office refused to allow a telephone company to start business, but arrogated to itself the right to supply telephone service to the public. In Great Britain, France, Austria-Hungary, Belgium, Holland, Switzerland, Russia, Denmark, Norway, and Sweden, telephone companies were organised and started work in the early 'eighties with the crude apparatus which was then available. Even in Germany a beginning was made in organising a telephone company, but the German Post Office compensated the enterprising organisers and took over the nucleus of the system before it was established as a business.

In most of the licences or concessions granted to the early telephone companies the various Governments reserved the right to buy out the companies at stated intervals, generally at periods from five to seven years apart, and in almost all cases the Government reserved to itself the privilege of building and operating the trunk or long-distance lines between towns. In a large proportion of cases the Government option to purchase the telephone systems was exercised at a very early date, as soon as it had become evident that the telephone service was a practical and remunerative business, and the telephone became a Government monopoly in France in 1889, in Belgium and Switzerland about the same time, and in Austria and Hungary a few years later. In other countries the misfortune of a complete Government monopoly has not yet overtaken the telephone business, but the strings of Government control have been drawn tighter, and it has become increasingly difficult for those telephone companies which still exist to develop their business along normal lines.

CHAPTER IV.

THE TELEPHONIC HISTORY OF GREAT BRITAIN.

The repressive policy of the British Post Office. How the existence of the State Telegraph Monopoly has resulted in checking the development of the Telephone in Britain. Post Office jealousy of telephonic progress. The transfer of the trunk lines.

GREAT BRITAIN has had perhaps the most interesting telephonic history of any European country, and the vicissitudes of the telephone business in this country well illustrate both the evils of excessive political interference with a technical industry and the wonderful vitality of the telephone business, which has enabled it to develop on a relatively large scale in spite of continuous and powerful efforts to hamper and restrict that development. No sooner did the first telephone companies begin to do business in England than suit was brought against them by the Post Office for infringement of the Postmaster-General's legal monopoly in telegraphy.

This monopoly was conferred on the Postmaster-General by the Telegraph Act of 1869, which empowered the Post Office to buy up the various telegraph companies then existing, and to organise a Post Office telegraph service. The desire for a legal monopoly in telegraphy had been expressly disclaimed by the Post Office witnesses who appeared before the Select Committee which examined the Telegraph Bill, and the Select Committee reported against a monopoly being granted, but the monopoly clause was inserted—for safety's sake, no doubt—as the Bill was going through Parliament, and became law. Ten years later it was used to assert proprietorship over the telephone, and although the telephone had not been invented until six years after the passage of the Act, the definition of a telegraph was found to be so widely drawn as to include the telephone or any other means of sending and receiving electric signals, with or without wires.

Consequently the decision went against the telephone, and an arrangement was come to between the telephone companies and the Postmaster-General under which the companies agreed not to appeal against the decision, and the Postmaster-General agreed to grant licences to the telephone companies to continue their business, on certain terms.

These terms were sufficiently onerous, the licence being limited to 31 years from the end of 1880, and the licencee being obliged to pay the Post Office a royalty of 10 per cent. of the gross receipts from the telephone exchange business. At first licences were granted for certain small areas only, but after some years the Postmaster-General, impelled by strong criticism in Press and Parliament of the obstructive attitude of the Post Office towards the telephone, changed his policy, and granted licences for the whole Kingdom. Later still it became the policy of the Post Office to grant no further telephone licences, either for the whole country or for restricted areas, until the passing of the Telegraph Act of 1899 gave the municipalities powers to establish telephone systems, and the issue of licences to those municipalities which applied for them became compulsory on the Post Office.

The relations between the Post Office and the telephone companies having been settled by the licence of 1880, and the Post Office having secured by the royalty clause a substantial share of all future revenue from the telephone, it might have been thought that the attitude of the Post Office and the Government towards the telephone would have become one of benevolent encouragement. But, unhappily, the Government officials have never been able to regard the telephone companies otherwise than as chartered infringers of the luckless telegraph monopoly of the Post Office, which, great organisation as it is, has never paid its way, and to-day costs the taxpayer about £1,000,000 a year to maintain. The Post Office have watched the growth of the telephone with a sleepless jealousy, and all the forces of the bureaucracy—almost irresistible forces, even in this free country—have been continuously used to harass and restrict the activities of the telephone companies.

One of the most powerful hindrances to a sound telephone development has been the refusal of statutory powers to the telephone companies to run their wires. In the early days the telephone companies on various occasions promoted Bills in Parliament to obtain these powers—which the Post Office of course enjoys—but the whole influence of the Government of the day was exerted against these Bills, and they naturally failed to pass. Consequently the telephone companies have been dependent on private wayleaves, and on private agreements with the various local authorities which control the streets, for powers to build their line plant. In some cases the local authorities have pursued a policy of deliberate obstruction and have refused to allow the construction of modern underground systems of telephone distribution, thereby hampering the development of the telephone and restricting the efficiency of the service—to the natural disadvantage of the public.

Among the worst offenders in this respect have been the London County Council and the City Corporation, which have long obstructed the sound development of the telephone in London by refusing to allow the telephone company to build underground work. The inconsistency as well as the injustice of such an attitude is evident when it is considered that the moment the Post Office is empowered to build a telephone service in London it takes up the streets in all directions without saying with your leave, or by your leave. But the municipalities naturally take their cue from the central Government, and it having been the policy of the Government from the first struggle with the early telephone companies persistently to harass the telephone industry, many of the local authorities have done their best along the same lines.

Even working under these very serious difficulties, however, the telephone made rapid progress in Great Britain, though how much more rapid progress would have been, had the conditions under which the industry was conducted been as encouraging as they were discouraging, it is difficult to estimate. Companies were formed to develop different sections of the country, and after a period of educating the public in the value of the new means of communication (by the simple expedient of supplying prominent business people with free service for a time), the telephone service began to take a firm hold on the business community. The Post

Office at first tried to compete with the telephone companies in different places, but as a Government Department can never compete successfully with a commercial concern these attempts were failures, and the company systems grew while those of the Post Office remained stagnant or disappeared. The Post Office still maintains a few provincial telephone systems, either relics of these early days of competition or serving areas where there is no company service ; but, like the Post Office telegraph service, these systems are conducted at a substantial loss.

After some ten or twelve years of development under diversified management, the various telephone interests amalgamated, between 1889 and 1892, into the National Telephone Company, and numerous improvements in organisation and in general methods resulted from this amalgamation. The rates charged to the public were reduced in a number of towns from £20 to £10 for flat rate service, and plans were made for adopting underground construction in place of overhead in those towns where wayleaves for laying underground cables could be obtained. About this time the expiry of the original Bell telephone patent in England encouraged schemes for starting competing telephone services in a few places. A small company was organised in Manchester and installed a crude telephone system on a very small scale, which gave service for a few months and was then sold out to the established company. A new telephone company on a larger scale was organised by the late Duke of Marlborough to establish a competing service in London, but this company also amalgamated with the National Telephone Company before beginning active operations.

The Long-Distance Service : The Transfer of the Trunk Lines.

In 1892, a fresh attack was made on the telephone business by the Post Office. During the previous seven or eight years the telephone companies had established trunk lines between various towns, and had built up quite a successful long-distance telephone service, which was largely used by the business public. The Post Office found that this long-distance telephone communication had a certain effect on the telegraph traffic ; as the telegraph business of the Post Office had for many years been conducted at a loss, it might be considered that the Post Office officials would be glad to see their telegraph traffic contract rather than expand, and so reduce the loss ; but the officials took the opposite view and raised an agitation for the transfer of the trunk telephone lines to the Post Office. An agreement was eventually come to between the Post Office and the National Telephone Company, involving the limitation of the operations of the National Telephone Company to certain defined areas and the transfer of all the trunk lines to the Post Office. According to the view of the Telephone Company an important part of the understanding arrived at, though not expressed in the agreement, was that the National Telephone Company should be protected from competition in the local service during the remainder of the licence ; but the existence of this understanding was afterwards denied by official witnesses at the Parliamentary Telephone Inquiry of 1898, and there the matter must be left. In giving up the trunk lines, however, it is fairly clear that the Telephone Company gave up that which would have made their position in the future impregnable against any attempt at competition in the local service, since for business purposes a purely local telephone service—without long-distance facilities—has a relatively small value.

The trunk telephone lines were finally transferred to the Post Office in 1896, and the Post Office adopted a new scale of rates for trunk calls and put in hand the construction of a number of lines to extend the trunk system to many new places. Incidentally, it may be mentioned here that by making rates for trunk telephone calls of 3d. for 25 miles, and 6d. for 50 miles, the Post Office officials introduced trunk telephone rates which compete very powerfully with their own telegraph rate, and in the hands of the Post Office the trunk telephone service became definitely a cut-throat competitor of the telegraph service. The effect of the extremely low trunk telephone charges for short distances is evident in the financial returns of the Post Office, which show that the average value of a conversation on the trunk lines is only 5·9 pence. As the minimum charge for a telegram is 6d., it is clear that the trunk telephone service competes most effectively with the telegraph service.

CHAPTER V.

THE TELEPHONIC HISTORY OF GREAT BRITAIN—(*continued*).

Incidents of the policy of official obstruction. The Glasgow Inquiry of 1897; the Parliamentary Inquiry of 1898; the Telegraph Act of 1899. The London Agreement of 1901. How political schemes miscarried. Municipal competition in Telephony and its failure. The costly experiment of the Glasgow Municipality.

THE opposition to the development of the telephone which had been the steadfast policy in official circles provoked a long-drawn-out Parliamentary Inquiry in 1895. Before this Select Committee the proposal to establish competing municipal exchanges was exhaustively discussed, and was condemned by all competent expert witnesses. Owing to the dissolution of Parliament this Committee made no report, but it has been stated that the draft report recommended the purchase of the entire telephone system of the country by the Government.

Municipal Telephony.

For some years past one or two of the municipalities, notably that of Glasgow, had been agitating for permission to establish municipal telephone systems. The persistent agitation of the Glasgow Corporation finally resulted in the Treasury Inquiry at Glasgow in 1897, to inquire into the efficiency and adequacy of the telephone service supplied by the National Telephone Company in Glasgow. This inquiry lasted about ten days, and the desire of the Glasgow Corporation to establish a municipal telephone service was fully ventilated. The report of the Commissioner appointed by the Treasury to conduct the inquiry was entirely adverse to the Glasgow Corporation and in favour of the Telephone Company. The Commissioner freely condemned the tactics of the Corporation, and pointed out that the Company had been prevented from improving the efficiency of its service by the refusal of the Corporation to allow the Company to put down a modern underground plant in Glasgow.

This victory of common-sense over organised agitation and obstruction was not permanent, however, as continued agitation against the sound development of the telephone resulted in a further Parliamentary Inquiry in 1898. The Select Committee was presided over by the late Mr. R. W. Hanbury, then Secretary to the Treasury, and the representative of the Post Office in the House of Commons (the then Postmaster-General being the Duke of Norfolk), and the inquiry was conducted in a manner markedly hostile to the telephone interests. The inquiry was a very extended one, and expert witnesses both from the Post Office and from the National Telephone Company were bullied and brow-beaten in a manner quite unusual in Parliamentary Inquiries. The result was a triumph for municipal and official obstruction, as the Committee reported in favour of active competition in the telephone business, and on the strength of the Committee's report Mr. Hanbury introduced a Bill into Parliament which became the Telegraph Act of 1899, and empowered the Post Office to grant telephone licences to the municipalities, and granted £2,000,000 to the Post Office to establish a competing telephone service in London.

The main object of this measure was to encourage such active competition in telephony as would destroy, or at least largely diminish the value of the National Telephone Company's undertaking against the time when the Post Office would succeed to the telephone business at the expiry of the original licence at the end of 1911. In practice, however, matters worked out differently. With the exception of Glasgow, where the Town Council had for several years nourished a persistent agitation for the privilege of establishing a municipal telephone service, and of four or five other towns where promoters and agitators succeeded in arousing the interest of the Town Council in telephone competition, the Telegraph Act of 1899 was practically disregarded by the municipalities. A few obtained telephone licences from the Post Office, but in the end only six municipal telephone systems were started, and after a very short experience of the difficulties of the telephone business, four municipalities were glad to sell their systems to the Post Office and to the National Telephone Company. Within six years after the passage of the Telegraph Act of 1899 municipal telephony had lived and died!

The Telephone Development of London.

With regard to the competition in London, Mr. Hanbury's Act similarly miscarried. Mr. Hanbury had advocated a strenuous competition between the Post Office and the National Telephone Company in supplying telephone service to Londoners; but when the question was approached in a practical way by practical men, the difficulties of starting a new system in a great city to undermine, by means of competitive rates, the position of a large and well-established system, were considered to be too great to warrant the expenditure of public money on such a wasteful and unpractical form of competition. In the end an agreement was come to in 1901, between the Post Office and the National Telephone Company, under which the two concerns virtually agreed to share the telephoning of London, to charge the same rates, and to give an intercommunicating service.

When this agreement was made public, there was a great hue and cry in the Press, as the politicians had planned for and had promised a cut-throat competition which was to annihilate the Telephone Company; but experience has proved that it was a sound and practical measure. Instead of throwing away public money on a competition which would certainly have been unsuccessful, and instead of inflicting on Londoners the annoyance and inconvenience of two separate telephone systems, the Post Office has assisted in the telephone development of London, and the expenditure made on the Post Office London system has gone to increase the facilities of the general telephone service, instead of being spent on a separate and isolated system which would have been of little practical value.

The combined systems of the National Telephone Company and Post Office in London now contain over 160,000 stations; as far as the telephone-using public is concerned all these telephones belong to one system, and so great has been the increase in telephonic facilities, due to the agreement for intercommunication, that it now seems ludicrous that in 1901 the agreement should have been styled "The Telephone Betrayal" by the hysterical section of the Press. If the large capital expended by the Post Office in London had been spent by the National Telephone Company in extending the original system along uniform lines, the results would have been still better and far more economical, as a certain amount of duplication of plant and organisation would have been avoided; but the intercommunication agreement has been fully justified both by the experience gained in London and by that of Glasgow, where the opposite policy of cut-throat competition was adopted by the municipality and quickly ended in failure.

The value of intercommunication is also shown by the case of the Hull municipal system, which is simply a small annex of the National Company's Hull exchange. By the threat of forcing the National Company to remove its underground plant the Hull Town Council obtained free intercommunication, and the Hull municipal telephone system virtually lives on the much larger and older-established business of its neighbour. Without intercommunication it would have died the death long ago, or, by a quiet political deal, would have been sold to the Post Office.

The Glasgow Town Council spent over £360,000 between 1901 and 1906 in building an opposition system to that of the National Telephone Company in Glasgow, and then, having got into both practical and financial difficulties, sold the system to the Post Office for £305,000. The waste of capital is not by any means measured by the amount lost by the Glasgow Corporation, as a large amount of the plant taken over by the Post Office was of such a character as to require complete replacement by modern equipment, and the duplication of plant at Glasgow caused by the strenuous competition of four or five years was naturally on a large scale. It is probable that the waste of capital through the Glasgow telephone competition, due to duplication of facilities and to the adoption, for the sake of cheapness, by the Glasgow Corporation of plant which was not up to the standards of recognised telephone practice at the time, will amount in the long run to not less than £250,000, probably to a good deal more.

Municipal Telephone Difficulties.

The municipal telephone campaign was accompanied by a great deal of noisy agitation, and as the municipalities exert such powerful political influence, they first obtained full scope to indulge in experiments in telephone competition, and later were relieved, at the taxpayer's expense, of the financial loss which those experiments involved. But the municipal incursion into the telephone field had little effect on the general development of the telephone in Great Britain, and served only to demonstrate the wastefulness and futility of competition in the telephone service. The telephone service differs from other means of communication in that the value of the service to each user depends on the scope and uniformity of the system as a whole. Two competing systems in the same district simply split the inhabitants of that district into two isolated telephonic camps, so to speak.

This fundamental feature of the telephone service—that its value to the user, the public, lies in the extent of the facilities for general intercommunication—was ignored by the promoters of municipal telephony. But the bearing of this fundamental fact is that in a telephone competition the two competitors engage in a life and death struggle to obtain, or retain, the larger share of the public patronage, and so obtain the more valuable system of the two and that which in the long run must attract all important customers and become the survivor.

It was in such a struggle, with a well-established and experienced adversary, that the few municipalities which started competitive telephone systems embarked, but the incapacity of a municipal department to compete successfully with a well-organised commercial concern made it impossible for any of them to overtake the National Telephone Company or seriously to shake its position ; the business public, not being governed either by sentimental considerations or by political influence, naturally clung to the system which gave the greatest facilities and the most efficient service.

Therefore, within a very short period municipal telephony, from which so much was expected by its political and other sponsors, had shot its bolt and ceased to be a factor in the active development of the telephone. Had a greater number of municipalities taken up the work than actually did, and had the municipal telephone systems—notably that of Glasgow, which was the pioneer and was generally looked to as an example—been better organised and conducted more in accordance with the technical requirements of the telephone business, there might be a different story to tell. As it was, municipal telephony in Great Britain was a failure, politically, financially, and practically, and has become a mere passing incident in British telephone history.

CHAPTER VI.

THE TELEPHONIC HISTORY OF GREAT BRITAIN—(*continued*).

The problem of the expiry of the National Telephone Company's licence. The Agreement of 1905. The strangling effect of the Agreement on present and future telephone development. The Post Office telegraph monopoly and its future telephone monopoly compared. The Government's disregard of the essential characteristics of the telephone business.

THE failure of the municipal telephone movement became evident within a very few years after the passage of the Telegraph Act of 1899. By 1904 it was clear that effective competition on a general scale was not to be expected from the municipalities, and as Mr. Hanbury's suggested active competition by the Post Office had been negatived by the Post Office officials themselves, there remained no further hope of destroying the National Telephone Company's business, and it became incumbent on the politicians to find some solution of the problem of 1911.

The problem is this: When the original licence was drafted neither side to the transaction looked forward to the time when it would expire, and no provision was made for the continuance of the service after the expiry of the licence on December 31st, 1911. As matters stood, on that date the Telephone Company would cease to have the right to conduct the telephone business, and that was all. The politicians and officials have aroused an unfriendly sentiment in Parliament towards the Telephone Company, and although the great majority of Members of Parliament are wholly unacquainted with either the practical or the technical features of the telephone business, and complacently ignore the deplorable financial results of the Post Office telegraph monopoly, it is understood that Parliament would not favour a renewal of the licence, but prefers the conversion of the telephone business into a Post Office monopoly.

This, of course, has always been the natural desire of the Post Office officials, and doubtless the transfer would have taken place long ago had not the rapid expansion of the telephone business caused the capital concerned to become so quickly an imposing sum. Therefore the purchase of the whole business, often contemplated, has always been postponed, in the hope of something turning up to the advantage of the Post Office. The last hope was Mr. Hanbury's scheme of ruthless competition, and this having signally failed, it became necessary to consider in reasonable time the situation which would arise after 1911, when the whole telephone service of the country would stop, unless the licence were renewed or the plant were purchased by the Post Office and continued in operation.

Under the agreement of 1901 respecting the joint operation of the London service, the Post Office undertook to buy the London plant of the National Telephone Company at the end of the licence; taking this agreement as a model, an agreement was finally arrived at, in 1905, extending the scope of the London agreement to the telephone plant of the whole country. This agreement was referred to a Select Committee which sat through the summer of 1905, and, with some modifications, it was approved by Parliament in August of that year. Strong efforts were made by the municipal politicians to upset the agreement, and the proposal that the Post Office should build duplicate telephone systems throughout the country, so as to avoid the necessity of purchasing the National Telephone Company's plant, was seriously argued before the Select Committee.

It is this agreement which now governs the situation of the telephone business of Great Britain, dooming it to become a Government monopoly. The agreement is already exercising a strangling effect on telephone development, and, by checking plans and construction for future requirements, creating a situation which will cause many difficulties in the future. The governing feature of the agreement is that the Post Office is to purchase the plant at its value as working plant *in situ*, without allowance for goodwill or for past or future profits. Regard is also to be had to the suitability of the plant for the purposes of the telephone service of the Post

Office. Under these terms it is evident that the National Telephone Company, which has built up a great business under very difficult and discouraging conditions, and has organised and developed a great public service to a high pitch of scope and efficiency, is treated with discouraging harshness and parsimony. The politicians and officials may be within the letter of the licence and of the law in dictating such terms, but such a policy towards those who have provided the capital to develop a great technical industry does not encourage the enterprise which is necessary for the development of new industries and is really the vital force of any industrial country.

The Telegraph Transfer of 1870 and the Proposed Telephone Transfer Compared.

The great financial failure made by the Post Office with the telegraph monopoly has always been a source of friction between the Treasury and the Post Office, and doubtless the political authorities were anxious that the results of the proposed telephone transfer should be more favourable to the Government than those of the purchase of the telegraph systems; so advantage was taken of the fixed termination of the telephone licence to dictate the lowest possible terms for the purchase of the telephone plant, and the Telephone Company, faced by a legal sentence of death due on a fixed date, had no recourse but to accept the terms.

The conditions of the proposed telephone transfer and those of the telegraph transfer of 1870 are widely different. When the Post Office took over the telegraph companies telegraphy was in an early stage of development—still in its experimental period—and several of the systems bought out were poorly organised and equipped. Undoubtedly the telegraph companies received favourable terms—they were not anxious to sell, and the Post Office officials were feverishly anxious to buy and to silence opposition to the scheme for a Government telegraph monopoly. But the price paid for the telegraphs is not in any way the cause of the financial failure of the Post Office telegraph monopoly. The Post Office has never conducted the telegraph business on commercial lines; it keeps no telegraph capital account and pays no interest even on the original purchase money. The telegraph capital account was closed in 1879, and the interest is paid out of the Consolidated Fund, so that the original cost of the telegraphs to the Post Office has no bearing whatever on current financial results. The reason why the Post Office telegraphs lose money is that a Government Department, subject to political control, cannot conduct a technical business on commercial lines.

In the case of the proposed telephone transfer we have an industry which, unlike the telegraph in 1870, has long ago emerged from its experimental period and has been developed to a high pitch of organisation and efficiency. The principal problems have all been solved ; we know what kind of plant to use to give an efficient service, what tariffs to adopt to secure a large development ; and the general organisation of staff and the commercial management of the telephone service have become highly specialised subjects. Therefore in taking over the telephone system the Government will be taking over a highly organised concern, working on lines that are the result of experience gained from a long period of experiment and development and conducting a thoroughly sound and remunerative business. Moreover, given enterprise and skill in the general management, the telephone business is capable of almost indefinite expansion. This is clear from the record of the telephone in America ; and it is only a question of capital expenditure and administrative skill for the development of the telephone in Great Britain to be increased tenfold within a few years.

In view of these facts—that the telephone service is a safe, profitable, and growing business, with, given adequate management, a limitless future before it—the Government could well afford to compensate reasonably those who have risked their capital during the trying and obscure period of experiment and early development. That the contrary policy, of giving the minimum possible price for the plant without any allowance for goodwill, has been followed in framing the purchase agreement is relatively unjust, and certainly illiberal ; and it is already producing results which will seriously handicap both the efficiency and the development of the service when it becomes a Government possession.

One of the essential features of sound telephone development, and one of the essential requirements for high general efficiency of the service, is constant provision of plant for the future. It is a common mistake to regard a telephone system as a plant having a fixed and limited capacity. On the contrary, the telephone system of any town is (or should be) in a constant state of change and growth. It is necessary to study for long periods in advance what the telephonic requirements of the district will be, and to provide from time to time certain sections of the plant—buildings, switchboards, conduits and cables—on a scale considerably greater than the actual requirements of the moment call for. Unless this is done systematically there will arise congestion of the plant, lack of facilities, and consequent loss of efficiency in the service. It is not only to supply the wants of future subscribers that this margin of spare facilities must be provided, but also to maintain the day-to-day efficiency of the service. Existing subscribers frequently require changes in their telephone installations, increasing or altering their equipment, or because of removal from place to place. The traffic tends constantly to increase, and unless there is a certain elasticity in the plant obtained by the systematic provision of spare facilities, inconvenience and inefficiency are sure to result. The policy of the Government towards the Telephone Company ignores this important feature of telephone development, and the agreement of 1905, by limiting the purchase terms to working plant at its value as working plant, expressly discourages the construction of plant with a large margin of spare capacity, and so discourages work which is absolutely necessary for the proper development of the telephone and for the maintenance of a high standard of efficiency.

Under ordinary circumstances the Telephone Company would now (1910) be spending capital in plant all over the country, much of which would not come into use until long after 1911, and therefore would not be revenue earning during the period of the licence. Under the terms of the agreement the Post Office would be entitled to try, and undoubtedly would try, to buy this plant at less than its cost price; so clearly it is not business for the company to expend capital on such future requirements. If the management of the telephone system were to be continuous, or if the purchase terms had been drawn with a due regard to the real conditions of the telephone business, this future plant would be constructed, in accordance with regular practice, well in advance of the requirements of the moment. But the harsh terms of the purchase agreement necessarily put the brakes on advance capital expenditure, and the normal development of the telephone system all over the country undergoes a check for several years.

Thus it is seen how the repressive policy of Government monopoly casts its shadow before, and even long in advance of its actual realisation exercises a strangling effect on a business where active development and an enterprising policy of providing for the future are essential factors for satisfactorily meeting the requirements of the public. When the Post Office succeeds to the telephone service it will succeed to a congested plant, without proper facilities for expansion and without proper development studies to govern plans for future construction. The result will naturally be a long period of embarrassment and bad service, and probably an expenditure, to cope with the situation, much greater than would have been involved had the situation been faced in an intelligent manner beforehand and the purchase terms arranged in accordance with the natural requirements of the telephone business.

It is at this point that the development of the telephone in the United Kingdom stands at the moment of writing. While there is yet time for some solution to be found which will avert the overhanging evil of Government monopoly, it does not seem likely, having regard to the apathy and prejudice with which technical industrial questions are regarded in political circles, that any endeavour will be made in that direction. That a Government telephone monopoly, even in England, will be progressive, efficient, and financially successful nobody with any knowledge of the subject believes ; but, equally, nobody with any knowledge of the subject believes it possible to convert the political authorities to a sound view of the evils of Government monopoly.

CHAPTER VII.

SOME REASONS FOR POPULAR PREJUDICE AGAINST THE TELEPHONE.

Unenterprising management in the early years of the business. The direct use of the telephone by the public accentuates difficulties. Telephone rates. The education of the public. General inefficiency of Government monopolies. The causes of the low efficiency of the telephone service in Europe.

REVIEWING the history of the telephone service in England, and in Europe generally, it must be candidly admitted that the early telephone companies did not appreciate the importance of their mission, and did not make the most of their opportunities. The service was necessarily bad in the early days, because the appliances were extremely defective. But in America more reliable apparatus was rapidly evolved, and the general organisation of the service was studied and improved with greater effect than on this side of the Atlantic. The European telephone companies in general pursued a cautious and conservative policy, which ill-accorded with the requirements of a business which in America was soon found to be capable, under enterprising management, of almost indefinite expansion.

Somewhat arbitrary methods of dealing with customers, and a policy of extreme reticence—of avoiding the circulation of any information regarding the difficulties and peculiarities of the telephone business—added to the unpopularity which the early telephone companies incurred. It is a curious fact that telephone administrations are unpopular all the world over, as far as unpopularity can be gauged from the flouts and gibes of the newspapers, which are generally supposed to represent public opinion.

This unpopularity partly arises, I am convinced, fromthe exasperation caused by the failure of a telephone call, which happens occasionally from one cause or another in the best organised systems, and, being a direct personal experience of an individual, always causes a deep impression. In other services a failure may occur—as for example, a lost letter—unnoticed ; but the public uses the telephone personally, and therefore every failure is immediately and directly felt by some individual. This is a fundamental and ever-present reason why the highest attainable efficiency is indispensable in the telephone service ; and the failure to recognise this by Government telephone administrations is the main cause of general dissatisfaction with the telephone service throughout Europe.

The rates charged for telephone service have also in the past given rise to much public complaint. There was a tendency in the early days of telephony to regard the telephone as a luxury, and to charge for it accordingly, and European telephone companies in general were slow to follow the example of the American companies in adopting the practical measured rate tariff, which gives low prices for small users, charges each customer in proportion to service rendered and has been a powerful factor in extending the development of the telephone. Given all these difficulties—defective plant and service, an unpractical tariff, somewhat arbitrary methods of dealing with the public and a policy of extreme reticence—it is not to be wondered at that the early telephone companies achieved a substantial measure of unpopularity.

Of late years these methods have all been radically altered. American experience has largely been availed of, and the standard of equipment of the modern telephone system in England is as high as that of America ; in fact, the plant used is exactly similar. The tariff has been greatly revised, and a system of measured rates introduced, and for a long time past the National Telephone Company, by means of trained commercial departments, have pursued an enlightened policy of educating the public in the advantages of the telephone service, and in its proper use.

The Evils of Government Monopoly.

If I were asked to summarise the progress of the telephone in Europe in the past thirty years, I should say that the telephone companies have learned their business, while the Government telephone administrations have not. Moreover, it seems hopeless ever to expect Government administrations really to learn the telephone business, because efficiency and enterprise are the two essential features of telephone work. Government officials are the first to admit that high efficiency cannot be expected of the mass of Government employees, owing to the permanency of their employment; and it is common knowledge that the efficiency of a Government department is only about 50 per cent. of that of a well-organised commercial concern. Enterprise is out of the question in a Government department, especially in a Government telephone department, which is merely an inferior branch of the Post Office, which in turn, in most countries, is a branch of some division of the great machine of State.

The whole tendency of Government methods and of the routine of Government departments is to repress individual initiative and to stamp out any willingness to take responsibility. It is not to be imagined for a moment that the ordinary civil servant, however intelligent and however scientifically accomplished he may be, can conduct a Government telephone administration with the efficiency and progressiveness, or with the economy, which would be required of the head of a commercial organisation. Even if he had the will and the expert knowledge to enable him to do so, he would not be allowed, under Government methods, to exercise the necessary power and responsibility. Therefore the fundamental characteristics of Government management are all against a Government telephone monopoly being either efficient or progressive, or being conducted in accordance with commercial principles. Linked to the unprogressive Government telegraph monopoly, the telephone business, from being the highly specialised industry which it really is, sinks to a mere sub-department of the Government service. In this position, which it has long occupied in many European States, it becomes relatively stagnant, and both the real requirements of the telephone service and its possibilities are unheeded and uncared for.

Apart from the fact that the Government telegraph departments which have controlled the telephone have not been willing—despite the wonderful development obtained in America—to recognise the possibilities of telephone development, they have always made the fundamental error of giving greater importance to a popular rate than to efficiency of service. High efficiency is the whole story in telephony. As has been said above, the public becomes temporarily a part of the telephone system during every call, and feels directly and resents acutely every difficulty, however small ; but there is another point which insistently demands constant and ever-increasing efficiency, which is that unless a telephone call can be completed immediately its object and its purpose are often gone, and the whole transaction becomes a failure. With a written message, once written and dispatched, the sender has to trust to luck whether it reaches its destination quickly or slowly, mutilated or intact. But a telephone call is a spoken message, and the reply is to be obtained on the spot ; the speaker is waiting to speak it from the moment he reaches for the telephone, and seconds seem like minutes when one is waiting at the telephone ; moreover, if the seconds swell into minutes, and the minutes into half-hours or hours (as happens daily in every European long-distance service) the delay in many cases entirely destroys the value of the call, and makes it purposeless or useless.

The Causes of Telephonic Inefficiency in Europe.

It is generally admitted that a Government department cannot obtain as high efficiency from its staff as a commercial organisation, because the discipline and training of the staff in Government employ cannot be as effective as with a commercial concern. All the world over, Government employees are wont to consider themselves superior persons, and in no service do the manners and mannerisms of superior persons create so much friction or so lower the general efficiency as in the telephone service. But the Government telephone administrations, in dealing with the long-distance service, have entirely lost sight of the extreme rapidity which is the whole value of telephone communication, and have made cheap rates, without regard to the real financial requirements of the business, and then have practically left the service to take care of itself; with the result that in hardly a single European country is there a long-distance telephone service worthy of the name. In fairness it must be stated that this policy of fixing uncommercial rates is generally due to the political authorities, and is often contrary to the views of their technical subordinates.

In America the suburban and long-distance telephone services always strike Europeans as phenomenally rapid. In point of fact, the suburban service in America is almost as rapid as the local service, and the delay in "getting through" to points within a radius of 30 or 40 miles of the large cities is always to be measured in seconds; in the long-distance service an average delay of a few minutes is the standard maintained, and it is rare for a call between any two of the large cities to be delayed over ten minutes. So accustomed is the American public to this standard of service, that a delay of over ten minutes in a call, say, from Boston to New York, or New York to Philadelphia, Washington or Chicago, would give rise to complaint. This is a real telephone service, and it is the kind of service that the public of every country really demands, because the public, in the long run, always puts efficiency first.

The Government telephone administrations, partly through lack of understanding of the telephone business, partly through political considerations, have put cheap rates first and efficiency second. It is impossible to give a high-speed and reliable long-distance telephone service at the rates currently in force in Europe, and make the service pay its way, because the American standard of service is only attainable by elaborate organisation of plant and staff, and specially by keeping the capacity of the plant in advance of the requirements of the traffic. This is expensive work, the requirements of the traffic being largely determined by the busiest hour of the day, and the busiest hour represents a large proportion of the total traffic of the 24 hours. In European telephone administrations it is the general practice to let the traffic constantly overwhelm the facilities, so that the usual daily condition of every long-distance line is that there is a long list of people waiting to use it, and some calls are delayed for hours. Even with this habitual delay, the long-distance telephone service finds many users, because the cheapness of the rates and the volume of communication obtained in a three-minute telephone call make the telephone compete favourably with the telegraph; but this does not constitute a real telephone service, and it cannot be doubted that the business public would willingly pay higher rates were a standard of service maintained which would enable the user always to rely on a long-distance call being completed within a few minutes.

Although the long-distance telephone service is better equipped in England than in most of the Continental countries, the general efficiency of the service, due to defective organisation and lack of understanding of the real requirements of the telephone service, is very inferior. With a "trunk call" in Great Britain it is a pure matter of luck whether one obtains a result in minutes or hours. Consequently the service is not a tenth as useful to the business public as it would be if it were organised and conducted according to the standards set in America.

It is common knowledge among those familiar with the subject that the great majority of complaints made by business people in England against the telephone are due to the inefficiency and unreliability of the trunk service. The Post Office, in their trunk telephone business, have made the common mistake of telegraph administrations of fixing too low rates and leaving the service to take care of itself, or to be taken care of by telegraph people, who do not as a rule understand the telephone business; the low average rate per trunk telephone call in England—5·9 pence—is clear evidence that the chief public use of the trunk service is for the short distances, where the price of a telephone call undercuts the price of a telegram. For the average value of the call to be 5·9d. the great majority of calls must be 3d. and 6d. calls, which are very much cheaper than sixpenny twelve-word telegrams—so it is evident that the public puts up with the relatively slow service for the sake of cheapness.

CHAPTER VIII.

THE ESSENTIAL FEATURES OF TELEPHONE POLICY.

Efficiency the prime requisite. Reasonable rates ensure constant development. Development requires continuous capital expenditure. British Post Office telegraph and telephone finance. The present loss on the telegraphs and probable future loss on the telephones. The technical progress of British telephony. Reasons for more rapid progress in America. A summary of the development of the telephone service and of telephone policy.

THE essence of the telephone service is efficiency—and here efficiency means rapidity combined with accuracy and reliability. High efficiency is vastly more important to the business public—indeed to all sections of the public—than apparent cheapness, and low rates accompanied by relative inefficiency are only apparently cheap. Until Government telephone administrators make this maxim their guiding principle, and also succeed in sufficiently convincing their high political chiefs of its soundness, so that the latter shall have the courage of their convictions when defending the telephone administration against agitation and political pressure, there will be no hope of a Government telephone service becoming the highly efficient public servant it is capable of becoming. Reasonable rates, capable of yielding a satisfactory return on the capital after providing adequately for working expenses, are necessary for the proper development of the telephone, because this development—to which no end can be seen—calls continuously for fresh capital expenditure. Neither private enterprise nor Government can afford to provide this constant flow of capital unless it is to be remuneratively employed.

The indifference with which Parliament and public regard the large annual loss on the Post Office telegraphs is altogether indefensible in a business community. The hopeless condition of Post Office telegraph finance has led to the general adoption of an unreasonable theory that the telegraph system ought to be maintained by the taxpayer for the public convenience. But there is no sound basis for this view, as most telegrams are sent by those who can well afford to pay for them. Moreover, it is a view absolutely inconsistent with the promises of the Post Office when the telegraph monopoly was established, which forecast great economies under Government working and sufficient profits to pay off rapidly the purchase capital ; and it is a view inconsistent with the general financial policy of the Post Office, under which strenuous resistance is made to every sacrifice of revenue demanded by reforms in postal charges. To extend the unsound finance of the telegraph monopoly to the telephone business and to teach the public to regard the telephone as well as the telegraph as a system to be maintained for the general convenience regardless of commercial principles, would be a certain means of checking the development of the telephone, because the large sums of capital continually required would not be forthcoming if they were to be unremuneratively employed.

The richest country could not afford to throw million after million into a bottomless pit such as Government management has made of the Post Office telegraphs. Therefore commercial rates for telephone service are a necessary factor in the sound development of the telephone ; and inadequate study of the important question of telephone tariffs and lack of conviction and courage in applying and maintaining correct rates are contributory factors in the general failure of Government monopolies to give the telephone service the development and usefulness of which it is capable.

It should be said here that the British Post Office, thanks to the discussion of their telegraph and telephone finance in Parliament, have adopted a form of accounts for their telephone systems which precludes the Post Office telephone work from becoming such a financial bottomless pit as the telegraphs have become. In the telephone accounts a capital account is shown, and as the accounts are presented annually to Parliament and all new capital expenditure must be approved by Parliament, there is little danger of telephone finance and general management getting totally beyond control, as telegraph finance and general management did in the early days of the telegraph monopoly. But these accounts show very clearly how much more costly is Government management than commercial management. It appears that whereas the working expenses of the National Telephone Company are only 56·6 per cent. of the gross receipts, those of the Post Office local telephone services are 73·5 per cent. of the gross receipts. The Post Office accounts are favoured by the fact that rates and taxes are not paid by the Post Office and rents are not charged in respect of buildings occupied by telephone services, whereas these charges, and doubtless others not included in the Post Office accounts, have to be met by the Telephone Company; so that if a strictly accurate comparison could be made the economy of commercial working would be still more emphatically demonstrated. Even as they stand, however, the Post Office telephone accounts are a serious warning of the economic disadvantage of transferring the whole telephone business of the country to Government management.

The Technical Progress of British Telephony.

As regards the technical progress of telephony in Great Britain, it may be said that the National Telephone Company and the Post Office have for some years past been occupied in transforming their plant from the old-fashioned magneto system to the modern central battery system. This transformation was effected on a general scale much earlier in America, where central battery plant has been standard practice for about ten years. Here again we have an illustration of the evil effects of political interference and obstruction on a technical industry. When the central battery system was perfected, over ten years ago, the advantages it gave in automatic signalling, increased accuracy and rapidity of working, and more economical and stable maintenance caused the existing magneto system at once to become obsolete. The American telephone companies, having an unlimited future before them, and having in all cases full underground wayleaves, were able at once to undertake the complete reconstruction of their plants, discarding the magneto system and adopting uniform central battery plant.

Such a wholesale reconstruction, for example, involving the rebuilding of every exchange and the re-equipment of every station, was carried out in New York in three years, and completed in 1901. This measure naturally involved large capital expenditure and heavy depreciation, since much working plant was scrapped long before its natural period of renewal; but it put the telephone service on a new footing of high efficiency and complete uniformity of working, and therefore was a measure justified to those who were animated by the single motive of sound development of the telephone service and who were able to work untrammelled by external influence. But the position of the National Telephone Company has been very different; the limited tenure, the uncertain future of the business, the frequent changes of Government telephone policy, the obstruction of the local authorities with regard to underground work, have all been restraining influences which rendered impossible the rapid and wholesale reconstruction of plant which was so desirable, and forced on the Company a more cautious policy of capital expenditure than that followed under the freer conditions of the telephone business in America.

Consequently the technical progress of the telephone service has been materially slower on this side of the Atlantic than on the other, and this has naturally reacted on the development of the system and on its general usefulness to the public; efficiency is the best possible advertiser, and high efficiency in the telephone service is not practicable without a uniform plant of modern type. If it had been possible to complete at an earlier date the general reconstruction of the telephone systems which is still in progress in the United Kingdom, both the development of the British telephone service and its usefulness to the public would have reached long ago a much higher plane.

A Summary of Telephone Policy.

In the preceding sketch of the development of the telephone in the United Kingdom, the general principles which govern the sound development of the telephone service have been referred to here and there, but in order to judge of the progress made in the rest of Europe it will be as well here, before describing the telephone conditions in Continental Europe, to recapitulate these general principles, as it is by them that telephone progress in Europe must be measured.

The sound development of the telephone service requires a combination of enlightened technical policy and enlightened commercial policy. The former includes all engineering matters, some of which in telephone work, notably the study of telephonic transmission, are of a highly scientific nature. The latter includes such questions as the adoption of correct tariffs, the education of the general public in the advantages of the telephone, and a constant study of the requirements of the telephone-using public. The commercial policy of telephony is almost as technical in its way as the engineering branch of the industry and splits itself up into a variety of branches. To secure rapid expansion of the telephone system, which simply means to give the telephone service its greatest possible usefulness to the public, it is necessary that the whole organisation should be backed by adequate capital resources, and for this flow of capital to be maintained the capital must be able to earn its living. These are in brief the general principles by which the development of the telephone is governed, and it is by them that telephone progress in Europe must be judged.

The technical progress of telephony has been relatively rapid when it is considered that when the telephone was invented there were no appliances in existence by which a telephone exchange service could be conducted. The earliest exchanges were constructed by means of adapting to the work crude appliances borrowed from the telegraph, but it was necessary to devise the whole telephone exchange system from the beginning. The early switchboard systems were based on the use of electro-magnetic indicators for signals, actuated by a magneto generator at the subscriber's instrument, and this system is currently known as the "magneto system." The lines were at first single wire with earth return. To overcome the interference caused by induction, the single wire line was later abandoned in favour of the metallic circuit; and with the introduction of metallic circuit working material improvements were made in the switchboards.

A distinct step in advance was the branching switchboard with self-restoring indicators; this gave better electrical conditions and quicker operating. In 1898, or within twenty years of the establishment of the first commercial telephone exchanges, the central battery system of exchange working was evolved, and this effected a revolution in telephone exchange equipment. The central battery system abolished the magneto generators and the local transmitter batteries from the subscribers' stations and rendered the signalling for calling and disconnection automatic operations, effected simultaneously with the lifting and replacement of the receiver; this, by making the disconnection signal no longer dependent on the memory of the subscriber, removed one of the greatest sources of inaccuracy and inefficiency in the working of the service.

The central battery system, with its central supply of energy for operating signals automatically, has given rise to a host of improvements making for more accurate and more rapid working. This great advance in equipment has been accompanied by a great improvement in traffic organisation; and this is a vital feature in telephone management, for just as the efficiency of the battleship depends in the long run on the efficiency of the man behind the gun, so does the efficiency of the telephone system depend on the efficiency of the girl behind the plug. On the selection, training and supervision of telephone operators, a great amount of attention is now concentrated; and the traffic staff of a modern telephone system is a highly organised body of special workers, governed by the most careful instructions, checked and supervised by the most minute and searching tests and guided by the most elaborate statistics.

To the mind of the engineer it is perhaps a weak point in the telephone system that in the long run the efficiency of the entire expensive plant and elaborate organisation is dependent on the young woman operator, but those who have given the subject the closest study are convinced that the young woman operator—that is, a human brain and set of fingers (and the young woman is found to possess the best combination for the purpose)—is a necessary feature to cope with the great range of selection required in the telephone service. The automatic telephone exchange, of which several highly ingenious varieties have been devised by different inventors, does not meet the conditions under which the telephone service has to be operated to-day. The full automatic exchange transfers the working of the service from the expert operator, who can be trained and supervised, to the inexpert member of the public who cannot be trained or supervised.

Moreover, in great city telephone systems there is such an immense variety of calls made daily, many of which for various reasons require special handling, that the automatic system breaks down economically, as a large proportion of the calls would in any case have to be handled by human operators. It is for these reasons that the automatic exchange, although technically practical and operative, has made little headway and has been adopted only to a limited extent in isolated systems. The present trend of technical progress is in the direction of employing automatic apparatus for the completion of the connection, a method which has the advantage of retaining a simple telephone set in the hands of the public, and of preserving oral communication between the user and the operator.

Reverting for a moment to the commercial policy required for sound telephone development, too much stress cannot be laid on this factor. Good commercial policy in telephony involves not only a correct rate scheme but proper education of the public, so as to extend steadily the use of the telephone and so as to ensure correct use of the service. The general efficiency of the service is largely dependent on good commercial organisation of the telephone administration. If telephone subscribers are left to go their own way, a large proportion of them will produce inefficient service for themselves and for others by misuse of their telephone equipment; and it is necessary continuously to educate subscribers so that they shall both obtain themselves and give to others an efficient service. The commercial policy of the telephone administration thus largely influences the engineering work, since the engineering work is necessarily affected by the rate of development of the system and by the efficient or inefficient use of the service by the public.

CHAPTER IX.

A SURVEY OF TELEPHONE DEVELOPMENT ON THE CONTINENT OF EUROPE.

Spain, Portugal and the Near East. Switzerland. Belgium. Holland.

IN summarising the progress of the telephone in Continental Europe that progress has to be judged by the standards and principles above referred to, and it is unquestionably true that the development of the telephone service, its efficiency, and its general usefulness to the public are in the long run dependent on the application of the principles of technical and commercial policy which the experience gained in the development of the telephone in America has proved to be correct.

Spain, Portugal, and the Near East.

Climatic conditions and the natural and political characteristics of the people have assuredly their effect on the development of any technical industry, and one would not expect relatively backward countries like Spain, Portugal, Greece, and the Balkan States to have done much with the telephone ; and, as a matter of fact, those countries have done practically nothing. Actual Government monopoly and excessive Government restrictions have extinguished the possibility of private enterprise undertaking any active development. In Portugal an English company works a small system in Lisbon and one in Oporto ; in Spain various small French and Spanish companies and firms work, under rigorous restrictions and excessive royalties, a handful of small systems, of which the largest, in Barcelona, contains about 3,000 stations.

In Greece, Servia, Bulgaria, and Roumania, the telephone is a Government monopoly and is not developed at all. Roumania, the most progressive of the Balkan countries, has between 8,000 and 9,000 telephones for a population of 6,000,000. One or two of the exchanges in this group of countries have recently been equipped with central battery switchboards, but with these exceptions the telephone plant in these countries belongs entirely to the experimental period of telephony, and the commercial policy is equally antiquated. It is quite true that in these countries the public is not so ready to take advantage of modern facilities as in the more active industrial countries, but nevertheless, if private enterprise were allowed fair opportunities, the development of the telephone in all of this group of countries might easily be increased tenfold within a very short time.

Switzerland.

In the telephonic survey of Europe I should next take a group of countries having some, although not too close a resemblance in general conditions and population, namely, Switzerland, Holland, and Belgium. Belgium and Holland have the advantage over Switzerland in the possession of great cities and in industrial wealth, but Switzerland has achieved the greater telephone development, having 20 telephones per thousand of population as against nine for Holland and six for Belgium. The telephonic history of Switzerland is interesting in several respects. The telephone became a Government monopoly practically from the beginning, as very little company work was done in Switzerland; but the Swiss Telegraph Department was fortunate in having men at its head who took an intelligent interest in the scientific side of telephony and realised the possibilities of the telephone.

The late Dr. Wietlisbach, Chief Engineer of the Swiss Telegraph Administration, was an accomplished student of telephone problems, and his scientific work—unhappily cut short by his early death—did much to give the telephone a good start in Switzerland, and to awaken for it more sympathetic attention from the Telegraph Administration as a whole than it has received in most other Continental countries. Unfortunately the telephone also attracted the attention of the politicians during its early days in Switzerland, and the Swiss Parliament forced on the Telegraph Administration telephone tariffs of such an unpractical character as seriously to impair the earning capacity of the telephone system.

In considering all European telephone tariffs it must be

borne in mind that the rates originally fixed were adopted when the telephone service was in its earliest stage of development, and these rates were based on supplying single wire service to very small groups of subscribers concentrated within a small radius of the exchange. Since the number of subscribers was small the traffic was also small. With the development of the telephone the circumstances have entirely changed. The whole plant is much more complicated than the crude single wire plants of the early days, the systems have spread out over much greater areas, and the average use of the service has greatly increased. The rates fixed in the early days of the business are therefore no criterion by which to judge telephone tariffs for the conditions of to-day.

In Switzerland the tariff was originally, as everywhere else, a flat rate tariff, but in 1889 a measured rate tariff, giving 800 free calls for the minimum rate, was introduced. At that time there were only 9,000 subscribers in the whole of Switzerland. The effect of the measured rate tariff was to cause the subscribers to cut down their use from an average of 1,310 calls a year to an average of 718, and a year later, when the measured rate tariff was completely in force, the average annual use of the service per subscriber fell to 546 calls. In 1896 a further drastic revision of the tariff was made and the 800 free calls were abolished, every call being charged for at the rate of 5 centimes. The subscriber pays extremely low maintenance rates, the first year of the contract the rate being 100 fr., the second year 70 fr., and the third and subsequent years 40 fr. Although the total number of subscribers in the Swiss local systems has risen from 20,000 to over 60,000 since this tariff was introduced, the average annual use per subscriber has only risen to 610 calls a year, or an average of about two a working day

The chief causes of the extremely low telephone traffic in Switzerland are that there are no large cities in our sense of the term, and that the Swiss are an extremely frugal people. Distances are short, and the general conditions of life are simple. The charge for each individual call causes under these circumstances a greater check on the traffic than it would in other places. It is interesting to note, as already mentioned above, that the telephone has been developed in

Switzerland to such an extent that the telephone traffic and revenue now far exceed those of the telegraph. For the past year the telephone revenue in Switzerland was over 8,800,000 fr., as compared with 3,445,000 fr. from the telegraph; and whereas there were 1,572.000 internal telegrams there were 45,000,000 telephone calls, of which 8,000,000 were interurban calls, so that the interurban calls alone were five times as numerous as the telegrams. This is partly due to the fact that the Swiss Government have made telephone rates which compete with their own telegraph rates, as over 90 per cent. of the interurban calls were for the short distances, where the charge for a telephone call is less than that for a ten-word telegram.

The rates charged for the telephone service in Switzerland produce a revenue insufficient to permit the Administration to adopt improved plant with great rapidity, and except those of Berne and Geneva, recently converted to common battery, the systems throughout Switzerland have magneto equipment, and to a large extent single wire lines. The chief demand upon the long distance service is for connections between neighbouring places, and real long distance talking, of course, does not occur in Switzerland, so that the interurban plant is also comparatively simple. The fact that all the towns are of small population and area enables them to be served by single exchange systems, and the extremely low traffic makes the working of the exchanges very simple, so that it is easy to maintain a sufficiently high standard of service to satisfy the very moderate requirements of the Swiss public.

Belgium.

In Belgium the telephone business was started in 1879–80 by companies, but only short term licences were granted and almost all the company systems were taken over by the Government in 1893. Since the establishment of a Government monopoly the progress of the telephone in Belgium has been very slow; and as many old-fashioned features of telephone practice, such as flat rates, extra charges for metallic circuits, and single-exchange systems with mileage charges beyond a limited zone, have been retained, the utility and general development of the telephone have been artificially restricted. In 1894 there were 8,500 subscribers in Belgium, and by 1910 the total has increased only to 42,540, whereas if the telephone service were conducted according to modern methods, Brussels or Antwerp alone would easily use as many telephones as the whole country uses now. Of recent years a number of local systems have been converted to common battery working, and the overhead single lines have been replaced by metallic circuit cables, but these reforms have been carried out very slowly, and antiquated plant has been retained in service long after it had become obsolete.

As with all Government monopolies, there is a total absence of commercial policy and of efforts to educate the public or to accelerate the development of the system; such growth as occurs is the normal increase due to people finding out for themselves that they want the telephone service. The adoption of measured rates has been discussed for some years in Belgium, but so far no active steps have been taken. This is always one of the great difficulties of Government monopolies—that proposals of the technical officials, who are versed in the real needs of the industry, are subject to review by politicians who have no expert knowledge whatever and are more often influenced by "political" motives than by practical and common sense reasons. It is well known that measured rates make both for the development and efficiency of the telephone service, but in all countries where flat rates have been retained up to now there is a sufficient body of large users (who consider they have a bargain in the flat rate) to raise a noisy agitation against a measured rate tariff; and in all countries politicians are more prone to be influenced by agitation than by truthful and practical reasoning.

Holland.

In Holland the early telephone work, both local and long distance, was undertaken by companies, principally by the Netherlands Bell Telephone Company, which is still in existence but on a much diminished scale. In later years it became the policy of the Dutch Government to operate the long distance service itself and to grant concessions for operating the local service to companies or to the municipalities.

Holland offers an example of a feature of the history of the telephone previously referred to. The early telephone companies failed to realise their opportunities and did not develop the business with sufficient activity and did not either keep their plant up to date or adopt modern tariffs and commercial policy. The result was that in the large cities the concessions to the companies were cancelled at the end of their original term and new concessions were granted to the municipalities of Amsterdam, Rotterdam, and the Hague; the municipalities built new systems and started the telephone business afresh, so to speak, in Amsterdam and Rotterdam in 1896, and in the Hague in 1902.

The Dutch Government has shown unusual liberality and common sense in these concessions, as no royalty is exacted from the telephone undertakings. Most other Governments have taxed the use of the telephone heavily by exacting a large contribution from the telephone undertaking in the form of a stiff percentage of the gross receipts. As in the long run this must be paid by the user of the service, the royalty becomes a tax on telephonic communication. This was apparently recognised in Holland, and no telephone royalty was demanded by the Government. The concessions were nevertheless surrounded by numerous regulations and restrictions, a most peculiar rule being that the area of each concession is confined to a circle having a radius of five kilometres (3·1 miles) from a point in the centre of the town. This results in the telephone undertakings having to refuse service to would-be subscribers situated in the outskirts of the towns.

Since the establishment of the new municipal systems the development of the telephone in Holland has been more rapid than previously, and the relative development is now substantially higher than that of its nearest neighbour, Belgium, being 9·3 telephones per thousand inhabitants for Holland, as against 6·1 per thousand inhabitants for Belgium. It is curious to observe, however, that considerable differences of practice occur in the three municipal systems of Amsterdam, Rotterdam, and the Hague in almost all respects—in technical methods, in the tariffs, and in general practice—although the three towns are only a short distance apart and must have a constant interchange of population. This lack of uniformity in telephone practice is a serious inconvenience from the public point of view, though it frequently is observed where the general direction of the telephone service is not in expert hands. Up to the present no measured rate tariff has been introduced in Holland; it has been strongly advocated, but the semi-political control of municipal management favours the retention of flat rates, which, superficially at any rate, benefit the large user at the expense of the small. Of recent years common battery working has been adopted in the three large Dutch cities and the service has been materially improved. In Rotterdam this change was made only a year or so ago, and it was then found, so crudely had the original magneto plant been constructed, that it was necessary entirely to reconstruct the line plant, although the municipal system had been at work only about ten years.

There is a steady growth of the telephone in Holland, amounting to about 4,000 stations a year, or an annual increase of nearly 10 per cent., but it cannot be doubted that if the possibilities of the telephone were fully realised and if modern commercial practice, as well as a higher standard of technical practice, were adopted, a rich industrial country like Holland would make a far greater use of the telephone than it does at present.

CHAPTER X.

A SURVEY OF TELEPHONE DEVELOPMENT ON THE CONTINENT OF EUROPE—(*continued*).

France and Germany. The effects of complete Government monopoly. Refusal to learn from the experience of others. Neglect of commercial principles and practice.

WE next come to the three great industrial countries of Continental Europe, France, Germany, and Austria-Hungary, and in each of them the telephone has had a sad history, considering the enormous opportunities these rich and largely populated countries, with their many great cities, offer for telephone development.

In France telephony is a slough of despond. The French bureaucracy, which, according to so keen an observer as Lord Rosebery, is "strangling France," has effectively strangled the French telephone service. The telephone was made a Government monopoly in France in 1889, and twenty years later it is described by most Frenchmen, and notably by those who have investigated it the most closely, as a national disgrace. The development is extremely low, the plant is largely antiquated, or where it has been renewed its efficiency is destroyed by bad organisation, the tariff is the tariff of the eighties, and the commercial practice is the rigid and cumbersome routine of the most bureaucratic of bureaucracies. The service, notably in Paris, which contains about a third of all the telephones in France, is a by-word with all Frenchmen. It is best to draw a charitable veil over the deplorable state of telephony in France, because in every respect telephone practice in France is the reverse of what it should be. At the moment of writing this, an announcement appears in the papers that the new Minister of Public Works is preparing a complete scheme of reorganisation of the French telephone service, which will be presented to Parliament next session, and it may be taken that progress with the telephone in France is all in the future, as nothing short of a complete reorganisation will place the telephone in France on even a moderate footing of efficiency.

Up to the present, France has been the most conspicuous example of the evil results of Government monopoly of the telephone service, because the country itself, with its many great cities and its great industrial activity, offers one of the finest fields in Europe, if not in the world, for the development of the telephone service. If the telephone in France were operated on commercial lines the present development could easily be multiplied tenfold. Paris, by reason of its peculiar area, and the wealth and activity of its population, comes next to New York as a favourable field for the development of the telephone. With a well-equipped and well-organised system, operated on commercial lines, Paris might easily have a system of from 200,000 to 300,000 telephones. It would be difficult to calculate the financial loss which the inhabitants of Paris daily suffer from the lack of adequate and efficient telephonic facilities.

Germany.

In Germany the telephone has always been a Government monopoly, and, as in other branches of industrial effort, the German Government has encouraged the development of the telephone on purely German lines without regard to the experience of other countries and without relying on external aid of any sort. The German Post Office has had its own rate schemes and its own ideas of technical practice in telephony, and has usually refused to adopt improvements devised in other countries until German manufacturers have been able to supply them, or imitations of them.

This tenacious adherence to purely domestic ideas and methods has resulted in seriously restricting the technical progress of telephony in Germany, and consequently in restricting the efficiency of the service. Single wire overhead line plant was largely maintained in service in Germany long after metallic circuit working and underground cable distribution had been generally adopted in other countries. Even to-day many of the German telephone systems consist largely of single wire overhead plant. When underground cables were introduced in some of the large cities they were constructed single wire instead of metallic circuit, and numbers of cables were laid in a mass in iron pipes of large section instead of each cable being laid in a separate duct, which is correctly the standard practice of other countries. Technical mistakes of an equally far-reaching character were made in switchboard practice, and at the subscriber's end of the system a fundamental error was made in allowing subscribers themselves to install extension stations of all kinds and connect them to the exchange lines.

These various errors in technical policy have naturally resulted in a very low standard of efficiency in the German telephone service, and have made it necessary, as modern telephone practice has tardily penetrated into Germany, for the Administration to reconstruct entirely the large city telephone systems. It has been necessary not only to substitute underground metallic circuit cables for the overhead single wires, as was done in America and in England many years ago, but in some cases to substitute conduit systems and new metallic circuit cables for the old underground systems of large iron pipes with their masses of single wire cables. As magneto switchboards have gradually been replaced by common battery switchboards it has been necessary, of course, to remodel the subscribers' installations, and enormous difficulties have arisen from the heterogeneous collection of extension stations with which many subscribers had provided themselves under the lax practice of the Administration in allowing the public to attach their own telephones to the exchange system.

This process of reconstruction is still in progress in the large cities of Germany, and it will be many years before the German telephone service is equipped with a uniform style of plant, such as has for a long time existed in the United States. Under these circumstances the standard of efficiency of the telephone service in Germany is necessarily low, and the telephone does not occupy nearly so important a place in the business life of the country as it is capable of occupying.

In tariff questions and in general commercial policy the practice of the German Post Office has not been notably superior to its technical practice. For many years there was a universal flat rate, the same for the small village as for Berlin or Hamburg. This was fixed in the early days, before the growth of the telephone was foreseen, and being based on the crude conditions of the overhead single wire telephone systems of the eighties, serving very small groups of subscribers, it was fixed at a comparatively low price. The result was that the service was very cheap in the large cities and comparatively dear in the small places, and a large development occurred in the cities, and, comparatively speaking, no development at all in the small places. In the cities the development was aided by the fact that no attention was paid to providing public telephone stations, and every shopkeeper and restaurant keeper was allowed to give his customers free use of the telephone. A revision of the tariff was made about ten years ago and a message rate tariff was introduced, giving much more moderate rates for small places and small users than had previously existed, and the flat rate was raised in some of the large cities.

Under this tariff considerable development took place in the smaller towns throughout Germany; but the German Administration in time discovered a feature of commercial telephony which has long been patent to all experienced telephone men, namely, that a measured rate tariff and a low flat rate tariff cannot be worked effectively side by side. The inevitable result is to divide telephone users into two classes, the extremely small users and people who want the telephone mainly for receiving purposes, and the moderate and large users. The former take the message rate and yield a very small revenue, and the latter take the flat rate, because it offers them a great bargain, and they do not yield a revenue corresponding to their use of the service. After several years' experience of the unsatisfactory results of a combined message rate and flat rate tariff the German Post Office finally resolved to abandon the flat rate altogether and to introduce a message rate tariff throughout Germany. A Bill was prepared to this effect and presented to Parliament, but has aroused so much opposition from the large users of the telephone service, who found themselves threatened with the abolition of the great bargain they have enjoyed for so many years under the flat rate, that it has not yet passed, and its fate is still uncertain.

It is interesting to contrast the results of complete Government monopoly of the telephone service in two such countries as France and Germany. In both countries the officials have refused to avail themselves of the experience of those who have most effectively developed the telephone both scientifically and commercially, namely, the Americans. From the point of view of both American and British telephone practice (the latter being largely modelled on the former) almost every technical error of omission and commission has been made in the telephone service of France and Germany ; but there is this to be said for the German Administration, that from the beginning they have made greater efforts to develop the telephone and paid greater attention to its possibilities than the French Administration, where the jealousy of the telegraph officials and the greater natural apathy of a French Government department have prevented serious and sustained efforts to develop the telephone service on a broad scale.

The result of the two different policies is that the relative development of the telephone in Germany in proportion to population is nearly three times higher than that of France. This is to some extent to be accounted for by the large number of private extension stations in the German systems previously referred to, but even allowing for these, the German telephone development is more than double that of France. This superior development is chiefly due to the introduction of measured rates in Germany, and to the much larger capital provided by the German Post Office for telephone development. In France the telephone system has been starved of capital, whether through political reasons or whether because, as is often publicly alleged, the French Chamber has lost confidence in the administration of the Post Office, it is not for an outside observer to determine.

In both countries the guiding principle of the telephone business—that high efficiency of service is the prime requisite—has been completely lost sight of. In neither country is the attention paid to telephone organisation, and especially to the organisation of the traffic branch, which the subject requires. As a result the efficiency of the service is naturally on a low plane, and in the long distance service it is accepted as a matter of course by the public that calls between important towns should habitually require several hours to complete.

CHAPTER XI.

A SURVEY OF TELEPHONE DEVELOPMENT ON THE CONTINENT OF EUROPE—(*continued*).

Austria-Hungary, Italy and Russia.

IN Austria-Hungary the development of the telephone is even lower than in France, and a beginning has not yet been made in the Dual Monarchy in supplying a modern and well-developed telephone service. The difficulties which have obstructed the telephone in Austria and Hungary, both countries which offer a great field for the most rapid means of communication, have been, as in other European countries, those arising from political control. A beginning was made by private companies, chiefly of English origin, but by 1896 the last of these had disappeared and the telephone service had been converted into a complete Government monopoly. Since then the antiquated tariffs adopted and the frequent political deadlocks which arise in Austria-Hungary have prevented any sound and rapid development of the telephone.

In Austria, for all places except Vienna, a tariff was adopted under which the subscriber had to pay the whole cost of building his line and then an annual rate for maintenance and for service. This effectively checked development, and the growth of the systems in the various Austrian towns soon almost ceased. In Vienna, on the other hand, the flat rate established in the early 'eighties, when the system was extremely small, was maintained long after it had become unprofitable; but owing to Parliamentary deadlocks the State Telegraph Administration was unable to obtain capital with which to develop the telephone system, so that the increase of subscribers which would naturally have occurred under a low flat rate could not be dealt with, and the progress of the Vienna system was extremely slow, the demand for service far exceeding the available facilities. This is a situation which frequently arises under Government monopolies; the telephone service is denied to a large portion of the public owing to capital not being forthcoming to provide for the development of the telephone system even at a sufficient rate to cope with the ordinary public demand, unstimulated by any active commercial policy.

In recent years more enlightened efforts have been made to develop the telephone in Austria by the adoption of graduated rates and the abolition of the old practice of charging the subscriber with the capital cost of his line. Arrangements have also been made for the annual supply of a certain amount of capital for new telephone construction. By these measures the checks on the development of the telephone in Austria have been somewhat lessened, and more rapid progress is now being made ; but this is purely relative to the previous stagnation, and it will be many years before the telephone in Austria reaches even a moderate degree of development. The measured rate tariff proposed by the Austrian Administration a few years ago was at first strenuously resisted by the Press and by the section of the public which had so long enjoyed bargain rates for unlimited use, but a measured rate tariff, with some modifications of the original scheme, was eventually adopted and has had a marked effect in increasing the rate of development.

In technical practice the Austrian Administration has, like most Government monopolies, lagged several steps behind modern methods. When the Vienna system was rebuilt, a little over ten years ago, the common battery system was available, but was not adopted ; and although two spacious new telephone buildings were constructed and a new underground system on the drawing-in principle was laid down, the magneto system of working was adhered to and the equipment adopted for the exchanges was generally of an old-fashioned type. During recent years the Austrian Administration has been experimenting with the automatic telephone exchange, and it is proposed that a number of small towns shall be equipped with automatic apparatus, with the idea that if these installations successfully meet the requirements of the public the automatic system will eventually be adopted in the larger cities and in Vienna itself.

Hungary has its separate telegraph Administration, and the telephone branch has followed its own policy, and not imitated that of Austria. Outside of Budapest there is little development of the telephone, and the development of the whole country is somewhat lower than that of Austria, and only slightly exceeds that of Italy, where the telephone is practically non-existent. In Budapest the telephone service was originally started by a company, but the State purchased the system after a few years and assumed the monopoly of telephone work. The system was reconstructed in 1903, and a large new telephone building put up, which is equipped with a common battery plant of the best American type. The low development and the slow progress of the telephone in Hungary are principally due to lack of capital, absence of commercial policy, and to the retention of the old flat rate tariff. For many years efforts have been made by the telephone officials to obtain the sanction of the Government to the adoption of a measured rate tariff, but so far the proposal has remained pigeon-holed.

Italy.

In Italy the telephone has hardly begun its career, there being only 62,000 telephones among a population of 33,500,000. The Italian Government has recently assumed an entire monopoly of the service, and has organised a State Telephone Department, separate from the Telegraph Administration—a plan which might well be copied by other Government monopolies, since the telephone industry undoubtedly suffers greatly from being under the control of telegraph officials. The telephone in Italy has its future all before it; efforts are being made to introduce modern technical practice and modern rate practice, and the future will show whether the new Italian State telephone department is capable of successfully developing the great field which undoubtedly exists for the telephone service in the many large cities of Italy.

Russia.

Russia is the last of the great European countries where the development of the telephone is extremely low, barely reaching one telephone per 1,000 inhabitants. Of recent years, however, very rapid progress has been made, notably in the three great cities, Moscow, St. Petersburg, and Warsaw. These three Russian cities offer another illustration of the fact already noted, that many of the early European telephone companies failed to realise either their opportunities or their responsibilities. Telephone companies were established under licence from the Government in Moscow, St. Petersburg, and Warsaw in the early days, but the progress made was extremely slow, and as the end of the concession approached it was realised in all three places that the telephone companies had failed in their mission. The rates were high, the service was bad, and the development low. So congested was the system in St. Petersburg, for example, that merchants were willing to pay a high premium for the right to take over a telephone line from a subscriber who was giving up the service.

These conditions may have arisen through onerous terms of the licences which prevented the companies from expending capital for development as the expiry of the licence approached; but since such conditions existed in the late nineties, when the possibilities of the telephone were well understood, and telephone practice, both technical and commercial, had been completely revolutionised, it looks as if the Russian companies had really failed to realise both their opportunities and their responsibilities towards the public. When the expiry of the original licences approached, the Russian Government, with a view to securing greater development, lower rates, and a higher efficiency of service, put up new concessions to competition, the competition to hinge upon the offer of low rates and of proper guarantees of an efficient service. In St. Petersburg the concession was obtained by the municipality, and in Moscow and Warsaw by companies organised by a group of Swedish, Danish and Russian capitalists.

These new concessions were granted in 1901, and work was begun in all three cities on the construction of entirely modern telephone plants, the common battery system being employed, with some modification in methods of operating due to local ideas. In all three cities the new systems came into operation between four and five years ago, and in each city the number of subscribers has increased about five-fold since the starting of the new services. Similar concessions were granted for Riga and Odessa, and new systems were constructed in those cities. The Russian Government, while hedging round these concessions with numerous restrictions and retaining the right to purchase the telephone systems at the end of the licence, has adopted the broad-minded attitude of encouraging the development of the telephone for the benefit of the public and exacts from the concessionaires a royalty of only three per cent. on the gross receipts, which compares very favourably with the high royalties exacted by some other Governments.

The long-distance service in Russia remains in the hands of the Government, and is very undeveloped, Moscow and St. Petersburg being practically the only great cities connected by long-distance telephone lines. The distances between cities in Russia are very great, and the maintenance of long distance lines is admittedly very difficult, as the sparsely inhabited character of the country is very favourable to the operations of copper wire thieves ; the lines between Moscow and St. Petersburg are interrupted with great frequency from this cause.

CHAPTER XII.

A SURVEY OF TELEPHONE DEVELOPMENT ON THE CONTINENT OF EUROPE—(*continued*).

Sweden, Denmark and Norway. The scope allowed to private enterprise and the deferment of State monopoly have permitted higher telephone development in the Scandinavian countries than in the rest of Europe. State competition with private enterprise in Stockholm. Conclusion. The ineradicable defects of Government monopoly.

IT is in the Scandinavian countries that the telephone has had the freest and most rapid development in Europe, as already shown by the statistics given relating to Sweden, Denmark and Norway. This special position of the telephone in the three small Northern countries of Europe is due to several causes, the principal being, perhaps, that in those countries the Government had no legal monopoly in the telegraph, and therefore was unable to exercise the same amount of repressive control over private telephone enterprise at the beginning as was exercised in less fortunate lands. Private enterprise had freer scope than elsewhere, notably in Sweden and Denmark, and the telephone became well-established and its value thoroughly appreciated by the public before new legislation was enacted to increase the slender measure of control which the Government Telegraph Administration at first possessed.

Sweden, Denmark and Norway have not been complete exceptions to the usual story of political interference and unsympathetic control which have so grievously restricted the progress of the telephone throughout Europe, and the tendency now in all three countries is either for the State to obtain an actual monopoly, or so completely to regulate the operations of the remaining telephone companies as to cause them to become virtually State-managed concerns. In Sweden the State now has a monopoly outside of Stockholm, in Norway the State has acquired the company systems in the principal towns and is acquiring others, and in Denmark the State, while leaving the active management still to the companies, dictates rates and regulations to such an extent that the companies have to adapt their policy to the exactions of the Government rather than to the requirements of the telephone business and of the public.

The comparative freedom from political control enjoyed by the telephone in Scandinavia in the beginning enabled many telephone companies, both large and small, to be started in all three countries, and the telephone seems to have "caught on" at once. Doubtless the simplicity and directness of telephonic communication appealed to such a practical people as the Scandinavians, while the relative scarcity of other means of communication, and the general conditions, such as small town areas, absence of other electrical construction and the cheapness of labour and material, were all factors favourable to rapid telephone development. The rates for telephone service in Scandinavia have always been cheap, and necessarily so, since the value of money there is high, and everything else is cheap when judged by English or American standards.

Many of the early Scandinavian telephone companies were small co-operative companies, in which the users were the principal shareholders. Numerous co-operative companies still exist in Norway and Denmark, but in Sweden they have all been absorbed either by the State telephone system or its great rival, the Stockholm Telephone Company. In the co-operative systems it has been usual for the subscribers to build and equip their own lines, generally using, for the sake of cheapness, iron wire and second-hand instruments, and to pay a small annual rate for exchange service and maintenance. In this way an extremely cheap telephone service has been obtained; the benefits of telephonic communication have been extended even to the most rural and sparsely settled parts, and the "telephone habit" has entered into the life of the Scandinavian peoples to an extent unapproached in other Continental countries.

To a very large extent, however, Stockholm has set the pace in matters telephonic for the whole of Scandinavia; and Stockholm owes its conspicuous place in the world of telephony to the unbounded enthusiasm and the organising genius of the late Henry Cedergren, whose early death those who knew him are still lamenting. Mr. Cedergren founded the Stockholm General Telephone Company in 1882, and remained at its head until his death, which occurred in the spring of 1909. He was a telephone man of great enterprise and foresight, and of unlimited energy, and after having taken up telephone work he lived entirely for telephone work. He periodically visited America, and constantly imported into Sweden American telephone methods, and under his management Stockholm has become the best telephoned city of Europe. Had he been allowed free scope, Mr. Cedergren would have developed the telephone throughout Sweden in the same active and enlightened manner, but the State Telegraph Department, jealous for the telegraph revenue, opposed his plans for building long distance lines, and obtained from Parliament powers and money to enable this work to be done by the Telegraph Administration.

The Telegraph Administration has since gradually acquired all the local telephone systems in Sweden except that of the Stockholm General Telephone Company, and has restricted the area of that Company to a circular district having a radius of 42 miles from Stockholm. Within this area there is strenuous competition between the State system, started in 1889, and the Company, but the latter has by far the best of the struggle, as the subscribers to the Company system outnumber those of the State system by more than three to one. The State Telegraph Administration has conducted a cut-throat competition, as it supplies service in Stockholm at an admittedly artificial rate—a rate lower than that which it charges in other smaller towns in Sweden—but notwithstanding this the Company maintains its preponderating position and steadily obtains the larger share of the new business. This active competition, which has now lasted for twenty years, has of course greatly stimulated the development of the telephone in Stockholm, and the extraordinary prevalence of the telephone there strikes every traveller as a distinct feature of that very attractive city. Not only has one two telephones—one of each system—in one's hotel bedroom, but one sees telephones literally everywhere, even down to the open-air refreshment stalls in public squares and gardens, and the trestle-benches in the fish market, which are in use for only a few hours each morning.

While the Stockholm competition has undoubtedly to a certain extent produced an artificial development of the telephone there—artificial for various reasons, but from the public point of view because every business man would prefer one uniform and well-developed system to two separate ones —the high standard of service maintained by the Stockholm General Telephone Company, its enlightened technical and commercial policy and its thorough organisation, have set a high standard of telephone work which is reflected throughout Scandinavia and has had its effect even on the telephone work of the various State Telegraph Administrations, which is noticeably better organised and equipped than that of other Continental countries.

Conclusion.

Whatever the reader's feeling may be, one cannot write even this brief and imperfect sketch of the history of the telephone in Europe without such a survey of wasted opportunities causing a sensation of the most poignant regret. It has been my lot to observe at frequent intervals during the past fifteen years the working of the telephone service in all its aspects in every European country, and I have even spent some time in arguing over "the telephone question" with the high political authorities of a country where there is at present no telephone service. All my observations and all my studies of the subject, during a continuous connection with telephony in America and in Europe, which now extends to over twenty years, or two-thirds of the whole life of commercial telephony, lead irresistibly to the one conclusion—that Government monopoly has strangled the telephone in Europe, checked its development, restrained its efficiency, limited its usefulness to the people. Like the Angel of Death, to paraphrase John Bright's famous speech on the Crimean war, the Government telegraph monopoly has overshadowed telephone progress throughout the length and breadth of Europe, and killed or maimed every effort made to give the telephone the widespread usefulness to the people of which we know it to be capable.

To compute the loss which the peoples of Europe have suffered, and still daily suffer, through being deprived of the highest possible development of the most rapid means of communication, would be impossible; it is an immense, incalculable loss, just as the gain in saving of time and friction which the community blessed with an efficient and highly-developed telephone service enjoys is immense and incalculable.

From the experience of America we know that an efficient telephone service, in which the greatest possible rapidity and certainty of communication over both short and long distances is made the governing principle, is capable of almost indefinite expansion—so deeply does this instantaneous means of communication enter into both business and social life. It may be asked—Why should not similar results, similar efficiency, and similar development be obtained under Government management? The best answer is the results of Government management in Europe during the past thirty years. The history of the telephone in Europe clearly shows that the "vested interest" of the Governments in the telegraphs has been allowed to check the natural advance of the telephone—the progress of the fittest has been artificially restrained. The growth of the brilliant and healthy youngster has been warped and stunted from mistaken regard for the less capable and less efficient elder brother.

Even if under Government ownership the management of the telephone were entirely separated from that of the telegraph, so as to eliminate this repressive influence, those who are familiar with the conduct of great Government departments, always necessarily subject to political control and political influence, would be the last to claim that a Government department can be an efficient substitute for private enterprise in the conduct of an industry where high efficiency, enterprising commercial policy, advanced technical policy, and sound and economical financial management are prime requisites for success. These are all the very antitheses of accepted Government methods; and, above all, the outstanding fact that under State management there is no man responsible for financial success or otherwise is sufficient always to prevent really successful and efficient working of a commercial enterprise by Government management. Financial soundness in the end governs the whole thing—organisation, efficiency, commercial and technical policy—and the difference between the Government official and the business man is that the former need only produce accounts while the latter must produce the money. In the world of affairs, of iron and steel, brass and copper, steam and electricity, machinery, organisation and effort, it is not love that makes the world go round, but money.

TELECOMMUNICATIONS

An Arno Press Collection

Abramson, Albert. **Electronic Motion Pictures:** A History of the Television Camera. 1955

[Bell, Alexander Graham]. **The Bell Telephone:** The Deposition of Alexander Graham Bell in the Suit Brought By the United States to Annul the Bell Patents. 1908

Bennett, A. R. **The Telephone Systems of the Continent of Europe** and Webb, Herbert Laws, **The Development of the Telephone in Europe.** 1895/1910. Two vols. in one

Blake, George G. **History of Radio Telegraphy and Telephony.** 1928

Bright, Charles. **Submarine Telegraphs:** Their History, Construction and Working. 1898

Brown, J. Willard. **The Signal Corps U. S. A. in the War of the Rebellion.** With an Introduction by Paul J. Scheips. 1896

Chief Signal Officer, U. S. Signal Corps. **Report of the Chief Signal Officer: 1919.** 1920

Danlellan, N[oobar] R. **A. T. & T.:** The Story of Industrial Conquest. 1939

Du Moncel, Count [Theodore A. L.] **The Telephone, the Microphone, and the Phonograph.** 1879

Eckhardt, George H. **Electronic Television.** 1936

Eoyang, Thomas T. **An Economic Study of the Radio Industry in the United States of America.** 1936

Everson, George. **The Story of Television:** The Life of Philo T. Farnsworth. 1949

Eyewitness to Early American Telegraphy. 1974

Fahie, J[ohn] J. **A History of Electric Telegraphy to the Year 1837.** 1884

Federal Communications Commission. **Investigation of the Telephone Industry in the United States.** 1939

Federal Communications Commission. **Public Service Responsibility of Broadcast Licensees.** 1946

Federal Trade Commission. **Report of the Federal Trade Commission on the Radio Industry.** 1924

Fessenden, Helen M. **Fessenden:** Builder of Tomorrows. 1940

Hancock, Harry E. **Wireless at Sea:** The First Fifty Years. 1950

Hawks, Ellison. **Pioneers of Wireless.** 1927

Herring, James M. and Gerald C. Gross. **Telecommunications:** Economics and Regulations. 1936

Lodge, Oliver J. **Signalling Through Space Without Wires:** Being a Description of the Work of Hertz and His Successors. [1900]

McNicol, Donald. **Radio's Conquest of Space:** The Experimental Rise in Radio Communication. 1946

Plum, William R[attle]. **The Military Telegraph During the Civil War in the United States.** With an Introduction by Paul J. Scheips. Two vols. 1882

Prime, Samuel Irenaeus. **The Life of Samuel F. B. Morse,** L.L. D., Inventor of the Electro-Magnetic Recording Telegraph. 1875

The Radio Industry: The Story of Its Development. By Leaders of the Radio Industry. 1928

Reid, James D. **The Telegraph in America:** Its Founders, Promoters and Noted Men. 1879

Rhodes, Frederick Leland. **Beginnings of Telephony.** 1929

Smith, Willoughby. **The Rise and Extension of Submarine Telegraphy.** 1891

Special Reports on American Broadcasting: 1932-1947. 1974

Thompson, Silvanus P., **Philipp Reis:** Inventor of the Telephone; A Biographical Sketch. 1883

Tiltman, Ronald F., **Baird of Television:** The Life Story of John Logie Baird. 1933

Wile, Frederic William. **Emile Berliner:** Maker of the Microphone. 1926

Woods, David L., **A History of Tactical Communication Techniques.** 1965

(27)